Physiological Aspects of Plant Response to Pathogens and Abiotic Stress—2nd Edition

Physiological Aspects of Plant Response to Pathogens and Abiotic Stress—2nd Edition

Guest Editors

Violetta Katarzyna Macioszek
Iwona Ciereszko
Andrzej Kiejstut Kononowicz

MDPI

Basel • Beijing • Wuhan • Barcelona • Belgrade • Novi Sad • Cluj • Manchester

Guest Editors

Violetta Katarzyna Macioszek
Department of Biology and Plant Ecology
Faculty of Biology
University of Bialystok
Bialystok
Poland

Iwona Ciereszko
Department of Biology and Plant Ecology
Faculty of Biology
University of Bialystok
Bialystok
Poland

Andrzej Kiejstut Kononowicz
Department of Plant Ecophysiology
Faculty of Biology and Environmental Protection
University of Lodz
Lodz
Poland

Editorial Office
MDPI AG
Grosspeteranlage 5
4052 Basel, Switzerland

This is a reprint of the Special Issue, published open access by the journal *Plants* (ISSN 2223-7747), freely accessible at: https://www.mdpi.com/journal/plants/special_issues/physiological_plant_pathogens_abiotic_stress_II.

For citation purposes, cite each article independently as indicated on the article page online and as indicated below:

Lastname, A.A.; Lastname, B.B. Article Title. *Journal Name* **Year**, *Volume Number*, Page Range.

ISBN 978-3-7258-3889-9 (Hbk)
ISBN 978-3-7258-3890-5 (PDF)
https://doi.org/10.3390/books978-3-7258-3890-5

Contents

About the Editors

Violetta Katarzyna Macioszek

Violetta Katarzyna Macioszek is currently a member of the Department of Biology and Plant Ecology at the Faculty of Biology of the University of Bialystok in Poland. She received her MSc and PhD in Biology from the University of Lodz in Poland. She has more than 25 years of experience in plant biology research. She is a member of a few Polish scientific societies and was a member of their councils. Her research interests include photosynthesis, plant-pathogen interactions, fungal pathogenesis, plant stress biology, and phytoremediation.

Iwona Ciereszko

Iwona Ciereszko is the head of the Department of Biology and Plant Ecology at the Faculty of Biology of the University of Bialystok in Poland. She obtained her PhD in Plant Physiology from the Warsaw University (1997), and her full Professor degree from the University of Bialystok (2019). She was the Dean of the Faculty of Biology and Chemistry at the University of Bialystok (2012–2016), a member of many scientific councils at the Polish Academy of Sciences and University of Bialystok, and head of the Polish Society of Experimental Plant Biology (2015–2017). Her research interests include photosynthesis, sugar signaling and metabolism, plant growth, metabolism, and development under various stress conditions, including phosphorus and nitrogen deficiency.

Andrzej Kiejstut Kononowicz

Andrzej Kiejstut Kononowicz is a full Professor Emeritus of the Department of Plant Ecophysiology at the Faculty of Biology and Environmental Protection of the University of Lodz in Poland. He obtained all his scientific titles at the University of Lodz. Throughout his academic life, he was subsequently associated with the University of Lodz as the adjunct, professor, and Head of the Department of General Genetics, Plant Molecular Biology and Biotechnology. He also worked as a visiting professor at Purdue University and Auburn University in the USA for many years. He was the Head of many Polish scientific societies. His research interests include plant transgenesis, plant molecular genetics, and plant stress biology.

Preface

Plant physiology encompasses how plants function, including photosynthesis, respiration, water and nutrient uptake, and hormonal regulation. These physiological processes determine plant growth, development, and adaptation in diverse ecosystems. However, changes in abiotic and biotic environments shape a plant's physiological state. Abiotic conditions such as temperature, light, water availability, and soil nutrients influence metabolic pathways. However, stress responses induced by abiotic factors, e.g., heavy metals, disturb stable physiological conditions and redirect the plant's metabolism. Biotic interactions, such as challenges from pathogens and pests or symbioses with beneficial microbes, further re-program plant physiological processes.

This reprint covers physiological aspects of resistance, tolerance, and susceptible responses of plants induced by various pathogens, i.e., viruses, bacteria, and fungi-causing diseases, as well as abiotic unfavorable factors such as heavy metals, fluctuations in temperature, salinity, and drought.

Violetta Katarzyna Macioszek, Iwona Ciereszko, and Andrzej Kiejstut Kononowicz
Guest Editors

plants

Article

Silencing of Putative Plasmodesmata-Associated Genes *PDLP* and *SRC2* Reveals Their Differential Involvement during Plant Infection with Cucumber Mosaic Virus

Richita Saikia [1,2,3], Athanasios Kaldis [1], Carl Jonas Spetz [3], Basanta Kumar Borah [2] and Andreas Voloudakis [1,*]

1 Laboratory of Plant Breeding and Biometry, Faculty of Crop Science, Agricultural University of Athens, 11855 Athens, Greece; richasaikia2195@gmail.com (R.S.); akaldis2003@hotmail.com (A.K.)

2 Department of Agricultural Biotechnology, Assam Agricultural University, Jorhat 785013, Assam, India; basantabora@gmail.com

3 Division of Biotechnology and Plant Health, Norwegian Institute of Bioeconomy Research, 1433 Ås, Norway; carl.spetz@nibio.no

* Correspondence: avoloud@aua.gr

Citation: Saikia, R.; Kaldis, A.; Spetz, C.J.; Borah, B.K.; Voloudakis, A. Silencing of Putative Plasmodesmata-Associated Genes *PDLP* and *SRC2* Reveals Their Differential Involvement during Plant Infection with Cucumber Mosaic Virus. *Plants* **2025**, *14*, 495. https://doi.org/10.3390/plants14030495

Academic Editors: Violetta Katarzyna Macioszek, Iwona Ciereszko and Andrzej K. Kononowicz

Received: 30 October 2024
Revised: 3 February 2025
Accepted: 4 February 2025
Published: 6 February 2025

Citation: Saikia, R.; Kaldis, A.; Spetz, C.J.; Borah, B.K.; Voloudakis, A. Silencing of Putative Plasmodesmata-Associated Genes *PDLP* and *SRC2* Reveals Their Differential Involvement during Plant Infection with Cucumber Mosaic Virus. *Plants* **2025**, *14*, 495. https://doi.org/10.3390/plants14030495

Abstract: Plant viruses utilize a subset of host plasmodesmata-associated proteins to establish infection in plants. In the present study, we aimed to understand the role of two plant genes, one encoding a putative plasmodesma located protein (PDLP) and a homolog of soybean gene regulated by cold 2 protein (SRC2) during Cucumber mosaic virus (CMV) infection. Virus-induced gene silencing (VIGS) was used to silence *PDLP* and *SRC2* genes in *Nicotiana benthamiana* and in two related solanaceous plants, *N. tabacum* and *Capsicum chinense* Jacq. (Bhut Jolokia). Up to 50% downregulation in the expression of the *PDLP* gene using the TRV2-*PDLP* VIGS construct was observed in *N. benthamiana* and *N. tabacum* while, using the same gene construct, 30% downregulation of the target mRNA was observed in *C. chinense*. Similarly, using the TRV2-*SRC2* VIGS construct, a 60% downregulation of the *SRC2* mRNA was observed in *N. benthamiana*, *N. tabacum*, and a 40% downregulation in *C. chinense* as confirmed by qRT-PCR analysis. Downregulation of the *PDLP* gene in *N. benthamiana* resulted in delayed symptom appearance up to 7–12 days post inoculation with reduced CMV accumulation compared to the control plants expressing TRV2-*eGFP*. In contrast, *SRC2*-silenced plants showed enhanced susceptibility to CMV infection compared to the control plants. Our data suggest that the *PDLP* gene might facilitate infection of CMV, thus being a susceptibility factor, while the *SRC2* gene could play a role in resistance to CMV infection in *N. benthamiana*.

Keywords: tobacco rattle virus (TRV)-based gene silencing; plant–virus interaction; disease resistance; resistance and susceptibility gene; solanaceous plants

1. Introduction

To establish an infection, most plant viruses utilize specialized membrane-linked cytoplasmic channels known as plasmodesmata (PD) for intercellular trafficking of virus particles in the host [1]. They encode at least one dedicated protein termed movement protein (MP) that can interact with host proteins and modify the structure of PD [2,3]. Viral MPs can form tubules that assemble in PD and thereby manipulate the size exclusion limit of PD for the transport of viral complexes to the adjacent cells [4]. Viruses encoding such tubule-forming MPs belong to six major families: Badnaviridae, Bromoviridae, Caulimoviridae, Ilaviridae, Secoviridae, and Tospoviridae [5,6]. To date, only a few of the host proteins localized to PD have been shown to interact with viral MPs. These include proteins such as calreticulin [7], protein kinase PAPK1 [8], remorin [9], endoplasmic reticulum (ER)-embedded proteins including reticulons [10,11], plasmodesmata-associated class 1 reversibly glycosylated polypeptide [12], ankyrin-repeat containing protein (ANK) [13], synaptotagmin [14], and other plasmodesmata-located proteins (PDLPs) [5,15]. It is suggested that ANK together with β-glucanases could regulate the size exclusion limit of plasmodesmata [13]. ER-embedded proteins have been shown as positive regulators for viral movement and spread of cucumber mosaic virus (CMV) [10]. PDLPs have been shown as positive regulators for viral movement and spread of grapevine fan leaf virus (GFLV) and cauliflower mosaic virus (CaMV) [5]. Another viral protein of CaMV, named P6, colocalized with host PDLP1 and CaMV MP in the PD [16]. The study also showed that P6 physically interacts with a C2 calcium-dependent membrane-targeting protein, SRC2 (Soybean Response to Cold) from Arabidopsis, and this protein is capable of a close association with the tubule formation in the PD. Although the expression of pepper *CaSRC2-1* gene was shown to be upregulated during bacterial and viral infection in chilli (*Capsicum annum*) [17], little is known about the exact role of *SRC2* and *PDLP* genes during plant viral infection, particularly for broad host range viruses like CMV. CMV infects more than 1000 plant species [18,19]. These include some of the most valuable horticultural crops such as chilli, cucumber, okra, squash, and tomato [20]. CMV has a linear tripartite genome of positive-sense single-stranded (ss) RNA (RNA1/RNA2/RNA3) and belongs to the family Bromoviridae [21]. It is also known that the MP of CMV localizes to the PD like other viruses belonging to Bromoviridae family [22]; however, the exact role of host proteins during CMV infection is not well defined as of yet.

RNA interference (RNAi) or post-transcriptional gene silencing (PTGS) is a natural surveillance system adopted by plants that can regulate levels of endogenous or foreign transcripts by degrading them in a sequence-specific manner. The PTGS system becomes activated in response to the formation of double-stranded RNA (dsRNA), which involves sequence-specific recognition of target mRNA molecules based on sequence homology and subsequent degradation of these molecules by host ribonucleases and other proteins in the RNA induced silencing complex (RISC), thereby inhibiting their translation process [23]. The RNAi mechanism can be evoked using recombinant viral vectors that can efficiently deliver silencing inducer molecules to plant cell and study target gene function. This reverse genetic approach of studying plant genes using viral vectors is called virus-induced gene silencing (VIGS). Some of the recombinant viral vectors are derived from the genome of Potato virus X (PVX), Tobacco mosaic virus (TMV), and Tobacco rattle virus (TRV), of which TRV- and PVX- have been widely utilized [24,25]. The TRV-VIGS has been used to characterize genes in many plant species related to plant developmental processes, biotic and abiotic stress tolerance, symbiosis, metabolite synthesis, and plant evolution [25–28]. TRV-VIGS was first reported in *N. benthamiana* for silencing an endogenous gene, namely *phytoene desaturase* (*PDS*), and this vector system was further modified [29–31]. Since then, several disease resistance and susceptibility genes were identified against viral infection in plant species using the TRV-VIGS system [32–34].

In the present study, the role of two genes, one encoding putative plasmodesmata-localized protein PDLP and a protein homolog of SRC2 from *Nicotiana benthamiana* and *N. tabacum* were studied using the TRV-based VIGS system for their involvement in plant defense against or susceptibility to CMV.

2. Results

2.1. Homology Analysis of PDLP and SRC2 Genes in the Nicotiana and Capsicum Species

Using BLASTn and phylogenetic analysis with known *PDLP* genes from *Arabidopsis thaliana*, we selected a homolog of *PDLP* gene (XM_016610375.1) from *N. tabacum* which was identified in *N. benthamiana* (Niben101Scf03456g00018.1, 95.16% sequence identity with 99% query cover), *C. annum* (XM_016721975.2, 80.26% sequence identity with query cover 99%) and *C. chinensis* (MCIT02000006.1, 75.26% sequence identity with query cover 100%) (Figure 1A, Supplementary Figure S1). Similarly, a homolog of the *SRC2* gene (XM_016630487.1) from *N. tabacum* was selected which was identified in *N. benthamiana* (DQ465395.1, 93.26% sequence identity with 99% query cover), *C. annum* (NM_001324570.1, 76.98% sequence identity with 100% query cover), and *C. chinensis* (AB442165.1, 76.68% sequence identity with query cover 94%) (Figure 1B, Supplementary Figure S2) by using BLASTn. The homologous sequences of *PDLP* and *SRC2* were used to design primers for PCR analyses (RT-PCR and RT-qPCR) and for construction of the VIGS clones.

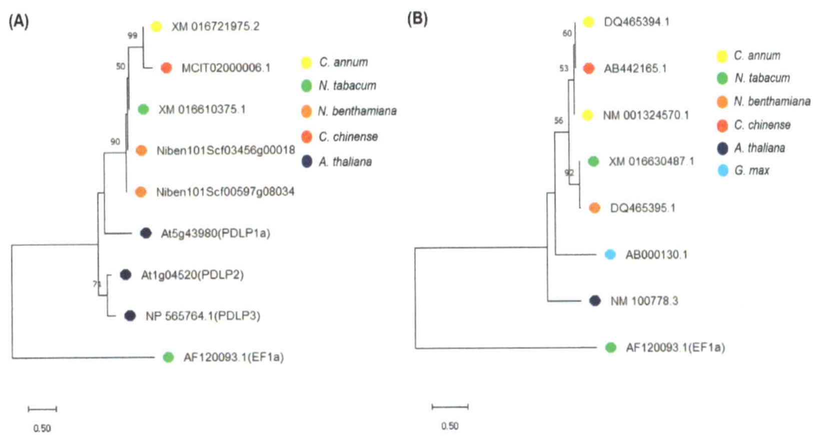

Figure 1. Phylogenetic analysis of (**A**) *PDLP*- and (**B**) *SRC2*-related genes from related plant species using the neighbor-joining (NJ) method. Each branch point is a representation of 10,000 bootstrap replicates. Bootstrap value higher than 50 is shown here. *Nicotiana tabacum* elongation factor 1-alpha gene sequence was used as an outgroup.

2.2. Relative Expression of PDLP and SRC2 Genes During CMV Infection in N. benthamiana

To investigate the expression pattern of *PDLP* and *SRC2* genes, *N. benthamiana* leaves were inoculated with CMV, phenotyped, and sampled at 2, 8, and 15 days post inoculation (dpi). Typical downward leaf curling and mosaic symptoms were observed at 8 dpi, but the symptoms were more prominent at 14 dpi (Figure 2). Using semi-quantitative RT-PCR, the CMV *CP* was detected only at 8 and 15 dpi in the inoculated leaves, but not in the 2 dpi leaf samples (Figure 3).

No CMV-specific band was detected in the systemic leaves of the plants treated with the phosphate buffer (mock) (Figure 3). The expression levels of the *PDLP* and *SRC2* genes were observed, and it was found that the *PDLP* gene expression was nearly the same between the mock and CMV inoculated plants, suggesting that the gene was constitutively expressed irrespective of the CMV infection. However, the expression of the *SRC2* gene was upregulated upon CMV inoculation, as observed at 8 and 15 dpi compared to the mock (Figure 3). Therefore, to understand the functional role of these two genes during viral infection, we decided to silence the genes in *N. benthamiana* employing a TRV-based VIGS strategy.

Figure 2. Phenotypic observation of *Nicotiana benthamiana* plants. (**A**) Cucumber mosaic virus (CMV)-infected plant and (**B**) mock (phosphate buffer)-treated plant at 14 days post treatment.

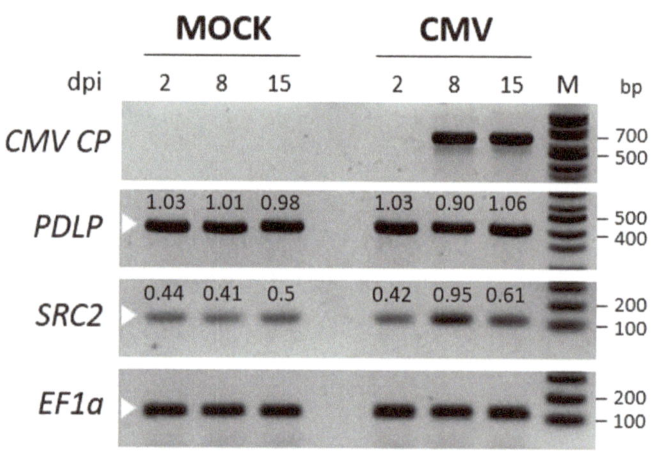

Figure 3. Time course expression profiles of *PDLP* and *SRC2* genes upon cucumber mosaic virus (CMV) infection in *Nicotiana benthamiana*. A semi-quantitative RT-PCR analysis is shown. A fragment of CMV coat protein gene (*CP*, 657 bp) and *N. benthamiana PDLP* (413 bp) and *SRC2* (123 bp) were amplified from cDNA obtained from systemic leaves at 2, 8, and 15 days post inoculation with CMV or phosphate buffer (mock). PCR was performed for 30 cycles and three biological replicates were pooled for each experimental group. The *elongation factor 1 alpha* (*EF1a*) gene fragment (116 bp) from *N. benthamiana* was used as an internal control. White color arrows indicate amplified target gene fragments. M indicates 100 bp ladder (New England BioLabs, NEB, Ipswich, MA, USA). The values above the bands represent the ratio of band intensity of normalized *PDLP* and *SRC2* genes in mock and CMV-infected *N. benthamiana* plants.

2.3. *VIGS of PDLP and SRC2 Genes in Tobacco and Bhut Jolokia Pepper*

2.3.1. Designing of VIGS Constructs and Silencing of *PDLP* and *SRC2* Genes in *N. benthamiana*, *N. tabacum*, and *C. chinense*

A conserved region of *PDLP* (383 bp) and *SRC2* (364 bp) genes from tobacco, *N. benthamiana*, *C. annum*, and *C. chinense* was identified using multiple sequence alignment by Clustal Omega (Figures S1 and S2). The conserved regions of *PDLP* and *SRC2* gene sequences were amplified using PCR with the respective primer pairs (Table 1) and cloned into the pTRV2 plasmid vector. Colony PCR of the transformed *E. coli* cells revealed positive bacterial colonies harboring the cloned constructs, pTRV2-*NtPDLP* and pTRV2-*NtSRC2*. Sanger sequencing of at least three positive clones from each gene fragment confirmed the presence of *NtPDLP* and *NtSRC2* gene fragments in pTRV2 in sense orientation. Alignment of the sequenced *PDLP* and *SRC2* gene fragments with *N. tabacum PDLP* (XM_016610375.1) and *SRC2* (XM_016630487.1) revealed 100% nucleotide identity.

Transient expression of *PDLP* and *SRC2* gene fragments cloned in the pTRV2 expression vector was carried out through agroinfiltration in *N. benthamiana*, *N. tabacum*, and *C. chinense* leaves (Supplementary Figure S3). The phytoene desaturase gene (*PDS*) from *N. benthamiana* cloned in pTRV2, which results in photobleaching effect upon gene silencing, was used as a positive control (Figure S4). The silencing experiment for *PDLP* and the *SRC2* gene in *N. benthamiana* was repeated two times, while the silencing experiment for these genes in *N. tabacum* and *C. chinense* plants was performed once. The silencing efficiency of the target genes (*PDLP* and *SRC2*) was evaluated using RT-qPCR and compared between the silenced and the control plants. In the *PDLP*-silenced group, the expression level of the *PDLP* gene was 0.4- to 0.6-fold lesser than the GFP control with the highest downregulation at 9 dpa (Figure 4). Similarly, the expression of the *SRC2* gene in the *SRC2*-silenced group was 0.3- to 0.5-fold lesser than the GFP control with highest downregulation observed at 21 dpa (Figure 4).

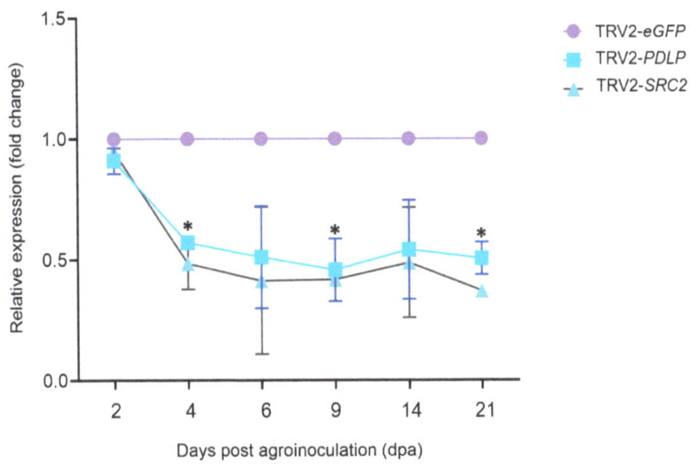

Figure 4. Time course expression analysis of *NtPDLP* and *NtSRC2* genes in silenced *Nicotiana benthamiana* plants. Relative quantification of *NtPDLP* and *NtSRC2* gene expression in silenced and control (TRV2-*eGFP*) plants at different time points. In addition, 50% and 60% percent downregulations of *PDLP* and *SRC2* genes were observed in *N. benthamiana* 14 days post agroinoculation. Purple color indicates control-treated (eGFP) group, blue color indicates pTRV2-*NtPDLP*-treated and green color indicates pTRV2-*NtSRC2*-treated plants. Asterisk (*) indicates significant differences between the control and silenced plants, p-value < 0.05 (Student's t-test); dpa = days post agroinoculation.

The silencing of *PDLP* and *SRC2* genes in *N. tabacum* and *C. chinense* Bhut Jolokia plants was analyzed at 14 dpa (Figure 5). Using the same VIGS constructs, a considerable level of *PDLP* and *SRC2* gene silencing was also observed in all species tested, despite having sequence polymorphisms among these species. The expression of the *PDLP* gene in *N. tabacum* and *C. chinense* was 0.5-fold and 0.7-fold lesser than the GFP control (a reduction in gene expression by 50% and 30%, respectively) (Figure 5). Similarly, in both *N. tabacum* and *C. chinense* Jacq., the expression of the *SRC2* gene was downregulated to approximately 0.4-fold and 0.6-fold, respectively, in comparison to the GFP control (downregulation up to 60% and 40% in *N. tabacum* and *C. chinense* Jacq., respectively) (Figure 5).

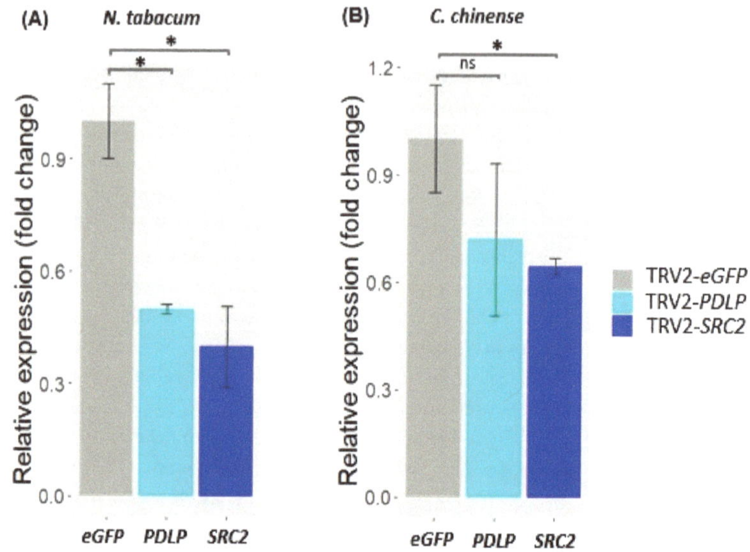

Figure 5. Expression analysis of *PDLP* and *SRC2* genes in silenced *Nicotiana tabacum* and *Capsicum chinense* plants at 14 days post agroinoculation (dpa). eGFP indicates the control-treated plants. Fifty and thirty percent downregulation of *PDLP* gene was observed in *N. tabacum* and *C. chinense* (Bhut Jolokia), respectively, whereas sixty and forty percent downregulation of *SRC2* gene was observed in *N. tabacum* and *C. chinense*, respectively. Student's *t*-test was performed for mean comparisons. Asterisk (*) indicates significant difference in gene expression level between *PDLP*-, *SRC2*-silenced plants, and control (*eGFP*) plants ($p < 0.05$, Student's *t*-test). ns: non-significant.

2.3.2. Disease Development in *PDLP*- and *SRC2*-Silenced *N. benthamiana* Plants Upon CMV Infection

The *PDLP*- and *SRC2*-silenced *N. benthamiana* plants were visually observed upon CMV infection and disease symptoms were recorded at different time points, i.e., 7, 12, and 19 dpi (Figure 6 and Supplementary Figure S5). In the *eGFP* control plants, the typical mosaic symptom of CMV was observed at 7 dpi and the symptoms gradually intensified with severe downward leaf curling at 19 and 21 dpi. In contrast, the *PDLP*-silenced plants displayed reduced viral symptoms in comparison to the *eGFP* control. In the *SRC2*-silenced plants, leaf curling similar to or even greater than the *eGFP* control plants was observed at 19 and 21 dpi, indicating that silencing of *SRC2* might have promoted CMV infection in *N. benthamiana* (Figure 6 and Supplementary Figure S5).

2.3.3. Relative Quantification of CMV Titer in *PDLP*- and *SRC2*-Silenced *N. benthamiana* Plants

Systemic (uninoculated) leaves were collected at different time points upon CMV infection, and the viral titer was evaluated using RT-qPCR analyses. Significant reduction in the CMV titer was observed at 7 and 12 dpi but not at 19 dpi in the *PDLP*-silenced plants in comparison to the control plants (i.e., eGFP) by RT-qPCR analysis (Figure 7). Similarly, it was found that the level of CMV was significantly higher in the *SRC2*-silenced plants when compared to the eGFP control plants at 12 and 19 dpi by using RT-qPCR (Figure 7).

Figure 6. Phenotypic observations of VIGS-silenced *Nicotiana benthamiana* plants at different time-points after cucumber mosaic virus (CMV) inoculation. Plants were first agroinfiltrated with VIGS constructs, followed by sap inoculation with CMV two days later to ensure the initiation of gene silencing. In the control (*eGFP*) plants, initial symptoms of CMV infection were observed at 7 days post infection (dpi). Subsequently, symptoms such as downward leaf curling and mosaic symptoms appeared more prominently at 12 and 19 dpi. The *PDLP*-silenced *N. benthamiana* plants showed reduced viral symptoms in comparison to the control and *SRC2*-silenced plants. Red arrows indicate typical disease symptoms. dpa = days post agroinoculation.

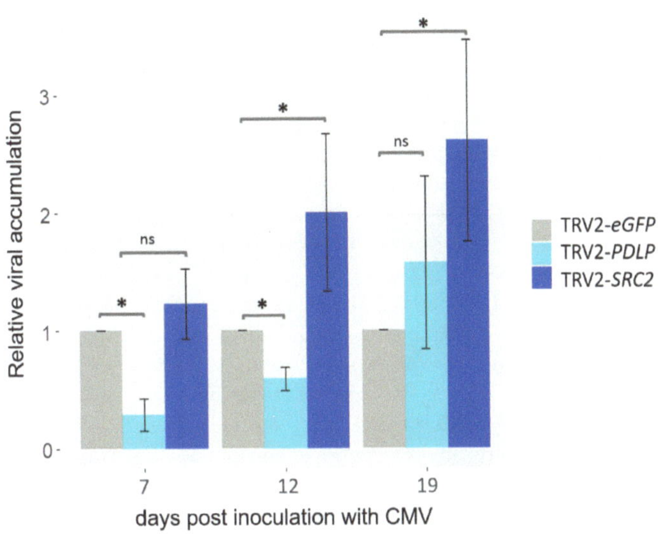

Figure 7. Relative abundance (RT-qPCR) of cucumber mosaic virus (CMV) titer in *PDLP*- and *SRC2*-silenced *Nicotiana benthamiana* plants at 7, 12, and 19 days post inoculation (dpi) with CMV. Plants agroinoculated with GV3101 strain carrying pTRV2-eGFP plasmids were used as control. The *elongation factor 1 alpha* (*EF1a*) gene fragment from *N. benthamiana* was used for internal control and normalization purposes. Student's t-test was performed for mean comparisons. Asterisk (*) indicates significant difference in relative viral titer between *PDLP*-, *SRC2*-silenced plants, and control (*eGFP*) plants ($p < 0.05$, Student's t-test). ns: non-significant.

2.3.4. Callose Deposition in the *PDLP*- and *SRC2*-Silenced *N. benthamiana* Plants Upon CMV Infection

To strengthen the RT-PCR and RT-qPCR results, a callose deposition assay was conducted. Callose deposits initiated at 7 dpi and progressively increased, with higher callose deposits observed at 19 dpi in the eGFP control plants after CMV infection. In contrast, lesser callose deposits were observed in the *PDLP*-silenced plants at 7, 12, and 19 dpi. In the *SRC2*-silenced plants, callose deposition exhibited a near-equivalent level to that of the control group at 7, 12, and 19 dpi (Supplementary Figure S6).

2.3.5. Accumulation of Reactive Oxygen Species (ROS) in the *PDLP*- and *SRC2*-Silenced *N. benthamiana* Plants Upon CMV Infection

The level of hydrogen peroxide (H_2O_2), a ROS molecule, was assessed in the *PDLP*- and *SRC2*-silenced plants following CMV infection. Remarkably, elevated levels of H_2O_2 were detected in systemic leaves of the control and *SRC2*-silenced groups at 7, 12, and 19 dpi (Supplementary Figure S7). In contrast, the concentrations of H_2O_2 in *PDLP*-silenced plants were notably lower in comparison to the control and *SRC2*-silenced groups upon CMV infection at 7, 12, and 19 dpi (Supplementary Figure S7).

3. Discussion

It is known that plasmodesmata-associated proteins and plasmodesmata (PD) are crucial for cell-to-cell movement of plant viruses [35]. *PDLP*s and *SRC2* are plasmodesmata-associated genes [5,16,35]. Some viruses recruit movement proteins that localize to the PD and interact with the PDLP protein to form a movement tubule facilitating, as a result, the transport of viral particles allowing systemic spread of the virus in the plant. The movement protein of CMV also localizes in the plasmodesmata [22]; however, the role of PDLPs in viral disease development is not well studied. The SRC2 protein is also a PD-associated membrane protein that colocalizes with *Arabidopsis* AtPDLP and interacts with the movement protein of cauliflower mosaic virus (CaMV) *in planta* [16]. The expression of pepper *CaSRC2-1* gene was upregulated during bacterial and viral infection in *Capsicum annum* [17]. Little is known about the expression of the *SRC2* gene of *Nicotiana* spp. during infection with CMV. Thus, these two genes, *PDLP* and *SRC2*, are promising candidates for promoting disease susceptibility or resistance to CMV in tobacco; therefore, they were functionally characterized in our study.

The expression of these two tobacco genes was transiently silenced utilizing a TRV-based VIGS strategy. The expression of *PDLP* upon CMV infection in *N. benthamiana* plants was consistent between mock and CMV inoculated samples, suggesting that the expression of *PDLP* is not significantly affected by CMV infection. On the other hand, the expression of *SRC2* was slightly upregulated at 8 dpi upon CMV infection, corresponding with the initiation of CMV expression at the same time point (Figure 3). Similar observations were made in pepper (*C. annum*) and soybean (*Glycine max*) [17,36]. Takahashi and Shimosaka (1997) found that *SRC2* expression in soybean was upregulated in response to cold stress and mechanical wounding [36]. Additionally, Kim et al. (2008) observed an induction of *SRC2* gene expression in pepper during bacterial and viral infection [17].

Through TRV-based VIGS, an average of about 80% downregulation can be achieved in the endogenous transcript levels of the targeted gene compared to vector control plants [25]. The initiation of silencing generally starts within 2–3 weeks after agroinfiltration [25]. The efficiency of silencing may vary due to various factors such as the VIGS-vector infection, multiplication, and systemic spread, the level of viral-vector titer in the host, the homology between the target gene fragment in the viral-vector and the endogenous target gene mRNA, and the environmental conditions during plant growth [25]. The present study revealed significant downregulation in both *PDLP* and *SRC2* at 4, 9, and 21 dpa (Figure 4). RT-qPCR analysis revealed a significant downregulation of the *PDLP* gene by 50% in *N. tabacum*, *N. benthamiana*, and 30% in *C. chinense* (Figures 4 and 5), along with the downregulation of the *SRC2* gene by 60% in *N. tabacum*, *N. benthamiana*, and 40% in *C. chinense* (Figures 3 and 4). Similar observations were reported in the knockdown lines of *PDLP-1*, *PDLP-2*, and *PDLP*-3 triple mutants in *Arabidopsis thaliana* which was resistant to powdery mildew infection [37]. Kim et al. (2008) proposed that *SRC2* from pepper (*C. annuum*) plays an essential role in inducing resistance to bacterial and viral pathogens by eliciting the hypersensitive response (HR) to restrict pathogen growth [17]. Liu et al. (2016) provided evidence that SRC2 recognizes INF1, an elicitor protein of *Phytophthora capsici*, and elicits HR in pepper [38]. They showed that silencing of the *SRC2* gene increased susceptibility of pepper plants to *P. capsici* infection. In addition, the silencing experiment did not reveal any developmental changes such as alteration in plant height in the *PDLP*- and *SRC2*-silenced plants when compared to the control plants (TRV2-*eGFP*). It is noteworthy that this is the first report of a TRV-based silencing of plant endogenous genes in *C. chinense* Bhut Jolokia. No developmental alterations were observed in *N. benthamiana* and *C. chinense* in any VIGS experiment, except for *N. tabacum* where development changes such as growth inhibition and minor leaf crumbling were observed at 14 dpa (Supplementary Figure S4). Photobleaching, indicative of *PDS* gene silencing, started at 9 dpa and became even more prominent at 14 and 21 dpa (Supplementary Figure S4), suggesting that silencing effect commenced approximately 1–2 weeks post agroinfiltration. This experiment allowed us to determine the silencing effect of plant endogenous genes with the TRV-based VIGS strategy, which was demonstrated within 14 dpa.

The current investigation unveiled that silencing of the *PDLP* in *N. benthamiana* resulted in a notable reduction in the accumulation of the CMV titer in the systemic leaves. The *PDLP* gene family comprises a set of multiple genes, specifically *PDLP1* to *PDLP8* in Arabidopsis, with predicted molecular weights ranging from 30 to 35 kDa [39]. The downregulation of the *PDLP* gene resulted in delayed symptoms of CMV infection up to 12 dpi but not until 19 dpi, suggesting that targeting multiple *PDLP* genes, if present, of *N. benthamiana* could potentially lead to further delay in viral symptom development. Another possibility is that CMV might exploit other plasmodesmata proteins to facilitate transport at later stages of systemic infection. Therefore, it is important to understand the role of other PDLPs in *N. benthamiana*. It can be speculated that the reduced H_2O_2 accumulation observed in the *PDLP*-silenced plants (Figure S7) is likely a result of decreased CMV accumulation compared to the *TRV2-eGFP*-treated plants. It is known that in a CMV–plant compatible interaction, there is an increase in H_2O_2 accumulation [40] that agrees with our observations. The reduced leaf mottling and mosaic symptoms in the *PDLP*-silenced plants correlate nicely with the reduced CMV titer and ROS accumulation (Figure 7 and Figure S7). Plants produce callose deposits in the cell periphery as a defense mechanism in response to pathogen invasion. Previous studies have demonstrated that overexpression of *PDLP1* and *PDLP5* enhances callose deposition at PD [37,41], which is consistent with the low callose deposition in our *PDLP*-silenced plants. Lim et al. (2016) demonstrated that the overexpression of *PDLP5* resulted in a reduction in the trafficking of azelaic acid (AzA) and glycerol-3-phosphate (G3P) through PD [42]. This impairment of PD trafficking was associated with a significant decline in Systemic Acquired Resistance (SAR) in plants. Similarly, Carella et al. (2015) observed that overexpressing of either *PDLP1* or *PDLP5* in Arabidopsis plants led to the loss of systemic movement of the crucial SAR protein, DEFECTIVE IN INDUCED RESISTANCE1 (DIR1), resulting in a compromised SAR [43]. The overexpression of *PDLPs* in *A. thaliana* correlates with an enhanced systemic spread of pathogens suggesting their role in promoting pathogen spread. We could hypothesize that *PDLP* silencing would augment SAR for the benefit of the plant and having negative effect for CMV. In addition, PDLPs are known to be involved in viral movement [5,6,15,16]. For example, for tubule-guided viral movement through plasmodesmata, the viral movement proteins (MPs) of GFLV and CaMV in Arabidopsis [5] and cowpea mosaic virus in *N. benthamiana* [15] interact with the host PDLPs. The interaction of viral MPs with host PDLPs results in the formation of movement tubules that allow cell-to-cell movement of plant viruses across adjacent cells [5]. CMV-MP also localizes in the plasmodesmata as it is a member of the '30K MP superfamily' group [22]. This finding is consistent with the results obtained from our *PDLP*-silenced *N. benthamiana* plants, where a reduction in *PDLP* expression led to decreased CMV accumulation. While the exact function of PDLPs in virus movement is still under investigation, future research can be prioritized to understand the structural interactions between tubule-forming MPs (e.g., CMV) and PDLPs. Overall, these findings suggest that *PDLP* silencing of *N. benthamiana* led to decreased CMV accumulation in systemic tissue, suggesting that PDLPs are possibly involved in facilitating the systemic spread of CMV and as a result increase the viral titer, so they should be considered as susceptibility factors. Generating knockout/knockdown lines of *PDLP* genes in Bhut Jolokia and *N. benthamiana* using the CRISPR/Cas9 system could aid in developing virus-resistant plants against CMV.

In contrast, silencing of *SRC2* resulted in increased susceptibility to CMV infection (Figure 7), indicating that SRC2 might be involved in the resistance to CMV in *N. benthamiana*. Kim et al. 2008 [17] reported that during both bacterial and viral infections in *C. annum*, the pepper encoding *CaSRC2*-1 (a single C2 domain-containing protein) gene was upregulated. Limited information exists regarding the expression patterns of the *SRC2* gene in *Nicotiana* spp. during the infection with CMV (our data indicate a slight *SRC2* upregulation upon CMV infection [Figure 3]). A recent study found that the silencing of *GmSRC2* in soya resulted in a significant enhancement of disease symptoms and an increase in the biomass of *Phytophthora sojae* [44]. A structural characteristic identified in AtSRC2.2 is its C2 domain, which serves as a lipid-binding domain found in numerous eukaryotic proteins [16]. The C2 domain, first identified in Ca^{2+}-dependent isoforms of protein kinase C, is integral to cellular signalling in plants and animals [17]. Plant proteins harboring C2 domains have been associated with facilitating the intercellular spread (cell-to-cell movement) of plant viruses [45]. An interesting characteristic of AtSRC2.2 is that its C2 domain shares 36% similarity with AtSYT4, a member of the synaptotagmin family, possessing two C2 domains (C_2A and C_2B) [16]. Not much is known about the involvement of AtSYT4 in virus movement, but there is another synaptotagmin, AtSYTA, which is involved in the movement of several viruses, such as cabbage leaf curl virus, tobacco mosaic virus, and squash leaf curl virus by interacting with their respective MPs [45]. It is noteworthy that AtSYTA is also associated with regulating freezing tolerance [46], a trait shared among SRC2 proteins. Several studies have suggested that the entry of Ca^{2+} is a key factor in initiating pattern-triggered immunity (PTI) in plant cells [44]. In the *SRC2*-silenced plants, a higher H_2O_2 accumulation was observed (Supplementary Figure S7) that could be the consequence of higher CMV load (Figure 7) in the compatible interaction with *N. benthamiana* agreeing with [40]. Therefore, further investigation is needed to understand the role of *SRC2* in plant defense response. The notion that enhancing resistance to CMV in plants could be achieved by overexpressing the *SRC2* gene is now a hypothesis to test.

4. Materials and Methods

4.1. Plant Materials and Growth Conditions

Nicotiana benthamiana, *N. tabacum*, and *Capsicum chinense* Jacq. (Bhut Jolokia) plants were grown in a greenhouse operating at 20 ± 3 °C and 70% humidity with a 16 h/8 h day/night cycle. Three- to four-week-old plants were selected for performing the VIGS experiment, and they were divided into four groups—TRV1 + TRV2-*NtPDLP*, TRV1 + TRV2-*NtSRC2*, TRV1 + TRV2-*eGFP* (negative control), and TRV1 + TRV2-*NbPDS* (positive control for silencing).

4.2. Development of TRV-Based Recombinant Constructs for VIGS

The TRV-based VIGS system, developed by Liu et al., 2002 [31], was used to generate the recombinant VIGS constructs for silencing of putative *PDLP* and *SRC2* genes in Solanaceous plants including *Nicotiana benthamiana*, *N. tabacum*, and *C. chinense*. TRV is a bipartite ss (+) RNA virus comprising two genomes, RNA1 and RNA2. Both TRV-RNA1 and TRV-RNA2 genomes were previously cloned in the binary plasmid vectors pBIN19 and pCambia0390, respectively [31]. The pTRV-RNA2 plasmid vector was digested by the *Bam*HI-HF restriction enzyme (New England Biolabs, Ipswich, MA, USA) to obtain a linearized plasmid DNA. The digested product was subjected to gel electrophoresis for confirmation along with the quantification of the digested pTRV2 and was stored at −20 °C for further use.

For the integration of the *PDLP* and *SRC2* gene fragments, the In-Fusion® HD seamless cloning system (Takara Bio Inc., Shiga, Japan) was employed. The nucleotide sequences corresponding to the putative *PDLP* (XM_016610375.1) and *SRC2* (XM_016630487.1) genes of *N. tabacum* were employed as queries in the VIGS tool provided by the Sol Genomics Network (SGN) to identify the optimal fragment for gene silencing purposes. Primers were designed for the amplification of the fragments of *PDLP* and *SRC2* gene using the primer3 tool (http://primer3.ut.ee/) (accessed on 15 February 2022). The designed primer had a 15-base extension at the 5′ end homologous to the linearized pTRV2 vector (Table 1). The amplification of *PDLP* and *SRC2* gene fragments was performed by RT-PCR utilizing RNA from *N. tabacum* leaves as a template. The PCR-amplified product underwent confirmation and quantification through 1.5% agarose gel electrophoresis and was subsequently cloned into the linearized pTRV2 vector as per the manufacturer's instructions (In-Fusion® HD cloning kit, Clontech, Mountain View, CA, USA). The recombinant plasmid constructs (pTRV2-*NtPDLP* and pTRV2-*NtSRC2*) were transformed into *E. coli* Stellar Competent Cells (Takara Bio Inc., Shiga, Japan), followed by a heat shock at 42 °C for 30 s. The bacterial cultures were spread over Luria–Bertani (LB) plates containing kanamycin (50 μg/mL) and were subjected to overnight incubation at 37 °C for colony development. Colony PCR was carried out to confirm the presence of recombinant plasmids in the transformed bacterial colonies, using specific primers (TRV2-1530F and TRV2-1809R) that bind on the TRV2 plasmid vector (Table 1). Two positive transformants for each of the recombinant plasmids were selected for plasmid isolation using NucleoSpin® Plasmid Kit (Machery-Nagel, Düren, Germany) and stored for further use. The confirmation of the successful integration of the *NtPDLP* and *NtSRC2* gene fragment into the pTRV2 vector was validated through Sanger sequencing (Azenta Life Sciences UK, Essex, UK).

Table 1. List of primers used in this study.

Name of Primer	Sequence (5′ to 3′) *	Product Size (bp)	Target Species	Function
Nt_PDLP-INF-378F Nt_PDLP-INF-760R	GCCTCCATGGGGATCGTGTACAAAGGCTGTGCTAA GCTCGGTACCGGATCGTCTCAAGCCTATCACTAAAC	413	*N. tabacum* *N. benthamiana*	Cloning of *PDLP* gene fragment to pTRV2
Nt_SRC2-INF-125F Nt_SRC2-INF-488R	GCCTCCATGGGGATCCATTAGATATCAAAGTTATTGC GCTCGGTACCGGATCGGTTTTCCGCTCTCTGTTAT	394	*N. tabacum* *N. benthamiana*	Cloning of *SRC2* gene fragment to pTRV2
eGFP-INF-286F eGFP-INF-646R	GCCTCCATGGGGATCGAGCGCACCATCTTCTTCAA GCTCGGTACCGGATCGCTTCT CGTTGGGGTCTTTG	391	-	Cloning of *eGFP* gene fragment to pTRV2
TRV2-1530F TRV2-1809R	GTTTTTATGTTCAGGCGGTTC TCAAGATCAGTCGAGAAT GTCA	280	*E. coli* *A. tumefaciens* agroinfiltrated plant species	Colony PCR, Detection of TRV2
TRV1-257F TRV1-491R	GCTGAGCAGAGGAGTCATTTC ACCCATGAACCATGTTTTTGT	235	*E. coli* *A. tumefaciens* agroinfiltrated plant species	Detection of TRV1
Nb_PDLP-RT-F Nb_PDLP-RT-R	GGGACTGTGTGAACTGTGTG ACTCCTAACCCCACACCAAC	210	*N. tabacum* *N. benthamiana*	Gene expression analysis for *PDLP*
Nb_SRC2-RT-F Nb_SRC2-RT-R	ATCCACCCGTACAACAACCT AGCAGCCATCTCTCCAACAT	164	*N. tabacum* *N. benthamiana*	Gene expression analysis for *SRC2*
CMV_CP_RT-F CMV_CP_RT-R	GAAGCTTGTTTCGCGCATTC TCCGGATGCTGCATACTGAT	175	CMV	Gene expression analysis for CMV *CP*
Nb_EF1a-F Nb_EF1a-R	AGCTTTACCTCCCAAGTCATC AGAACGCCTGTCAATCTTGG	116	*N. tabacum* *N. benthamiana*	Gene expression analysis for *EF1a* [47]

* The region that anneals to the TRV2 vector for the In-Fusion cloning is underlined.

To serve as a negative control in the VIGS assay, we introduced a 361 nt segment from the enhanced GFP (*eGFP*) gene (a gene that does not exist in plants) into the pTRV2 plasmid vector. The insert fragment was derived through PCR amplification using In-Fusion primers designed for this purpose, with a pBIN61-eGFP plasmid vector employed as the template. All subsequent steps for constructing the engineered pTRV2_eGFP plasmid were executed by the aforementioned procedures. The positive control in the VIGS assay involved the targeted silencing of a fragment of *N. benthamiana* phytoene desaturase gene (PDS) exhibiting a photobleaching phenotype. This control was employed to validate the efficiency of the gene silencing system.

4.3. Transformation and Mobilization of VIGS Constructs into Agrobacterium Tumefaciens (GV3101 Strain)

The recombinant plasmids (pTRV2-*NtPDLP* and pTRV2-*NtSRC2*) were transformed into *Agrobacterium tumefaciens* competent cells (GV3101 strain) using the freeze–thaw method with minor modifications [48]. The plasmids were added to the surface of frozen GV3101 competent cells and promptly placed in a heat block at 37 °C for 5 min. Following the heat treatment, the transformed GV3101 competent cells were mixed with LB and subjected to shaking at 28 °C for 4 h. The transformed bacterial cells were spread onto an LB agar plate containing 50 μg/mL kanamycin and 25 μg/mL rifampicin and incubated at 28 °C for 2–4 days for colony development. Positive colonies were further confirmed by colony PCR as mentioned earlier for *E. coli*. Similarly, the pTRV1 binary plasmid vector carrying the TRV-RNA1 genome was introduced into GV3101 competent cells.

4.4. Agrobacterium-Mediated Transient Expression and Virus Induced Gene Silencing

VIGS constructs were delivered to three- to four-week-old *N. benthamiana*, *N. tabacum*, and *C. chinense* plants via agroinfiltration. Primary and secondary cultures of the relevant Agrobacteria were cultured in LB supplemented with kanamycin (50 μg/mL) and rifampicin (25 μg/mL) at 28 °C in a shaker incubator overnight at 220 rpm. The secondary cultures were centrifuged at 3000× *g* for 10 min. The supernatant was discarded, and the resulting cell pellet was resuspended in a resuspension buffer (10 mM 2-(N-Morpholino) ethanesulfonic acid hydrate (MES-hydrate), pH 5.6 adjusted with KOH, 10 mM $MgCl_2$, and 200 μM acetosyringone). The final bacterial OD_{600} was adjusted to 0.5. The agroinfiltration mixtures were obtained by mixing the bacteria harboring TRV1 and TRV2-PDLP/TRV2-SRC2/TRV2-eGFP constructs in a 1:1 (*v*/*v*) ratio. The TRV2-eGFP was used as the negative control, whereas the TRV2-PDS was used as a positive control for silencing. The final mixtures were kept in the dark at room temperature for 3 h before agroinoculation. For the VIGS study, the plants were categorized into four distinct groups (three biological replicates per group), TRV1 + TRV2-*NtPDLP*, TRV1 + TRV2-*NtSRC2*, TRV1 + TRV2-*eGFP* (negative control), and TRV1 + TRV2-*NbPDS* (positive control for silencing). Two matured leaves per plant were agroinoculated using a needleless syringe on the abaxial side of the leaf. Subsequently, the plants were maintained in a growth chamber, during which tissue sampling was performed at different time points (2, 4, 6, 9, 14, and 21 days post agro-inoculation), and were subjected to RNA analysis, CMV inoculation, and disease symptom observation.

4.5. Mechanical Inoculation of CMV in Agroinfiltrated N. benthamiana Plants

Sap inoculation of CMV in plants was performed using carborundum powder (Supplementary Figure S3). The inoculating sap was prepared by collecting sap material from previously CMV-infected plants and gently rubbed on the adaxial side of the leaf using carborundum powder. For CMV inoculation in the silenced plants, leaves were initially agroinfiltrated with the VIGS constructs, and two days later, sap inoculation with CMV was performed to allow sufficient time for silencing to be initiated. Systemic leaf samples were collected at different time points (7, 12, and 19 days post inoculation (dpi) with CMV) and RNA was isolated to assess the viral titer in the silenced plants via quantitative RT-PCR (RT-qPCR).

4.6. PCR and RT-qPCR Analysis

PCR was performed to detect the viral titer of CMV using gene-specific primers (Table 1). The total reaction volume (20 μL) consisted of 2 μL of the 10X PCR Buffer (ThermoFisher Scientific, Waltham, MA, USA), 0.5 μL of 10 mM dNTPs, 0.5 μL of 10 μM of forward and reverse primers each (Table 1), and 0.1 μL of AmpliTaq DNA polymerase (250 U/μL) (ThermoFisher Scientific, Waltham, MA, USA). PCR was carried out with initial denaturation temperature at 95 °C for 5 min followed by 30 cycles consisting of denaturation at 95 °C for 30 s, annealing at 58 °C for 30 s, and extension at 72 °C for 30 s, and then final extension for 7 min at 72 °C.

For the validation of gene silencing, RT-qPCR was performed by extracting RNA from the systemic leaves of the silenced plants at different time points with the Plant/Fungi Total RNA Purification kit (NORGEN Biotek Corporation, Thorold, ON, Canada) according to the manufacturer's instructions. Three biological replicates were pooled for each experimental group. The investigation of pTRV2-*NtPDLP* and pTRV2-*NtSRC2* silencing was performed by sampling tissues at 2, 4, 6, 9, 14, and 21 dpa. Tissue sampling for the detection of CMV *CP* (proxy for CMV titer) was performed at 7, 12, and 19 days post inoculation (dpi) with CMV. The RNA samples were quantified, and their purity was determined spectrophotometrically with NanoDrop™ 2000 Spectrophotometer (ThermoFisher Scientific, USA). cDNA was synthesized using 200 ng total RNA in a 20 μL reaction volume, using the SuperScript VILO cDNA synthesis kit (Invitrogen, Waltham, MA, USA) as per the manufacturer's instructions. RNaseOUT™ Recombinant Ribonuclease Inhibitor safeguards against the degradation of target RNA due to ribonuclease contamination. The 5X VILO™ Reaction mix consists of random primers, $MgCl_2$, and dNTPs in a buffer formulation optimized for RT-qPCR. The RT reaction was performed in a thermocycler (BioRad, Hercules, CA, USA) at 25 °C for 10 min, followed by incubation at 42 °C for 60 min, and terminated at 85 °C for 5 min. The RT products were used as a template for amplifying the *PDLP* and *SRC2 genes* using the designed gene-specific primers (Table 1). PCR amplification for the detection of *NtPDLP, NtSRC2,* and *Elongation factor 1 alpha* (*EF1a*) was performed in a FastGene ULTRA Cycler (NIPPON Genetics EUROPE, Düren, Germany) using the FirePol DNA Polymerase (Solis BioDyne, Tartu, Estonia) and specific primers (Table 1). The total reaction volume consisted of 10X Buffer B (Solis BioDyne, Estonia), 10 mM dNTPs, 10 μM of forward and reverse (Table 1) primers each, 25 mM $MgCl_2$ and 0.2 μL (5 U/μL) *Taq* DNA polymerase (Solis BioDyne, Estonia). PCR for the *PDLP* and *EF1a* genes was carried out with initial denaturation temperature at 95 °C for 5 min followed by 35 cycles consisting of denaturation at 95 °C for 20 s, annealing at 58 °C for 20 s, and extension at 72 °C for 1 min, and then final extension continued for 7 min at 72 °C. PCR for the *SRC2* gene was carried out with initial denaturation temperature at 95 °C for 5 min followed by 35 cycles consisting of denaturation at 95 °C for 20 s, annealing at 58 °C for 15 s and extension at 72 °C for 1 min, and then final extension continued for 7 min at 72 °C. Primer specificity was verified by melting curve analysis. To obtain a semi-quantitative estimation of the time course expression profiles of *PDLP* and *SRC2* gene CMV infection, ImageJ software (v1.54m) (https://imagej.nih.gov/ij/) was employed to measure the band intensity of the gel image. In addition, band intensities were calculated for the housekeeping gene of mock and CMV-infected plants and used as control. The values for the target genes for mock and CMV-infected plants were normalized with the housekeeping gene (*EF1a*).

qPCR was performed in a CFX96^{TM} Touch Real-Time PCR Detection System (Bio-Rad, USA) to quantify the expression levels of the target genes (pTRV2-*PDLP* and pTRV2-*SRC2*) compared to the control group (plants inoculated with pTRV2-*eGFP*) and to check the viral titer (CMV) in the silenced and control plants using SYBR Green qPCR mix (ThermoFisher Scientific, USA). Expression of the target genes was normalized with the housekeeping gene *EF1a* that was found more stable to F-box based on *BestKeeper* housekeeping gene analysis [49], agreeing with [47]. The fold change was calculated using the delta CT ($2^{-\Delta\Delta Ct}$) method [50].

Statistical analyses were performed using Microsoft Office Excel (Microsoft Excel, Redmond, Washington, DC, USA). Student's *t*-test was used for significance analysis between two treatments. The standard error is shown as the standard deviation between the biological replicates.

4.7. Histochemical Analysis

4.7.1. Staining for Callose Deposition in Leaf Tissues

Aniline blue staining facilitates the detection and quantification of cell wall depositions of callose. Callose deposition assay using Aniline blue staining was performed following the procedure described in [51]. A total of three biological replicates were taken for the callose deposition test. Two infected leaf samples were harvested and transferred to 50 mL polypropylene tubes containing a destaining solution (1:3 acetic acid/ethanol). The tubes were incubated overnight with mild agitation. The destained leaves were washed with 150 mM K_2HPO_4 (washing solution) for 30 min. The plant materials were incubated in an aniline blue staining solution (1.3 g of K_2HPO_4 with 5 mg aniline blue in 50 mL H_2O) for at least 2 h in the polypropylene tubes wrapped in aluminium foil for light protection. The samples were embedded in 50% glycerol on a glass slide and the callose deposition was visualized using a fluorescent microscope equipped with a UV lamp. Glycerol ensures a prolonged observation time and reduces bubble formation. The callose deposits appear stained bright blue in color (Supplementary Figure S6).

4.7.2. Staining for H_2O_2 Detection in Leaf Tissues

The 3,3- diaminobenzidine (DAB) assay was performed following the procedure outlined in [52]. The DAB stain is mainly used to visualize the production of H_2O_2 in leaves and roots. For the detection of H_2O_2, 0.1 g of DAB was dissolved in 100 mL of ddH_2O (pH adjusted to 3.8 using HCl). The DAB solution was freshly prepared before starting the experiment. The DAB assay was conducted using three biological replicates for each experimental group. Systemic leaves were harvested and treated with the DAB solution and incubated overnight in the dark with mild agitation (50 rpm). The DAB solution was discarded, leaves were rinsed with ddH_2O to remove excessive staining solution and immersed in 96% ethanol. The samples were then incubated overnight to remove the chlorophyll. Leaves were washed with 75% ethanol and kept in 20% glycerol solution to prevent the leaves from breaking easily and enabling rehydration. Dark brown color precipitates indicated accumulation of H_2O_2. The images of the leaf samples were taken under a light microscope with a 5X objective lens (Supplementary Figure S7).

5. Conclusions

A better understanding of the functional roles of *PDLP* and *SRC2* in the plant's response to CMV infection is provided. This experimental study suggests that *PDLP* and *SRC2* are potential susceptibility and resistance genes during CMV infection in *Nicotiana benthamiana*, respectively. The VIGS method can be applied for screening candidate genes involved in viral resistance or disease susceptibility in *C. chinense* cv Bhut Jolokia and other crop plants.

Supplementary Materials: The following supporting information can be downloaded at: https://www.mdpi.com/article/10.3390/plants14030495/s1, Figure S1: Multiple sequence alignment of *PDLP* genes of *Nicotiana benthamiana* (Niben101Scf03456g00018.1), *Capsicum chinense* (MCIT02000006.1), *C. annum* (XM_016721975.2), and *N. tabacum* (XM_016610375.1). Red-colored lines indicate gene fragment (383 bp) used as VIGS targeted region for silencing of the *PDLP* gene. Sequence alignment was performed using Clustal Omega; Figure S2: Multiple sequence alignment of *SRC2* genes of *Nicotiana benthamiana* (DQ465395.1), *Capsicum chinense* (AB442165.1), *C. annum* (NM_001324570.1) and *N. tabacum* (XM_016630487.1). Red-colored lines indicate gene fragment (364 bp) used as VIGS targeted region for silencing of the *SRC2* gene. Sequence alignment was performed using Clustal Omega; Figure S3: Illustrative depiction of the methodology involved in the agroinfiltration of the VIGS constructs, cucumber mosaic virus (CMV) inoculation, and gene expression analysis in the experimental plants; Figure S4: Phenotypic observations of VIGS-silenced *Nicotiana benthamiana, Nicotiana tabacum, and Capsicum chinense* plants (A) at 21 days post agroinoculation (dpa). (B) Phytoene desaturase *(PDS)* gene was used as a positive control. Silencing of *PDLP* and *SRC2* genes did not alter the development in the *N. benthamiana* and *C. chinense;* in contrast, *N. tabacum* exhibited some leaf crumbling symptoms; Figure S5: Representative images of *N. benthamiana* plants upon CMV-inoculation at 19 dpi. dpi = days post CMV inoculation, CMV = cucumber mosaic virus, (-) = non-silenced plant with CMV infection, Mock = Phosphate buffer treated plant. TRV2-*SRC2*, TRV2-*eGFP* (control), and TRV2-*PDLP* are VIGS silenced plants observed upon CMV inoculation; Figure S6: Microscopic observation of callose deposits upon cucumber mosaic virus (CMV) in *PDLP*- and *SRC2*-silenced *Nicotiana benthamiana* plants. Callose deposition assay using Aniline blue staining was performed according to the procedure described by Schenk et al., 2015. The callose deposition was visualized using a fluorescent microscope with a UV lamp. White arrow indicates callose deposits. dpa = days post agroinoculation. dpi = days post infection. Figure S7: Macroscopic (A) and microscopic (B) detection of H_2O_2 by 3,3- diaminobenzidine (DAB) staining in VIGS-silenced *Nicotiana benthamiana* plants at different time-points after DAB staining. Control (*eGFP*) and *SRC2*-silenced plants exhibited elevated levels of H_2O_2 as compared to the *PDLP*-silenced group. Dark brown patches (indicated by green arrows) represent areas of reactive oxygen species (ROS) accumulation (see results of RT-qPCR in Figure 7), where darker shades indicate higher viral accumulation and lighter shades indicate lower viral accumulation. DAB assay was performed following the procedure outlined by Bach-Pages and Preston 2018. dpa = days post agroinoculation; dpi = days post CMV infection.

Author Contributions: R.S.: Writing—original draft, review, and editing, Methodology, Validation, Formal analysis, Data curation; A.K.: Writing—review, and editing, Methodology, Supervision, Data curation; C.J.S.: Writing—review, and editing, Supervision, Resources, Funding acquisition; B.K.B.: Writing—review, and editing, Supervision; A.V.: Conceptualization, Supervision, Project administration, Writing—review, and editing, Resources, Funding acquisition. All authors have read and agreed to the published version of the manuscript.

Funding: The funding to carry out the research in Norway was provided by the Norwegian Research Council Grant (287794). This research was partly funded by the Education, Audiovisual and Culture Executive Agency (EACEA) of the European Commission for the 3-year-long ERASMUS+ Capacity Building in Higher Education project (598797-EPP-1-2018-1-EL-EPPKA2-CBHE-JP) entitled 'AdaptNET—Strengthening education, research and innovation for climate-smart crops in India'.

Data Availability Statement: All data supporting this study are available within the paper and within the Supplementary Materials published online.

Acknowledgments: The authors acknowledge Supriya Chakraborty (Jawaharlal Nehru University, New Delhi, India) for proving the visual VIGS marker. The authors also acknowledge Dag-Ragnar Blystad (Norwegian Institute of Bioeconomy Research, Norway) for providing the CMV isolate to carry out the inoculation experiment on the VIGS silenced plants.

Conflicts of Interest: The authors declare no conflicts of interest.

References

1. Kumar, D.; Kumar, R.; Hyun, T.K.; Kim, J.-Y. Cell-to-Cell Movement of Viruses via Plasmodesmat. *J. Plant Res.* **2015**, *128*, 37–47. [CrossRef] [PubMed]
2. Schoelz, J.E.; Harries, P.A.; Nelson, R.S. Intracellular Transport of Plant Viruses: Finding the Door out of the Cell. *Mol. Plant* **2011**, *4*, 813–831. [CrossRef]
3. Lucas, W.J. Plant Viral Movement Proteins: Agents for Cell-to-Cell Trafficking of Viral Genomes. *Virology* **2006**, *344*, 169–184. [CrossRef]
4. Heinlein, M. Plasmodesmata: Channels for Viruses on the Move. *Methods Mol. Biol.* **2015**, *1217*, 25–52. [CrossRef] [PubMed]
5. Amari, K.; Boutant, E.; Hofmann, C.; Schmitt-Keichinger, C.; Fernandez-Calvino, L.; Didier, P.; Lerich, A.; Mutterer, J.; Thomas, C.L.; Heinlein, M.; et al. A Family of Plasmodesmal Proteins with Receptor-like Properties for Plant Viral Movement Proteins. *PLoS Pathog.* **2010**, *6*, e1001119. [CrossRef] [PubMed]
6. Ritzenthaler, C.; Hofmann, C. Tubule-Guided Movement of Plant Viruses. *Plant Cell Monogr.* **2007**, *7*, 64–83. [CrossRef]
7. Chen, M.-H.; Tian, G.-W.; Gafni, Y.; Citovsky, V. Effects of Calreticulin on Viral Cell-to-Cell Movement. *Plant Physiol.* **2005**, *138*, 1866–1876. [CrossRef]
8. Lee, J.-Y.; Taoka, K.; Yoo, B.-C.; Ben-Nissan, G.; Kim, D.-J.; Lucas, W.J. Plasmodesmal-Associated Protein Kinase in Tobacco and *Arabidopsis* Recognizes a Subset of Non-Cell-Autonomous Proteins. *Plant Cell* **2005**, *17*, 2817–2831. [CrossRef] [PubMed]
9. Raffaele, S.; Bayer, E.; Lafarge, D.; Cluzet, S.; German Retana, S.; Boubekeur, T.; Leborgne-Castel, N.; Carde, J.-P.; Lherminier, J.; Noirot, E.; et al. Remorin, a Solanaceae Protein Resident in Membrane Rafts and Plasmodesmata, Impairs Potato Virus X Movement. *Plant Cell* **2009**, *21*, 1541–1555. [CrossRef] [PubMed]
10. Ham, B.; Wang, X.; Toscano-Morales, R.; Lin, J.; Lucas, W. Plasmodesmal Endoplasmic Reticulum Proteins Regulate Intercellular Trafficking of Cucumber Mosaic Virus in Arabidopsis. *J. Exp. Bot.* **2023**, *74*, 4401–4414. [CrossRef]
11. Tilsner, J.; Kriechbaumer, V. Reticulons 3 and 6 Interact with Viral Movement Proteins. *Mol. Plant Pathol.* **2022**, *23*, 1807–1814. [CrossRef] [PubMed]
12. Zavaliev, R.; Sagi, G.; Gera, A.; Epel, B.L. The Constitutive Expression of *Arabidopsis* Plasmodesmal-Associated Class 1 Reversibly Glycosylated Polypeptide Impairs Plant Development and Virus Spread. *J. Exp. Bot.* **2010**, *61*, 131–142. [CrossRef]
13. Ueki, S.; Spektor, R.; Natale, D.M.; Citovsky, V. ANK, a Host Cytoplasmic Receptor for the Tobacco Mosaic Virus Cell-to-Cell Movement Protein, Facilitates Intercellular Transport through Plasmodesmata. *PLoS Pathog.* **2010**, *6*, e1001201. [CrossRef] [PubMed]
14. Uchiyama, A.; Shimada-Beltran, H.; Levy, A.; Zheng, J.Y.; Javia, P.A.; Lazarowitz, S.G. The Arabidopsis Synaptotagmin SYTA Regulates the Cell-to-Cell Movement of Diverse Plant Viruses. *Front. Plant Sci.* **2014**, *5*, 584. [CrossRef]
15. den Hollander, P.W.; Kieper, S.N.; Borst, J.W.; van Lent, J.W.M. The Role of Plasmodesma-Located Proteins in Tubule-Guided Virus Transport Is Limited to the Plasmodesmata. *Arch. Virol.* **2016**, *161*, 2431–2440. [CrossRef] [PubMed]
16. Rodriguez, A.; Angel, C.A.; Lutz, L.; Leisner, S.M.; Nelson, R.S.; Schoelz, J.E. Association of the P6 Protein of *Cauliflower Mosaic Virus* with Plasmodesmata and Plasmodesmal Proteins. *Plant Physiol.* **2014**, *166*, 1345–1358. [CrossRef] [PubMed]
17. Kim, Y.C.; Kim, S.Y.; Choi, D.; Ryu, C.M.; Park, J.M. Molecular Characterization of a Pepper C2 Domain-Containing SRC2 Protein Implicated in Resistance against Host and Non-Host Pathogens and Abiotic Stresses. *Planta* **2008**, *227*, 1169–1179. [CrossRef]
18. Roossinck, M.J. Evolutionary History of *Cucumber Mosaic Virus* Deduced by Phylogenetic Analyses. *J. Virol.* **2002**, *76*, 3382–3387. [CrossRef]
19. Edwardson, J.R.; Christie, R.G. *CRC Handbook of Viruses Infecting Legumes*; CRC Press: Boca Raton, FL, USA, 1991.
20. Ashwathappa, K.V.; Krishna Reddy, M.; Venkataravanappa, V.; Madhavi Reddy, K.; Hemachandra Reddy, P.; Lakshminarayana Reddy, C.N. Genome Characterization and Host Range Studies of *Cucumber Mosaic Virus* Belonging to the Subgroup IB Infecting Chilli in India and Screening of Chilli Genotypes for Identification of Resistance. *VirusDisease* **2021**, *32*, 535–547. [CrossRef] [PubMed]
21. Roossinck, M.J. *Cucumber Mosaic Virus*; a Model for RNA Virus Evolution. *Mol. Plant Pathol.* **2001**, *2*, 59–63. [CrossRef]
22. Sáray, R.; Fábián, A.; Palkovics, L.; Salánki, K. The 28 Ser Amino Acid of Cucumber Mosaic Virus Movement Protein Has a Role in Symptom Formation and Plasmodesmata Localization. *Viruses* **2021**, *13*, 222. [CrossRef] [PubMed]
23. Tenllado, F.; Llave, C.; Díaz-Ruíz, J.R. RNA Interference as a New Biotechnological Tool for the Control of Virus Diseases in Plants. *Virus Res.* **2004**, *102*, 85–96. [CrossRef] [PubMed]

24. Lange, M.; Yellina, A.L.; Orashakova, S.; Becker, A. Virus-Induced Gene Silencing (VIGS) in Plants: An Overview of Target Species and the Virus-Derived Vector Systems. *Methods Mol. Biol.* **2013**, *975*, 1–14. [CrossRef] [PubMed]
25. Senthil-Kumar, M.; Mysore, K.S. Tobacco Rattle Virus–Based Virus-Induced Gene Silencing in *Nicotiana Benthamiana. Nat. Protoc.* **2014**, *9*, 1549–1562. [CrossRef]
26. Purkayastha, A.; Dasgupta, I. Virus-Induced Gene Silencing: A Versatile Tool for Discovery of Gene Functions in Plants. *Plant Physiol. Biochem.* **2009**, *47*, 967–976. [CrossRef]
27. Becker, A.; Lange, M. VIGS--Genomics Goes Functional. *Trends Plant Sci.* **2010**, *15*, 1–4. [CrossRef] [PubMed]
28. Shi, G.; Hao, M.; Tian, B.; Cao, G.; Wei, F.; Xie, Z. A Methodological Advance of Tobacco Rattle Virus-Induced Gene Silencing for Functional Genomics in Plants. *Front. Plant Sci.* **2021**, *12*, 671091. [CrossRef]
29. Kumagai, M.H.; Donson, J.; Della-Cioppa, G.; Harvey, D.; Hanley, K.; Grill, L.K. Cytoplasmic Inhibition of Carotenoid Biosynthesis with Virus-Derived RNA. *Proc. Natl. Acad. Sci. USA* **1995**, *92*, 1679–1683. [CrossRef]
30. Ratcliff, F.; Martin-Hernandez, A.M.; Baulcombe, D.C. Technical Advance. Tobacco Rattle Virus as a Vector for Analysis of Gene Function by Silencing. *Plant J.* **2001**, *25*, 237–245. [CrossRef]
31. Liu, Y.; Schiff, M.; Dinesh-Kumar, S.P. Virus-Induced Gene Silencing in Tomato. *Plant J.* **2002**, *31*, 777–786. [CrossRef] [PubMed]
32. Verlaan, M.G.; Hutton, S.F.; Ibrahem, R.M.; Kormelink, R.; Visser, R.G.F.; Scott, J.W.; Edwards, J.D.; Bai, Y. The Tomato Yellow Leaf Curl Virus Resistance Genes Ty-1 and Ty-3 Are Allelic and Code for DFDGD-Class RNA–Dependent RNA Polymerases. *PLOS Genet.* **2013**, *9*, e1003399. [CrossRef] [PubMed]
33. Hayward, A.; Padmanabhan, M.; Dinesh-Kumar, S.P. Virus-Induced Gene Silencing in *Nicotiana Benthamiana* and Other Plant Species. *Methods Mol. Biol.* **2011**, *678*, 55–63. [CrossRef] [PubMed]
34. Liu, H.; Fu, D.; Zhu, B.; Yan, H.; Shen, X.; Zuo, J.; Zhu, Y.; Luo, Y. Virus-Induced Gene Silencing in Eggplant (*Solanum Melongena*). *J. Integr. Plant Biol.* **2012**, *54*, 422–429. [CrossRef] [PubMed]
35. Reagan, B.C.; Burch-Smith, T.M. Viruses Reveal the Secrets of Plasmodesmal Cell Biology. *Mol. Plant-Microbe Interact.* **2020**, *33*, 26–39. [CrossRef] [PubMed]
36. Takahashi, R.; Shimosaka, E. CDNA Sequence Analysis and Expression of Two Cold-Regulated Genes in Soybean. *Plant Sci.* **1997**, *123*, 93–104. [CrossRef]
37. Caillaud, M.-C.; Wirthmueller, L.; Sklenar, J.; Findlay, K.; Piquerez, S.J.M.; Jones, A.M.E.; Robatzek, S.; Jones, J.D.G.; Faulkner, C. The Plasmodesmal Protein PDLP1 Localises to Haustoria-Associated Membranes during Downy Mildew Infection and Regulates Callose Deposition. *PLOS Pathog.* **2014**, *10*, e1004496. [CrossRef]
38. Liu, Z.; Liu, Y.; Shi, L.; Yang, S.; Shen, L.; Yu, H.; Wang, R.; Wen, J.; Tang, Q.; Hussain, A.; et al. SGT1 Is Required in PcINF1/SRC2-1 Induced Pepper Defense Response by Interacting with SRC2-1. *Sci. Rep.* **2016**, *6*, 21651. [CrossRef]
39. Ye, Z.-W.; Chen, Q.-F.; Chye, M.-L. *Arabidopsis Thaliana* Acyl-CoA-Binding Protein ACBP6 Interacts with Plasmodesmata-Located Protein PDLP8. *Plant Signal. Behav.* **2017**, *12*, e1359365. [CrossRef] [PubMed]
40. Hernández, J.A.; Gullner, G.; Clemente-Moreno, M.J.; Künstler, A.; Juhász, C.; Díaz-Vivancos, P.; Király, L. Oxidative Stress and Antioxidative Responses in Plant–Virus Interactions. *Physiol. Mol. Plant Pathol.* **2016**, *94*, 134–148. [CrossRef]
41. Lee, J.Y.; Wang, X.; Cui, W.; Sager, R.; Modla, S.; Czymmek, K.; Zybaliov, B.; Van Wijk, K.; Zhang, C.; Lu, H.; et al. A Plasmodesmata-Localized Protein Mediates Crosstalk between Cell-to-Cell Communication and Innate Immunity in Arabidopsis. *Plant Cell* **2011**, *23*, 3353–3373. [CrossRef]
42. Lim, G.-H.; Shine, M.B.; de Lorenzo, L.; Yu, K.; Cui, W.; Navarre, D.; Hunt, A.G.; Lee, J.-Y.; Kachroo, A.; Kachroo, P. Plasmodesmata Localizing Proteins Regulate Transport and Signaling during Systemic Acquired Immunity in Plants. *Cell Host Microbe* **2016**, *19*, 541–549. [CrossRef] [PubMed]
43. Carella, P.; Isaacs, M.; Cameron, R.K. Plasmodesmata-Located Protein Overexpression Negatively Impacts the Manifestation of Systemic Acquired Resistance and the Long-Distance Movement of DEFECTIVE IN INDUCED RESISTANCE1 in *Arabidopsis*. *Plant Biol.* **2015**, *17*, 395–401. [CrossRef] [PubMed]
44. Deng, S.; Zhang, Y.; Fang, X.; Gou, H.; Sun, R.; Xuan, H.; Wang, H.; Zhao, J.; Xing, H.; Guo, N. Overexpression of *GmSRC2* Confers Resistance to *Phytophthora Sojae* in Soybean. *Preprint* **2023**. [CrossRef]
45. Lewis, J.D.; Lazarowitz, S.G. *Arabidopsis* Synaptotagmin SYTA Regulates Endocytosis and Virus Movement Protein Cell-to-Cell Transport. *Proc. Natl. Acad. Sci. USA* **2010**, *107*, 2491–2496. [CrossRef] [PubMed]
46. Yamazaki, T.; Kawamura, Y.; Minami, A.; Uemura, M. Calcium-Dependent Freezing Tolerance in *Arabidopsis* Involves Membrane Resealing via Synaptotagmin SYT1. *Plant Cell* **2008**, *20*, 3389–3404. [CrossRef]
47. Liu, D.; Shi, L.; Han, C.; Yu, J.; Li, D.; Zhang, Y. Validation of Reference Genes for Gene Expression Studies in Virus-Infected Nicotiana Benthamiana Using Quantitative Real-Time PCR. *PLoS ONE* **2012**, *7*, e46451. [CrossRef]
48. Weigel, D.; Glazebrook, J. Transformation of Agrobacterium Using Electroporation. *CSH Protoc.* **2006**, *2006*. [CrossRef] [PubMed]
49. Pfaffl, M.W.; Tichopad, A.; Prgomet, C.; Neuvians, T.P. Determination of Stable Housekeeping Genes, Differentially Regulated Target Genes and Sample Integrity: BestKeeper—Excel-Based Tool Using Pair-Wise Correlations. *Biotechnol. Lett.* **2004**, *26*, 509–515. [CrossRef] [PubMed]

50. Schmittgen, T.D.; Livak, K.J. Analyzing Real-Time PCR Data by the Comparative C(T) Method. *Nat. Protoc.* **2008**, *3*, 1101–1108. [CrossRef]
51. Schenk, S.T.; Schikora, A. Staining of Callose Depositions in Root and Leaf Tissues. *Bio-protocol* **2015**, *5*, e1429. [CrossRef]
52. Bach-Pages, M.; Preston, G.M. Methods to Quantify Biotic-Induced Stress in Plants. *Methods Mol. Biol.* **2018**, *1734*, 241–255. [CrossRef]

plants

MDPI

Article

Investigating the Role of Viruses in the Rapid Decline of Young Apple Trees in High-Density Orchards in New York

Anna O. Wunsch [1,*], Mario Miranda Sazo [2], Janet van Zoeren [2], Kurt H. Lamour [3], Oscar P. Hurtado-Gonzales [4], Awais Khan [1] and Marc Fuchs [1]

[1] Plant Pathology and Plant-Microbe Biology Section, School of Integrative Plant Science, Cornell University, Geneva, NY 14456, USA; awais.khan@cornell.edu (A.K.); marc.fuchs@cornell.edu (M.F.)
[2] Cornell Cooperative Extension Lake Ontario Fruit Program, Albion, NY 14411, USA; mrm67@cornell.edu (M.M.S.); jev67@cornell.edu (J.v.Z.)
[3] Department of Entomology and Plant Pathology, University of Tennessee, Knoxville, TN 37996, USA; klamour@utk.edu
[4] USDA-APHIS Plant Germplasm Quarantine Program, Beltsville, MD 20705, USA; oscar.hurtado-gonzales@usda.gov
* Correspondence: aw838@cornell.edu

Abstract: A sudden, unexplained decline and collapse of young apple trees on dwarfing and semi-dwarfing rootstocks has been reported across North America over the past decade. Although viruses have been detected in declining trees, no information is available on their potential causal role in the decline phenomenon. To this end, virus-inoculated apple trees were established in a high-density experimental orchard and monitored over five years. Tree decline was observed in year 4 (2022), resulting in 17% mortality, with declining trees exhibiting marked vascular tissue necrosis. However, none of the eight viruses and one viroid detected in the experimental orchard was significantly more prevalent in declining trees. Extreme temperature fluctuations in January 2022, followed by a severe water deficit in summer 2022, were recorded at the experimental orchard. Similar but distinct observations were made in a nearby commercial orchard with foliar nutrient imbalances documented in trees exhibiting symptoms of rapid decline. Together, our findings suggest that viruses are not primarily responsible for the rapid decline phenomenon and highlight the need for future work to investigate the roles of tree physiology and water stress in tree decline, as well as the potential efficacy of horticultural mitigation practices.

Keywords: latent apple virus; apple stem pitting virus; apple chlorotic leaf spot virus; apple stem grooving virus; high-density orchard; rapid apple decline

Citation: Wunsch, A.O.; Miranda Sazo, M.; van Zoeren, J.; Lamour, K.H.; Hurtado-Gonzales, O.P.; Khan, A.; Fuchs, M. Investigating the Role of Viruses in the Rapid Decline of Young Apple Trees in High-Density Orchards in New York. *Plants* **2024**, *13*, 2866. https://doi.org/10.3390/plants13202866

Academic Editors: Violetta Katarzyna Macioszek, Iwona Ciereszko and Andrzej K. Kononowicz

Received: 28 August 2024
Revised: 28 September 2024
Accepted: 8 October 2024
Published: 14 October 2024

1. Introduction

A sudden, unexplained, mid-season collapse of young apple trees (*Malus domestica* Borkh.) in high-density orchards, which has been referred to as "rapid apple decline" or "sudden apple decline", was first reported in the eastern United States in the 2010s [1–4]. Similar sudden, unexplained apple tree decline phenomena have since been reported in New York [3,5,6], North Carolina [4], Pennsylvania [1], Washington [7], and West Virginia [8] in the United States, as well as in British Columbia [9] in Canada.

These decline phenomena are characterized by the rapid collapse of young apple trees in high-density plantings over the course of several weeks, often resulting in tree mortality [1,2,8]. Tree collapse is sometimes preceded by a gradual decline of tree health over the course of a growing season, typified by chlorosis throughout the canopy and declining tree vigor [1,2,10,11]. Trees which collapse typically do so during the active growing season, often with a full crop load [1,2,4,6,11]. The underlying causal factors of the decline phenomena are yet to be identified, although many have been hypothesized, including fire blight of the rootstock caused by the bacterial pathogen *Erwinia amylovora*, herbicide injury, boring insects, winter injury, drought stress, fungal canker and wood rot fungi, and latent viruses [3,5,6,8,9,11,12]. This rapid mid-season collapse of young apple trees has not been attributed to a known disease, such as apple bacterial quick decline in Japan caused by *Dickeya dadantii* [13] or stem canker and dieback in Ontario caused by *Botryosphaeria dothidea* [14].

Several recent studies have attempted to assess the role of viruses and viroids in the decline phenomenon. In Pennsylvania, high-throughput sequencing revealed the presence of a novel luteovirus, i.e., apple luteovirus 1 (ALV1), in declining trees, as well as three common latent viruses of apple, i.e., apple chlorotic leaf spot virus (ACLSV), apple stem pitting virus (ASPV), and apple stem grooving virus (ASGV) [8]. Latent apple viruses generally do not elicit symptoms on commercial cultivars and rootstocks, often occur in mixed infections, and are endemic to cultivated apple trees [15]. A comprehensive study in a commercial orchard in New York detected two latent viruses, i.e., ACLSV and ASPV, in both declining and non-declining trees [6]. In Washington, high-throughput sequencing revealed the presence of several novel putative viruses, as well as nine known viruses and viroids, in both declining and non-declining trees in commercial orchards, but none was consistently associated with decline [12]. In British Columbia, high-throughput sequencing and molecular assays identified a novel ilarvirus, i.e., apple ilarvirus 2, as well as 21 known viruses and viroids, in both declining and non-declining trees in commercial orchards [9].

In addition to investigating the role of viruses and viroids in decline phenomena, studies conducted to date have also investigated plant and soil microbial community composition [6], weather patterns [6,16], tree water status [11], and root system architecture [10]. Singh et al. (2019) used high-throughput sequencing to identify the composition of bacterial and fungal communities in apple root, shoot, rhizosphere, and soil samples and documented no significant enrichment of any class of bacteria or fungi in declining trees in a commercial apple orchard in western New York [6]. However, this study identified unusually cold winter temperatures followed by summer drought in the vicinity of the declining orchard [6]. Similarly, Donahue & Elone (2021) documented uncharacteristically variable temperatures in their case study of a declining commercial orchard in eastern New York [16]. A comparison of the root system architectures of declining and non-declining trees in two commercial orchards in New York found that the root systems and scion trunk diameters of declining trees were significantly smaller than non-declining trees in one orchard but not the other [10]. Furthermore, a study of the water status of trees in a declining orchard in British Columbia found that tree decline and mortality were associated with disruption in xylem water transport and hydraulic failure [11].

All studies conducted to date which have investigated the role of viruses and viroids in the rapid decline of young apple trees have represented observational surveys of trees already in decline at a single point in time during the life of an orchard [6,8–10,12]. To our knowledge, no study to date has assessed the potential role of viruses and viroids in the decline phenomenon by monitoring declining trees for more than one growing season. The objectives of this study were to monitor the progress of decline in an experimental orchard plot beginning at orchard establishment, to reproduce the rapid decline phenomenon under experimental conditions, and to conclusively rule latent virus infections in or out as a primary cause while simultaneously monitoring the progress of decline in a nearby commercial orchard block. We hypothesized that concurrent infection by a greater number of viruses would result in higher rates of tree decline at both sites. Here, we summarize our efforts to characterize the rapid decline phenomenon in both a high-density experimental apple orchard and a high-density commercial apple orchard in the Lake Ontario fruit production region of New York.

2. Results

2.1. Observations of Rapid Tree Decline in an Experimental Apple Orchard

An experimental apple orchard was established in 2019 at Cornell AgriTech in Ontario County, NY, USA, in an effort to reproduce the rapid decline phenomenon. Trees representing six scion-rootstock combinations ('Baigent' Gala, 'Honeycrisp', and 'Royal Red Honeycrisp'™ (Willow Drive Nursery, WA, USA) on 'Malling 26' (M.26) or 'Geneva 935' (G.935) rootstocks) were custom-budded the year prior and then chip graft inoculated with either apple chlorotic leaf spot virus (ACLSV), apple stem pitting virus (ASPV), both viruses, or neither virus. Over the course of the five-year study, 65 of the 240 total trees established in the experimental orchard were lost for reasons unrelated to the tree decline phenomenon of interest, including unsuccessful budding, tree establishment failure, and fire blight caused by *Erwinia amylovora* infection. These trees were excluded from the study.

The decline severity of the remaining 175 trees (175 of 240; 73%) was visually evaluated according to a 0 to 4 severity scale from 2019 to 2023 (years 1–5). No tree declined (decline severity rating 0) in 2019, 2020, or 2021 (years 1–3). In 2022 (year 4), all 175 trees successfully emerged from dormancy, bloomed, and set fruit (decline severity rating 0) (Figure 1). Interestingly, some trees began showing initial signs of decline, corresponding to decline severity rating 1, in early June (approximately four weeks post-bloom) (Figure 1). By early August (approximately 12 weeks post-bloom), trees with symptoms corresponding to all five decline severity ratings were observed, including complete tree collapse (decline severity rating 4) in three trees (1.7%, 3 of 175) (Figures 1 and 2). By the onset of leaf senescence in mid-October (approximately 22 weeks post-bloom), the tree mortality rate (decline severity rating 4) reached 17% (30 of 175) (Figure 1). All three cultivars and both rootstocks evaluated were represented in the population of 30 trees which collapsed in 2022: 'Baigent'/G.935 (n = 2), 'Baigent"/M.26 (n = 15), 'Honeycrisp'/G.935 (n = 1), 'Honeycrisp'/M.26 (n = 11), and 'Royal Red Honeycrisp'™/G.935 (n = 1). No leaf drop or substantial trunk swelling or cracking at the graft union was apparent in any tree. However, cross-sectioning of the trunks of severely declining trees (decline severity rating 3) revealed extensive browning and necrosis of the vascular tissue (Figure 2). Together, this work documented the occurrence of the rapid decline phenomenon in the experimental orchard across all evaluated apple cultivars and rootstocks in year 4. The rapid decline was accompanied by vascular tissue necrosis.

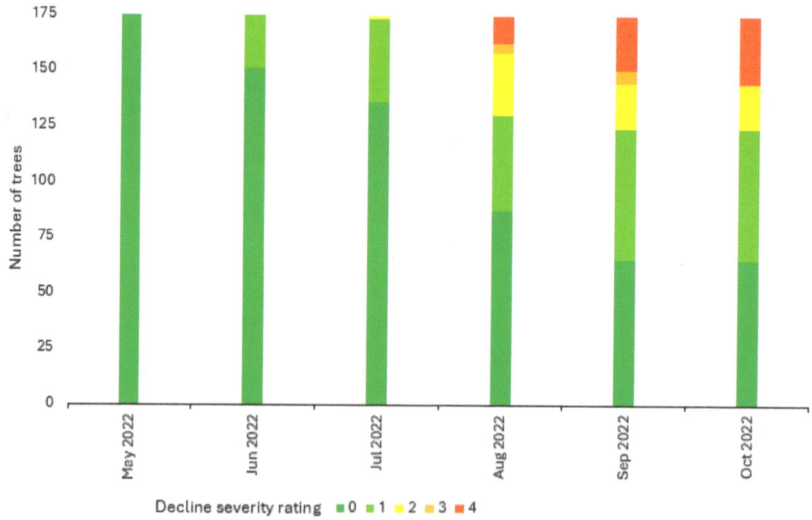

Figure 1. Decline severity ratings over time of three-year-old 'Baigent' Gala, 'Honeycrisp', and 'Royal Red Honeycrisp'™ apple trees on 'Malling 26' and 'Geneva 935' rootstocks in a high-density experimental orchard at Cornell AgriTech in Ontario County, NY, USA over the course of the 2022 growing season. Tree decline severity was visually assessed monthly from May to October using a scale from 0 to 4, with a rating of 0 indicating no decline symptoms and a rating of 4 indicating tree mortality. Ratings of 1 to 3 indicate increasing decline severity, with a rating of 1 corresponding to

chlorosis throughout <50% of the canopy, a rating of 2 corresponding to chlorosis throughout ≥50% of the canopy, and a rating of 3 corresponding to chlorosis throughout ≥50% of the canopy, leaf flagging, and poor tree vigor.

Figure 2. (**A**) Asymptomatic, declining, and collapsed three-year-old 'Baigent' Gala, 'Honeycrisp', and 'Royal Red Honeycrisp'™ apple trees on 'Malling 26' and 'Geneva 935' rootstocks in a high-density experimental orchard at Cornell AgriTech in Ontario County, NY, USA on 22 August 2022. Tree decline severity was visually assessed using a scale from 0 to 4, with a rating of 0 indicating no decline symptoms and a rating of 4 indicating tree mortality. From left to right, the eight trees pictured here represent decline severity ratings of 0, 3, 4, 0, 0, 4, 0, and 0, respectively; (**B**) A representative cross-section of the trunk of a severely declining tree (decline severity rating 3) showing extensive browning and necrosis of the vascular tissue approximately 30 cm above the graft

union. Cross-sections of 15 declining trees were evaluated (four with a decline severity rating of 3; 11 with a decline severity rating of 4), and all displayed similar tissue necrosis; (**C**) a representative cross-section of the trunk of a non-declining tree (decline severity rating 0) without browning or necrosis of the vascular tissue approximately 30 cm above the graft union. Cross-sections of two non-declining trees were evaluated (both had decline severity ratings of 0), and both displayed a similar lack of tissue necrosis.

Overall, the majority of the trees evaluated (62%, 109 of 175) experienced either mild (decline severity rating 1), moderate (decline severity rating 2) or severe decline (decline severity rating 3) during the 2022 growing season (Figure 1). The 30 trees that collapsed (decline severity rating 4) made up 28% (30 of 109) of those which showed symptoms of decline (Figure 1). Strikingly, none of the surviving 145 trees declined in 2023 (year 5), including those showing signs of mild or moderate decline at the end of the 2022 growing season (45%, 79 of 175).

2.2. Detection of Viruses and Viroids in the Experimental Apple Orchard

Diagnostic testing of 175 trees in the experimental orchard via multiplex polymerase chain reaction (PCR)-based amplicon sequencing revealed the presence of eight viruses and one viroid (Table 1). At least one virus or viroid was identified in every tree tested (Tables 1 and S1). The three most prevalent viruses were apple stem grooving virus (ASGV) (100%, 175 of 175), ACLSV (78%, 136 of 175), and ASPV (71%, 125 of 175) (Table 1). Five additional viruses, i.e., citrus concave gum-associated virus (CCGaV), apple green crinkle-associated virus (AGCaV), apple rubbery wood virus 2 (ARWV2), tobacco ringspot virus (TRSV), and tomato ringspot virus (ToRSV), as well as one viroid, apple hammerhead viroid (AHVd), were detected in smaller proportions of the trees tested (Tables 1 and S1). Ten of the viruses and viroids detectable by this methodology were not identified in the experimental orchard (Tables 1 and S1). This work indicated the presence of more viruses and viroids than expected based on the initial design of the experimental orchard that relied on chip graft inoculation of selected trees with either ACLSV, ASPV, or both viruses.

Table 1. Viruses and viroids detected via multiplex PCR-based amplicon sequencing of 175 three-year-old 'Baigent' Gala, 'Honeycrisp', and 'Royal Red Honeycrisp'™ apple trees on 'Malling 26' and 'Geneva 935' rootstocks in a high-density experimental orchard at Cornell AgriTech in Ontario County, NY, USA. Numbers of infected trees per virus or viroid are indicated.

	Rootstock/Scion						
Virus	G.935/Gala	G.935/HC	G.935/RRHC	M.26/Gala	M.26/HC	M.26/RRHC	Total
ACLSV	21	22	30	21	34	8	136
AGCaV	13	7	14	11	7	4	56
ALV1	0	0	0	0	0	0	0
ApMV	0	0	0	0	0	0	0
ARWV1	0	0	0	0	0	0	0
ARWV2	2	2	15	1	5	0	25
ASGV	31	34	30	33	39	8	175
ASPV	21	19	30	21	26	8	125
CCGaV	17	19	29	12	22	7	106
CiVA	0	0	0	0	0	0	0
PpPV2	0	0	0	0	0	0	0
PrVT	0	0	0	0	0	0	0
ToRSV	0	0	2	1	0	0	3
TRSV	0	0	0	0	1	0	1
Viroid	G.935/Gala	G.935/HC	G.935/RRHC	M.26/Gala	M.26/HC	M.26/RRHC	Total
ADFVd	0	0	0	0	0	0	0

Table 1. *Cont.*

	Rootstock/Scion						
AFCVd	0	0	0	0	0	0	0
AHVd	29	28	30	8	6	3	104
ASSVd	0	0	0	0	0	0	0
PBCVd	0	0	0	0	0	0	0

Abbreviations: ACLSV, apple chlorotic leaf spot virus; ADFVd, apple dimple fruit viroid; AFCVd, apple fruit crinkle viroid; AGCaV, apple green crinkle-associated virus; AHVd, apple hammerhead viroid; ALV1, apple luteovirus 1; ApMV, apple mosaic virus; ARWV1, apple rubbery wood virus 1; ARWV2, apple rubbery wood virus 2; ASGV, apple stem grooving virus; ASPV, apple stem pitting virus; ASSVd, apple scar skin viroid; CCGaV, citrus concave gum-associated virus; CiVA, citrus virus A; Gala, 'Baigent' Gala; G.935, 'Geneva 935'; HC, 'Honeycrisp'; M.26, 'Malling 26'; PBCVd, pear blister canker viroid; PpPV2, Pyrus pyrifolia partitivirus 2; PrVT, Prunus virus T; RRHC, 'Royal Red Honeycrisp'™; ToRSV, tomato ringspot virus; TRSV, tobacco ringspot virus.

2.3. The Relationship between Virus Infection and Decline Outcome in the Experimental Apple Orchard

The relationship between virus and viroid prevalence in collapsed trees (decline severity rating 4; n = 30) and in trees that did not collapse (decline severity ratings 0, 1, 2, or 3; n = 145) was evaluated for statistical significance (Figure 3A). Of the eight viruses and one viroid detected in the experimental orchard, none was found to be significantly more prevalent in trees that collapsed compared with trees that did not collapse (Figure 3A). Furthermore, AHVd prevalence was significantly higher in trees that did not collapse (69%, 100 of 145) than in trees that did collapse (13%, 4 of 30) ($p < 0.001$) (Figure 3A). Moreover, an analysis of variance revealed no statistically significant difference between the number of unique viruses and viroids detected in trees that collapsed ($\overline{x}$ = 3.63) and in trees that did not collapse ($\overline{x}$ = 4.29) (p = 0.062), revealing no association between tree mortality and the number of unique viruses and viroids detected.

Similarly, of the eight viruses and one viroid detected in the experimental orchard, none was found to be significantly more prevalent in declining trees (decline severity ratings 2–3; n = 20) or trees that had collapsed (decline severity rating 4; n = 30) compared with non-declining trees (decline severity ratings 0–1; n = 125) (Figure 3B). Indeed, the prevalences of both AHVd and AGCaV were significantly higher in non-declining trees (72% and 37%, respectively) than in declining trees (50% and 5%; $p < 0.001$ and p = 0.034, respectively) (Figure 3B). Together, these results revealed that the prevalence of common apple viruses and viroids was not associated with tree decline outcome in the experimental orchard.

Because the viruses and viroid detected in the experimental orchard were present in many unique combinations, the number of trees infected with any one combination was generally quite small, precluding a robust statistical analysis of correlation between virus combinations and decline outcome. Nevertheless, we observed no clear association between particular virus combinations and tree decline or collapse (Table S1).

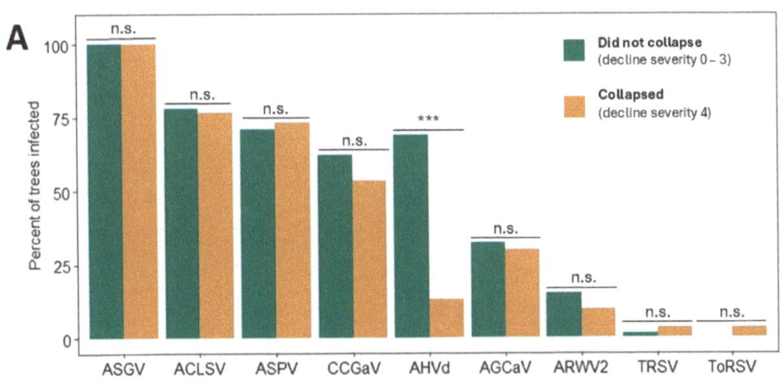

Figure 3. *Cont.*

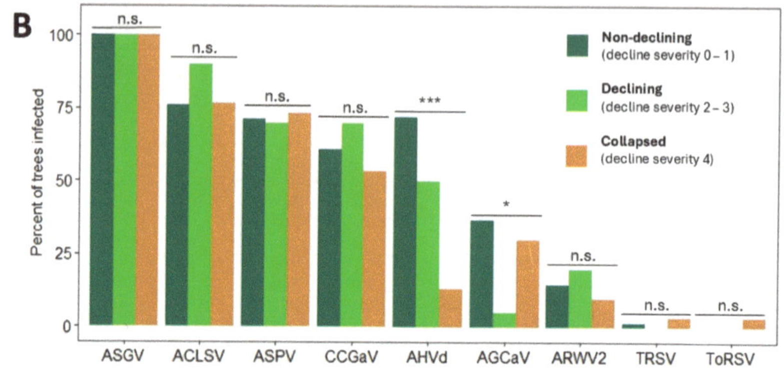

Figure 3. Prevalence of eight viruses and one viroid in 175 three-year-old 'Baigent' Gala, 'Honeycrisp', and 'Royal Red Honeycrisp'™ apple trees on 'Malling 26' and 'Geneva 935' rootstocks in a high-density experimental orchard at Cornell AgriTech in Ontario County, NY, USA that (**A**) did or did not collapse in 2022, or (**B**) collapsed, declined, or did not decline in 2022. Decline severity ratings were assigned as previously described. Statistical significance was determined by χ^2 tests with a p value threshold of 0.05. n.s. $p \geq 0.05$, * $p < 0.05$; *** $p < 0.001$. Abbreviations: ACLSV, apple chlorotic leaf spot virus; AGCaV, apple green crinkle-associated virus; AHVd, apple hammerhead viroid; ARWV2, apple rubbery wood virus 2; ASGV, apple stem grooving virus; ASPV, apple stem pitting virus; CCGaV, citrus concave gum-associated virus; ToRSV, tomato ringspot virus; TRSV, tobacco ringspot virus.

2.4. Root System Architecture of Trees in the Experimental Apple Orchard

Root system architecture parameters were compared between excavated root systems of trees that experienced substantial decline (decline severity ratings of 2, 3, or 4; $n = 25$) and those that did not experience substantial decline (decline severity ratings of 0 or 1; $n = 14$). Of the root system parameters evaluated, both scion diameter and rootstock diameter at the graft union differed significantly between scion cultivars ($p = 0.001$ and $p = 0.025$, respectively) (Table S2). These differences are attributable to known differences in trunk cross-sectional area and overall tree size between 'Gala' and 'Honeycrisp' cultivars, as previously described [17]. No other significant differences in the root system architecture parameters evaluated were observed based on cultivar or rootstock (Tables 2, S2 and S3). Although none of the root system parameters evaluated differed significantly between declining and non-declining trees when a p value threshold of 0.05 was applied, root system depth was significantly shallower in trees that experienced substantial decline than in trees that did not experience substantial decline when a p value threshold of 0.10 was applied instead ($p = 0.058$), possibly due to the high root system depth variability observed (Table 2). Although these data must be interpreted with caution, they may indicate reduced root depth in declining trees.

Table 2. Root system architecture traits of declining (decline severity ratings 2, 3, and 4) and non-declining (decline severity ratings 0 and 1) three-year-old 'Baigent' Gala, 'Honeycrisp', and 'Royal Red Honeycrisp'™ apple trees on 'Malling 26' and 'Geneva 935' rootstocks in a high-density experimental orchard at Cornell AgriTech in Ontario County, NY, USA.

	Declining		Non-Declining				
Number of Trees	25		14				
Parameter	**Mean [a]**	**SD**	**Mean [a]**	**SD**	**Decline *p* Value [b]**	**Decline × Cultivar *p* Value [c]**	**Decline × Rootstock *p* Value [c]**
ØScion (cm)	3.58 a	0.71	3.57 a	0.58	0.940	0.136	0.230
ØRootstock (cm)	4.72 a	0.83	4.61 a	0.82	0.667	0.179	0.472
ØR/ØS (cm)	1.33 a	0.15	1.30 a	0.18	0.552	0.997	0.140
RootstockUG (cm)	25.81 a	10.06	28.91 a	5.83	0.228	0.835	0.351

Table 2. *Cont.*

	Declining		Non-Declining				
Number of Trees	25		14				
Parameter	Mean [a]	SD	Mean [a]	SD	Decline *p* Value [b]	Decline × Cultivar *p* Value [c]	Decline × Rootstock *p* Value [c]
RSDepth (cm)	53.55 a	14.85	63.67 a	15.82	0.058 †	0.431	0.953
RSWidth (cm)	73.44 a	18.77	68.91 a	15.62	0.424	0.214	0.800
RSArea (cm^2)	4281.49 a	1487.92	4314.82 a	1715.49	0.952	0.716	0.867

Root system parameters evaluated were scion trunk diameter at the graft union (ØScion), rootstock trunk diameter at the graft union (ØRootstock), ratio of rootstock trunk diameter to scion trunk diameter (ØR/ØS), rootstock shank length below the soil level (RootstockUG), root system depth (RSDepth), root system width (RSWidth), and projected area of the root system (RSArea). SD indicates standard deviations. [a] Means followed by the same letter in both columns are not significantly different according to Tukey's Honest Significant Difference test with a *p* value threshold of 0.05. [b] *p* values indicate statistical significance of the interactions between root system trait and decline outcome according to one-way ANOVA with a *p* value threshold of 0.05. [c] *p* values indicate the statistical significance of the interactions between root system trait, decline status, and cultivar or rootstock according to two-way ANOVA with a *p* value threshold of 0.05. † $p < 0.10$.

2.5. Weather Patterns at the Experimental Apple Orchard Site

Weather data analysis indicated that between 2019 and 2023, the rainfall received at the experimental orchard site was below average for the region every year except for 2023, based on a 20-year rainfall average of 54.0 cm for these months (Figure 4) [18]. The site received significantly less rainfall in 2022 than in other years, with cumulative rainfall from April to September of only 29.0 cm (Figure 4). The rainfall deficit was particularly pronounced during July, August, and September 2022 (Figure 4). During these three months, the orchard site received only 7.4 cm of rainfall (compared with 24.1, 24.1, 32.3, and 22.4 cm during the same months in 2019, 2020, 2021, and 2023, respectively) (Figure 4).

Additionally, the orchard site experienced more extreme temperature fluctuations in January 2022 compared with the other years of the study, with the lowest minimum temperature of −24.5 °C (compared to the minimum January temperatures of −19.7, −13.4, −14.7, and −10.2 °C in 2019, 2020, 2021, and 2023, respectively), as well as the greatest temperature range, with a total difference between maximum and minimum temperatures of 34.8 °C in January 2022 (compared to January temperature ranges of 31.9, 30.5, 20.4, and 19.6 °C in 2019, 2020, 2021, and 2023, respectively) (Figure 4).

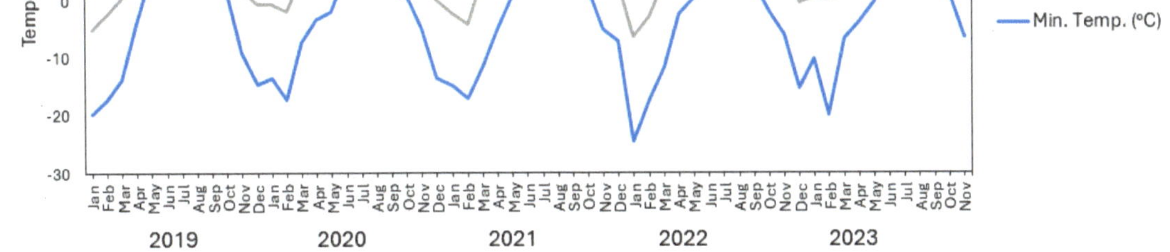

Figure 4. *Cont.*

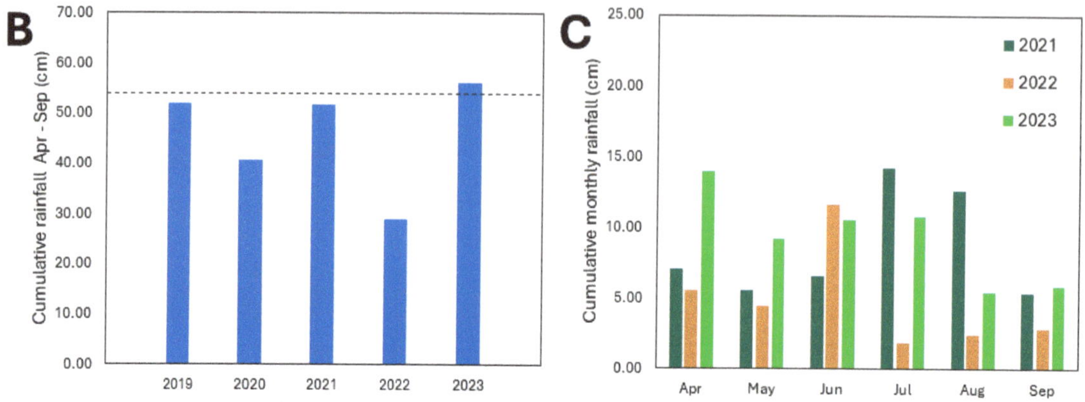

Figure 4. (**A**) Maximum, average, and minimum temperatures each month from January 2019 to November 2023; (**B**) cumulative rainfall during each growing season (April–September) from 2019 to 2023, where the dashed line indicates the twenty-year average rainfall (2003–2023) received during April–September in this region; (**C**) cumulative rainfall each month of the growing season (April–September) from 2021 to 2023 at the site of the experimental apple orchard at Cornell AgriTech in Ontario County, NY, USA.

2.6. Water Balance of Trees at the Experimental Apple Orchard Site

Modeling the water balance of trees in the experimental orchard indicated that the water demand of rainfed trees was higher than the rainfall received each growing season, with maximum negative water balances of −40,451, −33,999, −266,206, and −101,015 L per hectare in 2020, 2021, 2022, and 2023, respectively (Figure 5A). However, the drip irrigation provided was sufficient to satisfy orchard water demand every year of the study, with the exception of 2022 (Figure 5B). When accounting for irrigation, the experimental orchard had an estimated maximum negative water balance of −74,676 L per hectare on 4 September 2022. These analyses revealed a substantial water deficit of trees at the experimental apple orchard site when decline was documented in the summer of 2022.

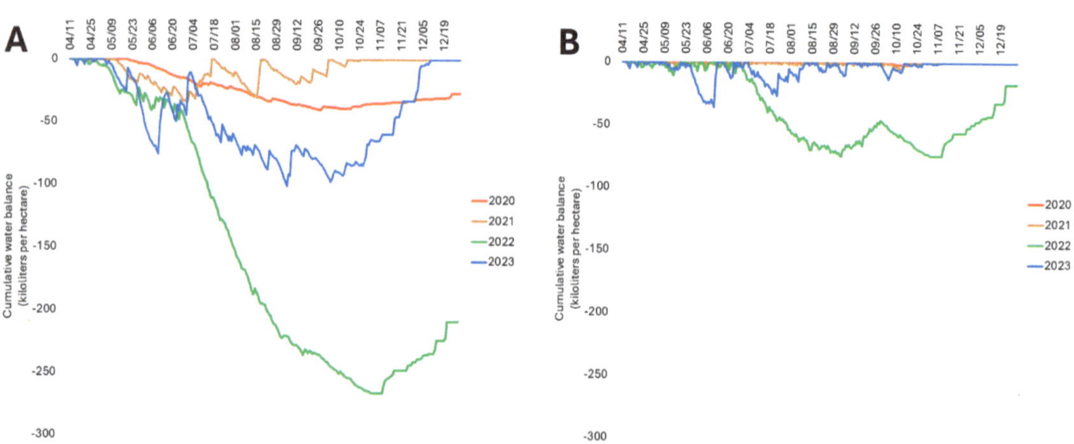

Figure 5. Cumulative orchard water balance in a high-density experimental apple orchard at Cornell AgriTech in Ontario County, NY, USA, from 2020 to 2023 (**A**) without accounting for supplemental irrigation; and (**B**) accounting for supplemental irrigation. The cumulative orchard water balance is expressed in kiloliters per hectare.

2.7. Observations of Rapid Tree Decline in a Commercial Apple Orchard

In addition to monitoring decline severity in the experimental orchard, approximately 600 'Honeycrisp'/'Malling 9 Nic29' trees planted in a commercial orchard in Wayne County, NY, USA, in 2018 were visually evaluated according to the same 0 to 4 decline severity scale during the 2021–2023 growing seasons (years 4–6). The commercial orchard site was located 35 km from the experimental orchard and was selected for this study after the orchardist noticed suddenly declining trees in 2020 (year 3).

In 2021 (year 4), all trees in the commercial orchard block successfully emerged from dormancy, bloomed, and set fruit (decline severity rating 0). At the first decline severity assessment in June (approximately seven weeks post-bloom), trees with symptoms corresponding to all five decline severity ratings were observed, including complete tree collapse (decline severity rating 4) in 20 trees (3.4%, 20 of 588) (Figure 6). By late September (approximately 20 weeks post-bloom), the tree mortality rate (decline severity rating 4) had reached 6.5% (38 of 588) (Figure 6). Trees that collapsed during the 2021 season were removed and replaced with new trees by the orchardist that autumn.

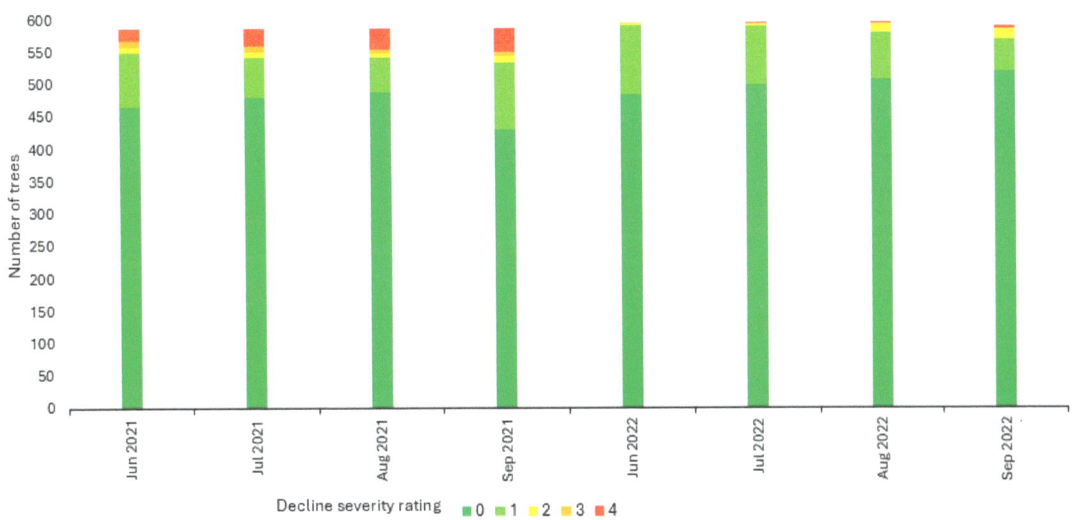

Figure 6. Decline severity over time of 'Honeycrisp' apple trees on 'Malling 9 Nic29' rootstocks established in 2018 in a high-density commercial orchard block in Wayne County, NY, USA, over the course of the 2021 and 2022 growing seasons (June–September). Tree decline severity was visually assessed using a scale from 0 to 4, with a rating of 0 indicating no decline symptoms and a rating of 4 indicating tree mortality. Ratings of 1 to 3 indicate increasing decline severity, with 1 corresponding to chlorosis throughout <50% of the canopy, 2 to chlorosis throughout ≥50% of the canopy, and 3 to chlorosis throughout ≥50% of the canopy, leaf flagging, and poor tree vigor.

The impact of rapid decline in the commercial orchard was significantly less severe in 2022 (year 5). All surviving trees successfully emerged from dormancy, bloomed, and set fruit (decline severity rating 0), including those which had exhibited mild to moderate decline severity the previous year. At the first decline severity assessment in June (approximately four weeks post-bloom), some trees exhibited decline symptoms (decline severity ratings 1, 2, and 3), but none had collapsed (decline severity rating 4) (Figure 6). By September (approximately 21 weeks post-bloom), the tree mortality rate reached only 0.7% (4 of 594), representing a ten-fold decrease in mortality compared to the previous year. Moreover, none of the replacement trees planted the previous year exhibited decline symptoms in 2022. The few trees that collapsed during the 2022 growing season were removed and replaced with new trees by the orchardist that autumn. In 2023 (year 6), no trees declined, including those replaced in 2021 and 2022. No leaf drop or significant trunk swelling or cracking at the graft union was apparent in any tree of the commercial orchard over the course of the study. Together, these results documented the occurrence of extensive rapid decline in the commercial orchard block in year 4, followed by a reduced decline incidence in year 5 and zero decline incidence in year 6.

2.8. Weather Patterns at the Commercial Apple Orchard Site and Water Balance of Trees

An analysis of weather patterns at the commercial orchard site was conducted using data obtained from a nearby weather station. The trees in this orchard block were not supplied with drip irrigation to supplement rainfall. Temperatures never rose above 30 °C in the summer or fell below −15 °C in the winter over the duration of the study (Figure 7), as is typical for this region due to the moderation of seasonal temperature extremes by the proximal Lake Ontario [18]. Monthly temperature ranges were also greatly moderated, with an average monthly temperature range of a mere 9.6 °C between monthly maximum and minimum temperatures at the commercial orchard site, as is typical for this region (Figure 7) [18].

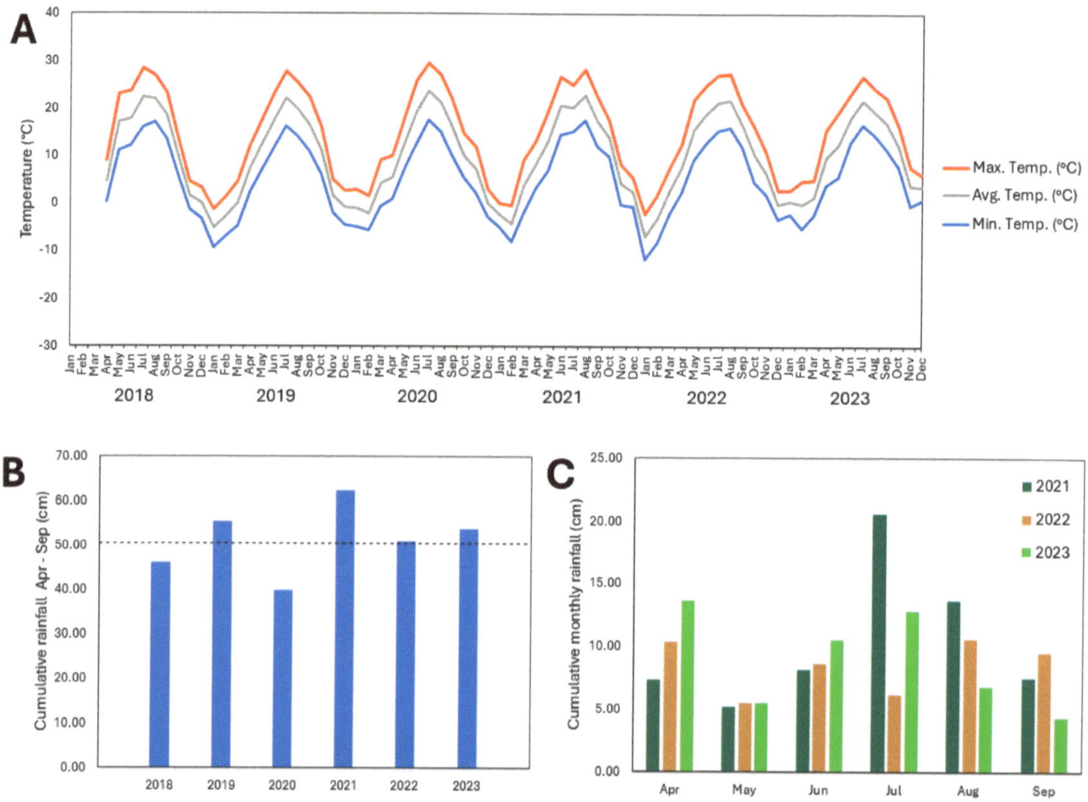

Figure 7. (**A**) Maximum, average, and minimum temperatures each month from March 2018 to December 2023; (**B**) cumulative rainfall during each growing season (April–September) 2018–2023, where the dashed line indicates the twenty-year average rainfall (2003–2023) received during April–September in this region; and (**C**) cumulative rainfall each month of the growing season (April–September) 2021–2023 near the site of a commercial apple orchard in Wayne County, NY, USA.

Between 2018 and 2023, the commercial orchard site received the most rainfall (62.4 cm) during the 2021 growing season and the least (39.9 cm) during the 2020 growing season

(Figure 7). The rainfall received during the growing season was below average for the region in 2018 and 2020 but was at or above average in 2019, 2021, 2022, and 2023, based on a 20-year rainfall average of 50.4 cm for these months (Figure 7) [18].

Modeling the water balance of trees at the commercial apple orchard site indicated that the water demand of rainfed trees was higher than the rainfall received each growing season, with maximum negative water balances of −19,342, −84,538, −75,039, −195,786, and −83,104 L per hectare in 2019, 2020, 2021, 2022, and 2023, respectively (Figure 8). Orchard water deficit was therefore greatest by far in 2022 and least in 2019 (Figure 8).

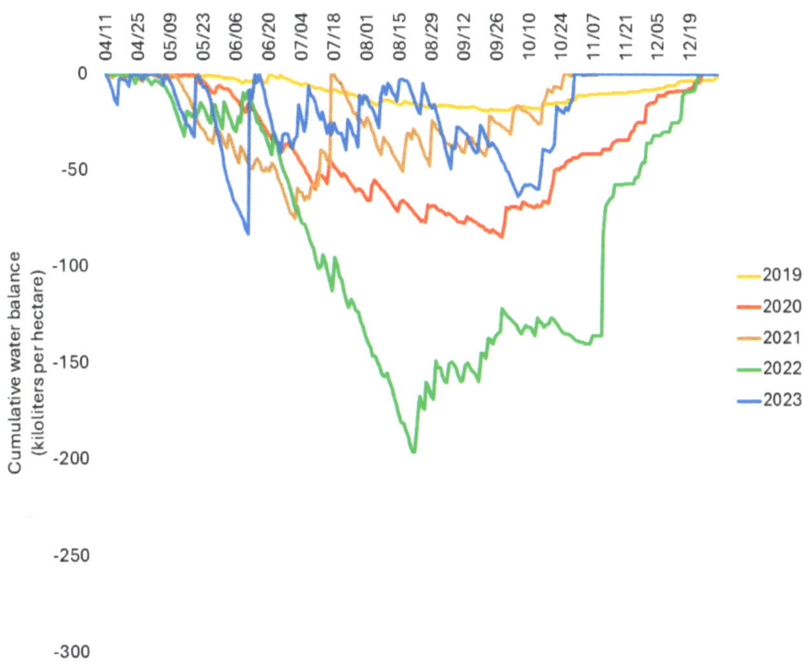

Figure 8. Cumulative orchard water balance in an unirrigated high-density commercial apple orchard in Wayne County, NY, USA, 2019–2023. The cumulative orchard water balance is expressed in kiloliters per hectare.

2.9. Foliar Nutrient Status of Trees in the Commercial Apple Orchard

Foliar nutrient analysis of trees in the commercial orchard in August 2021 (year 4) demonstrated that leaf elemental nitrogen, magnesium, manganese, iron, copper, and zinc concentrations were in the normal ranges for all four decline severity categories evaluated (decline severity ratings 0, 1, 2, and 3) (Figure 9A,D,F–H,J). However, leaf elemental calcium concentrations were in the low range for all four decline severity categories evaluated, possibly indicating an insufficient level of biologically available calcium at the orchard site (Figure 9E). Notably, the leaf elemental concentrations of potassium, phosphorous, and boron differed by decline severity category (Figure 9B,C,I). Both potassium and phosphorous were present in leaf tissue at concentrations in the normal ranges for non-declining trees (decline severity category 0) but at concentrations in the low ranges for mildly, moderately, and severely declining trees (decline severity ratings 1, 2, and 3) (Figure 9B,C). Conversely, boron was present in leaf tissue at concentrations in the low range for non-declining trees (decline severity category 0) and mildly and moderately declining trees (decline severity ratings 1 and 2) but at a concentration in the normal range for severely declining trees (decline severity rating 3) (Figure 9I). Together, these foliar nutrient analyses revealed nutrient imbalances in declining trees in comparison with non-declining trees.

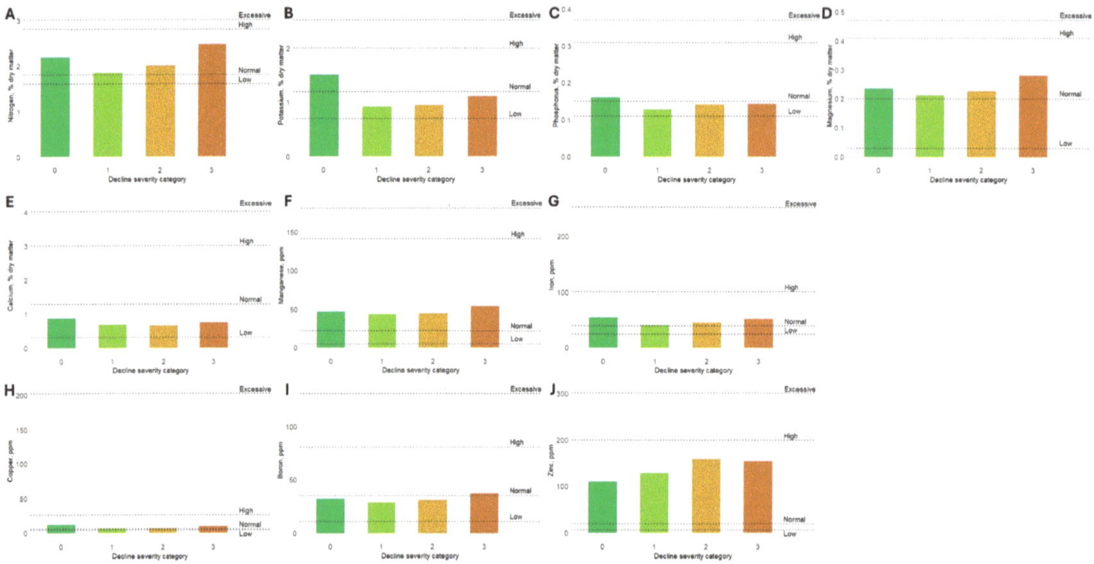

Figure 9. Foliar dry matter nutrient concentrations of (**A**) nitrogen, (**B**) potassium, (**C**) phosphorus, (**D**) magnesium, (**E**) calcium, (**F**) manganese, (**G**) iron, (**H**) copper, (**I**) boron, and (**J**) zinc in three-year-old 'Honeycrisp' apple trees on 'Malling 9 Nic29' rootstocks representing increasing decline severity. Thresholds indicating "Low", "Normal", "High", and "Excessive" mineral nutrient concentrations in cultivated apple according to Pennsylvania State University nutrient guidelines are indicated by dashed lines [19].

3. Discussion

To our knowledge, this study represents the first investigation in which the rapid decline of young apple trees was recapitulated under experimental conditions. Prior attempts to characterize the decline phenomenon have solely comprised observational studies in commercial orchards. In this study, tree decline occurred in the Cornell AgriTech experimental orchard only in 2022, during the fourth growing season following orchard establishment, when full fruiting capacity was achieved for the first time. The observed decline symptoms were consistent with those previously reported in commercial apple orchards, with initial tree decline symptoms developing between June and September (between 4 and 18 weeks post-bloom), while the collapse of severely declining trees occurred between August and October (between 12 and 22 weeks post-bloom). Although nearly one-fifth (17%) of the evaluated trees collapsed, the decline symptoms of many mildly and moderately declining trees appeared to stabilize, never advancing to severe decline and collapse. All of the evaluated trees in decline severity categories 1 or 2 at the end of the 2022 growing season successfully emerged from dormancy the following spring, apparently having recovered.

A similar phenomenon, in which a considerable proportion of declining trees appeared to stabilize at mild or moderate decline stages rather than proceeding to severe decline and collapse, was observed in a declining commercial Zestar!™ (cv. 'Minnewashta', University of Minnesota, Minneapolis, MN, USA) apple orchard in the Hudson Valley production region of New York [16]. The distinction between gradual, reversible tree decline and abrupt tree collapse was explored by Xu et al. (2023) in a multi-site study of declining orchards in British Columbia [11]. In that study, assessments of stem hydraulic characteristics, stomatal conductance, carbon isotope content, and fruit dry matter accumulation revealed an association between tree mortality and severe xylem water transport disruption, while trees with symptoms of gradual decline exhibited moderate hydraulic dysfunction followed either by continued decline or eventual recovery [11]. Although stem hydraulic characteristics were not assessed in this study, our observations corroborate the patterns previously reported [11,16]. The leaf flagging and vascular necrosis observed in declining trees in the experimental orchard (Figure 2) appear to signal a vascular transport disruption. Similar trunk necrosis was previously reported in the commercial orchard monitored in this study [10], as well as in other orchards in New York [6] and in British Columbia [11].

The tree decline phenomenon observed in the Wayne County commercial orchard progressed similarly to that described above in the Cornell AgriTech experimental orchard. Decline symptoms observed were consistent with those previously reported in other commercial apple orchards. Initial symptoms of tree decline primarily developed between June and September each year; however, a subset of trees had already progressed to advanced stages of decline by the time of our first visit in June 2021 (year 4), suggesting that initial symptom development had begun earlier in the season. A similar distinction between abrupt tree collapse and gradual tree decline followed by either symptom stabilization and recovery or further decline, as discussed above, was observed in the commercial orchard block. No severely declining tree (decline severity rating 3) survived past the end of the growing season. However, many declining trees appeared to stabilize at a mild or moderate decline (decline severity ratings 1 and 2) and, after successful emergence from dormancy the following spring, did not display decline symptoms in the following growing seasons. Tree decline in the commercial orchard was most prevalent during the fourth growing season following orchard establishment (2021) and much less prevalent the following year (2022). No decline symptoms were observed in 2023, including in trees planted in 2021 and 2022 to replace those which had collapsed.

Multiplex PCR-based amplicon sequencing revealed the presence of eight viruses and one viroid in frequent mixed infections in the experimental orchard, with up to seven unique viruses and viroids detected in a single tree. These results indicated the presence of more viruses and viroids than expected based on the initial design of the experimental orchard that relied on chip graft inoculation of selected trees with either ACLSV, ASPV, or both viruses. We suspect that this was a result of undetected virus and viroid infection of the rootstock liners or scionwood sourced for this study. Virus testing by multiplex PCR-based amplicon sequencing also revealed the lack of uninfected trees in the experimental orchard, signifying a major limitation of our work due to the absence of uninfected trees to be used as negative controls. However, of the eight viruses and one viroid detected, all but ToRSV were detected in both declining and non-declining trees, as well as in trees which did not collapse and those which did. None of the eight viruses or viroid detected was more prevalent in trees which declined or in trees which collapsed entirely, leading us to reject our hypothesis that one or multiple viruses or viroids would be significantly more prevalent in declining trees. Surprisingly, the prevalence of AHVd was significantly higher in non-declining than in declining trees, as well as in trees which did not collapse versus those which did. As AHVd was detected primarily in trees on G.935 rootstocks, we suspect that this viroid was present in the G.935 rootstock liners sourced for this study. Similarly, three common apple latent viruses, i.e., ACLSV, ASPV, and ASGV were detected at similar rates in both declining and non-declining trees in the Wayne County commercial orchard, as previously reported [10]. Together, our results corroborate the findings of previous studies in Pennsylvania [8], New York [6], Washington [12], and British Columbia [9] by demonstrating the occurrence of viruses in both declining and non-declining trees and, further, identify no association between any virus and decline. ALV1 was initially hypothesized to be implicated in the rapid decline phenomenon [8] but was not detected in this study. Additionally, no association was found between decline outcome and the number of unique viruses and viroids detected in the tree. Together, our findings suggest that individual apple viruses and viroids, or coinfections thereof, are not primarily responsible for the rapid decline of young apple trees on dwarfing and semi-dwarfing rootstocks in high-density orchards. It is important to note, however, that this study focused on evaluating the role of a known set of viruses in the decline phenomenon but not of other pathogens.

The incidence of tree decline varied dramatically from year to year in both orchards evaluated in this study. Over the course of five years in the experimental orchard, tree decline was observed only during the 2022 growing season (year 4), despite the implementation of identical orchard management practices each year (apart from the manual removal of fruitlets in the first two years to promote tree establishment). We therefore hypothesized that variations in weather conditions from year to year at the experimental orchard site would correlate with the year-to-year differences in tree decline incidence. Our analyses revealed a significant rainfall deficit in mid- to late summer and early autumn 2022 and a striking negative water balance in August, September, and October 2022 despite the supplemental water provided via drip irrigation. As our weather analyses are purely observational, we cannot conclude with certainty that this water deficit was responsible for tree decline or collapse in the experimental orchard. However, Xu et al. (2023) recently documented that tree mortality of rapidly declining trees at multiple sites in British Columbia was associated with severe disruption in xylem water transport and hydraulic failure, as evidenced by reduced xylem function, reduced stomatal conductance and stem water potential [11]. It is possible that a substantial water deficit, as observed at the experimental orchard site, may have contributed to mortality in trees already experiencing water transport disruption. Conversely, at the commercial orchard site, the greatest decline incidence was observed during the growing season (2021), with the greatest cumulative rainfall during the study. Although trees at the commercial orchard site were not irrigated, the estimated orchard water balance never fell below zero over the course of the entire study. These findings suggest that either orchard water balance is not primarily responsible for the decline phenomenon or the etiologies of the decline phenomena observed in the experimental and commercial orchards are distinct.

Weather data analyses additionally revealed that the experimental orchard site experienced more extreme temperature fluctuations and a lower minimum temperature in January 2022 compared with the other years of the study. It is possible that the affected trees may have been weakened due to cold temperatures and extreme temperature fluctuations. It should be noted that in the case of the experimental orchard, weather data were obtained from an on-farm weather station located less than three kilometers from the orchard site. However, in the case of the commercial orchard site, weather data were obtained from a weather station located approximately ten kilometers from the orchard site; therefore, microclimatic variations in weather conditions at the commercial orchard site may not be well represented in the data evaluated.

As previously reported for the Wayne County commercial apple orchard by Serrano et al. (2023), the root systems of declining trees significantly differed from the root systems of their non-declining counterparts [10]. Declining trees had significantly smaller scion trunk diameter, root system width, number of primary roots, total root length, root system surface area, and root system dry weight. Other root system traits did not differ significantly between the declining and non-declining trees evaluated, including rootstock trunk diameter, the ratio of scion to rootstock trunk diameter, root system depth, length of the rootstock shank below the soil level, projected root system area, total number of root tips, maximum root diameter, and root system volume [10]. Interestingly, no significant differences between the root systems of declining vs. non-declining trees in a second orchard in eastern New York were previously found [10]. A decrease in average root system depth was noted in declining trees in our experimental orchard, although this difference was not found to be statistically significant when a p value threshold of 0.05 was applied. These findings suggest that either root system architecture is not primarily responsible for the decline phenomenon or the etiologies of the decline phenomena observed in the experimental and commercial orchards are distinct.

A growing body of evidence is pointing toward vascular transport disruption as a cause of the gradual decline of apple trees, suggesting vascular failure as the immediate cause of tree collapse [10,11]. However, it remains unclear what factors may be initiating this vascular transport disruption. More work is needed to address these issues in experimental orchards and elucidate the specific triggers of the decline phenomenon. Ideally, uninfected trees should be used as negative controls in such studies. If the involvement of tree water balance, water stress response, crop load, or root system architecture in decline is confirmed, horticultural mitigation, such as minimizing water stress, would appear to be a compelling option for preventing apple decline. We hope that future studies will elucidate the specific triggers of the decline phenomenon to inform more precise management strategies.

4. Materials and Methods

4.1. Plant Material and Virus Inoculation

Standard nonfeathered nursery trees representing six scion-rootstock combinations ('Baigent' Gala, 'Honeycrisp', and 'Royal Red Honeycrisp'™ on 'Malling 26' or 'Geneva 935' rootstocks) were custom-budded in 2018 and chip graft inoculated with either apple chlorotic leaf spot virus (ACLSV), apple stem pitting virus (ASPV), both viruses, or neither virus. Briefly, a thin segment of bark was excised from the stem of the test plant, exposing the cambium, and then replaced with a segment of bark and cambium tissue of equivalent dimensions from an infected source tree selected based on previous virus diagnosis via double-antibody sandwich enzyme-linked immunosorbent assay (DAS-ELISA) with specific antibodies (BIOREBA, Reinach, Switzerland). Chip grafts were moistened with water, wrapped with grafting tape, and allowed to heal. The initial experimental design consisted of a full factorial design of three scion cultivars, two rootstock genotypes, and four virus treatments comprising 24 combinations. Ten replicates of each of the 24 combinations were produced for a total of 240 trees.

4.2. Experimental Orchard Establishment and Management

An experimental orchard comprising 240 trees was established in June 2019 at Cornell AgriTech in Ontario County, NY, USA, on a Lima series well-drained loam soil [20]. Following bareroot storage at 4 °C, trees were established at the orchard site in a tall spindle production system at 0.9 m × 3.4 m high-density spacing (equivalent to approximately 3260 trees per hectare) according to a randomized complete block design. Rainfall was supplemented with approximately 0.6 cm of water per week by drip irrigation throughout each growing season (May–September), and insecticide, fungicide, and antibiotic applications were made as needed throughout the course of the study for pest and disease management. Any observed fire blight cankers (shoot infections by *E. amylovora*) were pruned out at least 30 cm below the symptomatic tissue using pruning shears disinfected with 2.5% sodium hypochlorite between each pruning cut, and trees were removed if fire blight reached the central leader.

Trees were supported by a three-wire trellis (2.7 m), and the tall spindle production system was maintained by annual winter pruning to remove branches competing with the central leader and lateral branches with diameters greater than 2 cm. Leaders were not headed for the duration of the study. Fruitlets produced by trees during the first two years following orchard establishment (2019 and 2020) were manually removed to encourage successful tree establishment. For the remainder of the study (2021 and 2022), trees were allowed to crop, and fruitlets were manually thinned each spring to approximately two fruits per cluster according to standard practices for conventional tall spindle apple production in the region.

4.3. Commercial Orchard Site Selection

A high-density orchard block of 'Honeycrisp' apple trees on 'Malling 9 Nic29' rootstocks established in 2018 was selected for this study after the orchardist first noticed suddenly declining trees in 2020 (year 3). This commercial block was located in Wayne County, NY, USA, in the Lake Ontario apple production region. Trees were established at the orchard site in a tall spindle production system at 1.25 m × 3 m high-density spacing with trellis support. Conventional horticultural and pest management practices for apple orchards in the region were applied throughout the duration of this work. The orchard block was not irrigated.

4.4. Assessment of Tree Decline Severity

Trees in the experimental orchard at Cornell AgriTech were visually evaluated for decline during the 2020–2023 growing seasons using a scale from 0 to 4, with a rating of 0 indicating no decline symptoms and a rating of 4 indicating tree mortality. Ratings of 1 to 3 indicate increasing decline severity, with 1 corresponding to chlorosis throughout <50% of the canopy, 2 to chlorosis throughout ≥50% of the canopy, and 3 to chlorosis throughout ≥50% of the canopy, leaf flagging, and poor tree vigor. Prior to tree excavation for analysis of root system architecture in late autumn 2022, trees were grouped into two categories: a non-declining category comprised of trees with severity ratings of 0 and 1 and a declining category comprised of trees with severity ratings of 2, 3, and 4 as previously described [10].

Trees in the commercial orchard block in Wayne County were visually evaluated for decline during the 2021–2023 growing seasons using the same 0 to 4 rating.

4.5. Tissue Sampling and Virus Detection from Declining and Non-Declining Trees

Leaf and root tissue samples of declining and non-declining trees in the experimental orchard at Cornell AgriTech were screened for 19 viruses and viroids of pome fruit trees via multiplex PCR-based amplicon sequencing as previously described [21,22]. Briefly, a leaf sample consisting of 5–6 mature leaves from throughout the scaffold and a composite root tissue sample consisting of fine roots excavated from the top 15 cm of soil surrounding the trunk were collected from each tree in the summer of 2022. Then, approximately 100 mg of each tissue sample was homogenized in a guanidine thiocyanate and β-mercaptoethanol lysis buffer with two steel beads (4.5 mm diameter) using a Retsch MM400 mixer mill (Retsch, Haan, Germany) at 30 Hz for 70 s (leaf tissue) or 180 s (root tissue). Next, total RNA was isolated using the GenCatch Plant Total RNA Miniprep Kit (Epoch Life Science, Missouri City, TX, USA) following the manufacturers' guidelines, quantified using a Qubit 2.0 Fluorometer (ThermoFisher Scientific, Waltham, MA, USA), and stored at −80 °C prior to virus screening. Each sample was tested for the presence of 14 viruses and five viroids of pome fruit trees via multiplex PCR-based amplicon sequencing as described by Costa et al. [21]. Namely, each sample was tested for the presence of the alphapartitivirus Pyrus pyrifolia partitivirus 2 (PpPV2), the capillovirus ASGV, the coguviruses citrus concave gum-associated virus (CCGaV) and citrus virus A (CiVA), the foveaviruses apple green crinkle-associated virus (AGCaV) and ASPV, the ilarvirus apple mosaic virus (ApMV), the nepoviruses tobacco ringspot virus (TRSV) and tomato ringspot virus (ToRSV), the rubodviruses apple rubbery wood virus 1 (ARWV1) and apple rubbery wood virus 2 (ARWV2), the tepovirus Prunus virus T (PrVT), the tombusvirus apple luteovirus 1 (ALV1), the trichovirus ACLSV, the apscaviroids apple dimple fruit viroid (ADFVd), apple fruit crinkle viroid (AFCVd), apple scar skin viroid (ASSVd), and pear blister canker viroid (PBCVd), and the pelamoviroid apple hammerhead viroid (AHVd).

4.6. Foliar Nutrient Analysis of Declining and Non-Declining Trees

Foliar nutrient analysis was conducted for declining and non-declining trees in the Wayne County commercial orchard in August 2021. Briefly, approximately six leaves were collected from current season terminal shoots from throughout the canopies of each of ten trees corresponding to each decline severity category 0–3 described above. Leaf tissue samples were not collected from trees representing decline severity category 4 (tree mortality). Bulked leaf samples comprising 60–100 leaves were then prepared for each of the four decline severity rating categories and submitted to DairyOne Forage Laboratory (Ithaca, NY, USA) for nutrient analysis via microwave digestion and inductively coupled plasma optical emission spectroscopy. Nutrient concentration thresholds were assigned according to Pennsylvania State University nutrient guidelines for cultivated apple [19].

4.7. Analysis of Weather Data and Orchard Water Balance

Analysis of weather patterns at the experimental orchard site at Cornell AgriTech was conducted using data obtained from an on-farm weather station located less than three kilometers from the orchard site via Cornell University's Network for Environment and Weather Applications (NEWA). Analysis of weather patterns at the commercial orchard site in Wayne County was conducted using data obtained from a weather station located approximately ten kilometers from the orchard site via NEWA.

Water balance analyses of the experimental orchard and commercial orchard sites were conducted using the NEWA Apple Irrigation Model [23] using the appropriate green tip dates, orchard ages, and tree spacing parameters for each orchard.

4.8. Root System Sampling, Processing, and Analysis

Trees in the experimental orchard at Cornell AgriTech were assigned to one of two decline categories based on their decline severity ratings in October 2022 as described above declining (decline severity ratings 2, 3, and 4) and non-declining (decline severity ratings 0 and 1). Representative trees from each of the two categories were excavated in October 2022 to facilitate root system architecture analysis. The aboveground portion of each tree was cut approximately 30 cm above the graft union and removed. An excavator was then used to remove the soil from around the trunk of each tree in a radius equal to half the tree spacing (approximately 45 cm) and to a depth of approximately 35 cm. After removing loose soil, the lower portion of each tree, including the portion of the root system within the excavation zone, was removed and stored at 4 °C prior to processing and evaluation of root traits.

Each of the excavated root systems was cleaned with tap water and suspended from a horizontal metal rail by threading a screw into the trunk to facilitate imaging as previously described [10]. Four images of each root system were captured using a Nikon D850 camera (Nikon, Minato, Tokyo, Japan) by rotating the root system in 90° increments. Images were analyzed using ImageJ version 1.54d [24] as previously described [10]. Root system architecture parameters evaluated included scion trunk diameter at the graft union, rootstock trunk diameter at the graft union, ratio of rootstock trunk diameter to scion trunk diameter, rootstock shank length below the soil level, root system depth, root system width, and projected area of the root system.

Similarly, trees representative of both the declining and non-declining categories were excavated from the commercial orchard block in Wayne County in November 2021. Root system sampling, processing, and analysis were conducted as previously described [10].

4.9. Statistical Analysis

Statistical analyses were performed using R statistical software (v4.3.1; R Core Team 2023). Chi-squared tests of independence with Bonferroni adjustments were conducted to identify significant associations between decline outcome and virus presence. Type III analysis of variance tests were conducted to evaluate whether root system architecture traits differed by tree decline outcome using the 'car' statistical software package in R (v3.1-3) [25]. Tukey's Honest Significant Difference tests were conducted to assess the significance of differences between group means using the 'agricolae' statistical software package in R (v1.3-7) [26].

Supplementary Materials: The following supporting information can be downloaded at https://www.mdpi.com/article/10.3390/plants13202866/s1, Table S1: Virus and viroid status of 175 three-year-old 'Baigent' Gala, 'Honeycrisp', and 'Royal Red Honeycrisp'™ apple trees on 'Malling 26' and 'Geneva 935' rootstocks in a high-density experimental orchard at Cornell AgriTech in Ontario County, NY, USA; Table S2: Root system architecture traits by cultivar ('Baigent' Gala or 'Honeycrisp') of declining (decline severity ratings 2, 3, and 4) and non-declining (decline severity ratings 0 and 1) three-year-old apple trees in a high-density experimental orchard at Cornell AgriTech in Ontario County, NY, USA; Table S3: Root system architecture traits by rootstock ('Geneva 935' or 'Malling 26') of declining (decline severity ratings 2, 3, and 4) and non-declining (decline severity ratings 0 and 1) three-year-old apple trees in a high-density experimental orchard at Cornell AgriTech in Ontario County, NY, USA.

Author Contributions: Conceptualization, M.F., A.K. and A.O.W.; Methodology, M.F., O.P.H.-G., K.H.L., M.M.S., A.O.W. and J.v.Z.; Formal Analysis, O.P.H.-G. and A.O.W.; Investigation, M.F., O.P.H.-G., A.K., K.H.L., M.M.S., A.O.W. and J.v.Z.; Resources, M.F., O.P.H.-G., A.K., K.H.L., M.M.S. and J.v.Z.; Data Curation, O.P.H.-G. and A.O.W.; Writing—Original Draft Preparation, A.O.W.; Writing—Review and Editing, M.F. and A.O.W.; Visualization, A.O.W.; Supervision, M.F.; Project Administration, A.K. and M.F.; Funding Acquisition, A.K. and M.F. All authors have read and agreed to the published version of the manuscript.

Funding: This research was funded by the USDA NIFA Agriculture and Food Research Initiative Critical Agriculture Research and Extension program, grant number 2021-68008-34117, the New York Apple Research and Development Program, and Cornell AgriTech Venture Funds.

Data Availability Statement: The original contributions presented in the study are included in the article and the Supplementary Materials; further inquiries can be directed to the corresponding author.

Acknowledgments: The authors are grateful to Yu Yang for performing sequencing reactions, Xiaojun Hu for initial sequence data analysis, and Alicia Serrano Gómez for her assistance and advice regarding root system analysis. We also thank the New York orchardist and the fruit tree nursery with whom we were privileged to cooperate, and we especially thank Bill Pitts for his invaluable contributions to the research on tree decline.

Conflicts of Interest: The funders had no role in the design of the study; in the collection, analyses, or interpretation of data; in the writing of the manuscript; or in the decision to publish the results.

References

1. Peter, K.A. Apple Disease—Rapid Apple Decline. Penn State Extension. 2023. Available online: https://www.extension.psu.edu/apple-disease-rapid-apple-decline/ (accessed on 23 September 2024).
2. Prengaman, K. What's killing these trees? *Good Fruit Grower*, 29 March 2017. Available online: https://www.goodfruit.com/whats-killing-these-trees/ (accessed on 23 September 2024).
3. Rosenberger, D. Sudden apple decline: Trunk-related problems in apples. In Proceedings of the Empire State Producers Expo, Syracuse, NY, USA, 18 January 2017.
4. Villani, S. The Ambrosia Beetle Invasion of Western NC Apple Orchards. North Carolina State Extension Plant Pathology. 2017. Available online: https://plantpathology.ces.ncsu.edu/2017/06/the-ambrosia-beetle-invasion-of-western-nc-apple-orchards/ (accessed on 23 September 2024).
5. Rosenberger, D. Factors contributing to the death and decline of young apple trees. *Fruit Q.* **2019**, *27*, 5–8.
6. Singh, J.; Silva, K.J.P.; Fuchs, M.; Khan, A. Potential role of weather, soil and plant microbial communities in rapid decline of apple trees. *PLoS ONE* **2019**, *14*, e0213293. [CrossRef] [PubMed]
7. Harper, S.J. Apple Decline. Washington State University Tree Fruit. 2020. Available online: https://treefruit.wsu.edu/article/apple-decline/ (accessed on 23 September 2024).
8. Liu, H.; Wu, L.; Nikolaeva, E.; Peter, K.; Liu, Z.; Mollov, D.; Cao, M.; Li, R. Characterization of a new apple luteovirus identified by high-throughput sequencing. *Virol. J.* **2018**, *15*, 85. [CrossRef] [PubMed]
9. Xiao, H.; Hao, W.; Storoschuk, G.; MacDonald, J.L.; Sanfaçon, H. Characterizing the virome of apple orchards affected by rapid decline in the Okanagan and Similkameen valleys of British Columbia (Canada). *Pathogens* **2022**, *11*, 1231. [CrossRef] [PubMed]
10. Serrano, A.; Wunsch, A.; Sabety, J.; van Zoeren, J.; Basedow, M.; Miranda Sazo, M.; Fuchs, M.; Khan, A. The comparative root system architecture of declining and non-declining trees in two apple orchards in New York. *Plants* **2023**, *12*, 2644. [CrossRef] [PubMed]
11. Xu, H.; Hannam, K.D.; MacDonald, J.L.; Ediger, D. Field investigation into tree fates from recent apple tree decline: Abrupt hydraulic failure versus gradual hydraulic loss. *Stresses* **2023**, *3*, 256–269. [CrossRef]
12. Wright, A.A.; Cross, A.R.; Harper, S.J. A bushel of viruses: Identification of seventeen novel putative viruses by RNA-seq in six apple trees. *PLoS ONE* **2020**, *15*, e0227669. [CrossRef] [PubMed]
13. Fujikawa, T.; Ota, N.; Sasaki, M.; Nakamura, T.; Iwanami, T. Emergence of apple bacterial quick decline caused by *Dickeya dadantii* in Japan. *J. Gen. Plant Pathol.* **2019**, *85*, 314–319. [CrossRef]
14. Ilyukhin, E.; Schneider, K.; Ellouze, W. First report of *Botryosphaeria dothidea* causing stem canker and dieback of apple trees in Ontario, Canada. *Plant Dis.* **2022**, *106*, 2994. [CrossRef] [PubMed]
15. Hadidi, A.; Barba, M. Economic impact of pome and stone fruit viruses and viroids. In *Virus and Virus-Like Diseases of Pome and Stone Fruits*; Hadidi, A., Barba, M., Candresse, T., Jelkmann, W., Eds.; APS Press: St. Paul, MN, USA, 2011; pp. 1–7.
16. Donahue, D.J.; Elone, S.E. Case study of a declining apple orchard. *Fruit Q.* **2021**, *29*, 28–31.

17. Reig, G.; Lordan, J.; Miranda Sazo, M.; Hoying, S.; Fargione, M.; Reginato, G.; Donahue, D.J.; Francescatto, P.; Fazio, G.; Robinson, T. Long-term performance of 'Gala', 'Fuji' and 'Honeycrisp' apple trees grafted on Geneva® rootstocks and trained to four production systems under New York State climatic conditions. *Sci. Hortic.* **2019**, *244*, 277–293. [CrossRef]
18. Applied Climate Information System, National Oceanic and Atmospheric Administration. NOAA Online Weather Data. Available online: https://www.weather.gov/wrh/Climate (accessed on 8 August 2024).
19. Agricultural Analytical Services Lab, College of Agricultural Sciences, Pennsylvania State University. Interpretive Nutrient Levels for Plant Analysis. Available online: https://agsci.psu.edu/aasl/plant-analysis/plant-tissue-total-analysis/interpretive-nutrient-levels-for-plant-analysis/ (accessed on 23 September 2024).
20. Soil Survey Staff, Natural Resources Conservation Service, United States Department of Agriculture. Official Soil Series Descriptions. Available online: https://soilseries.sc.egov.usda.gov/ (accessed on 22 August 2023).
21. Costa, L.C.; Atha, B.; Hu, X.; Lamour, K.; Yang, Y.; O'Connell, M.; McFarland, C.; Foster, J.A.; Hurtado-Gonzales, O.P. High-throughput detection of a large set of viruses and viroids of pome and stone fruit trees by multiplex PCR-based amplicon sequencing. *Front. Plant Sci.* **2022**, *13*, 1072768. [CrossRef] [PubMed]
22. Wunsch, A.; Hoff, B.; Miranda Sazo, M.; van Zoeren, J.; Lamour, K.H.; Hurtado-Gonzales, O.P.; Fuchs, M. Viruses of apple are seedborne but likely not vertically transmitted. *Viruses* **2024**, *16*, 95. [CrossRef] [PubMed]
23. Robinson, T.L.; Lakso, A.N.; Lordan, J.; Francescatto, P.; Dragoni, D.; DeGaetano, A.T.; Eggleston, K. Precision irrigation management of apple with an apple-specific Penman-Monteith model. *Acta Hortic.* **2017**, *1150*, 245–250. [CrossRef]
24. Schneider, C.A.; Rasband, W.S.; Eliceiri, K.W. NIH Image to ImageJ: 25 years of image analysis. *Nat. Methods* **2012**, *9*, 671–675. [CrossRef] [PubMed]
25. Fox, J.; Weisberg, S. *An R Companion to Applied Regression*, 3rd ed.; Sage: Thousand Oaks, CA, USA, 2019. Available online: https://cran.r-project.org/web/packages/car/index.html (accessed on 23 September 2024).
26. de Mendiburu, F. agricolae: Statistical Procedures for Agricultural Research. R Package Version 1.3-7. 2023. Available online: https://CRAN.R-project.org/package=agricolae/ (accessed on 23 September 2024).

plants

Article

A Physiological and Molecular Focus on the Resistance of "Filippo Ceo" Almond Tree to *Xylella fastidiosa*

Mariarosaria De Pascali [1,2], Davide Greco [1], Marzia Vergine [1,*], Giambattista Carluccio [1], Luigi De Bellis [1,2] and Andrea Luvisi [1]

[1] Department of Biological and Environmental Sciences and Technologies, University of Salento, 73100 Lecce, Italy; mariarosaria.depascali@unisalento.it (M.D.P.); davide.greco@unisalento.it (D.G.); giambattista.carluccio@unisalento.it (G.C.); luigi.debellis@unisalento.it (L.D.B.); andrea.luvisi@unisalento.it (A.L.)

[2] National Biodiversity Future Center, 90133 Palermo, Italy

* Correspondence: marzia.vergine@unisalento.it

Abstract: The impact of *Xylella fastidiosa* (*Xf*) subsp. *pauca* on the environment and economy of Southern Italy has been devastating. To restore the landscape and support the local economy, introducing new crops is crucial for restoring destroyed olive groves, and the almond tree (*Prunus dulcis* Mill. D. A. Webb) could be a promising candidate. This work focused on the resistance of the cultivar "Filippo Ceo" to *Xf* and evaluated its physiological and molecular responses to individual stresses (drought or pathogen stress) and combined stress factors under field conditions over three seasons. Filippo Ceo showed a low pathogen concentration ($\approx 10^3$ CFU mL^{-1}) and a lack of almond leaf scorch symptoms. Physiologically, an excellent plant water status was observed (RWC 82–89%) regardless of the stress conditions, which was associated with an increased proline content compared to that of the control plants, particularly in response to *Xf* stress ($\approx$8-fold). The plant's response did not lead to a gene modulation that was specific to different stress factors but seemed more indistinct: upregulation of the *LEA* and *DHN* gene transcripts by *Xf* was observed, while the *PR* transcript was upregulated by drought stress. In addition, the genes encoding the transcription factors (TFs) were differentially induced by stress conditions. Filippo Ceo could be an excellent cultivar for coexistence with *Xf* subps. *pauca,* confirming its resistance to both water stress and the pathogen, although this similar health status was achieved differently due to transcriptional reprogramming that results in the modulation of genes directly or indirectly involved in defence strategies.

Keywords: climate change; combined stresses; drought; plant disease; transcription factor

Citation: De Pascali, M.; Greco, D.; Vergine, M.; Carluccio, G.; De Bellis, L.; Luvisi, A. A Physiological and Molecular Focus on the Resistance of "Filippo Ceo" Almond Tree to *Xylella fastidiosa*. *Plants* **2024**, *13*, 576. https://doi.org/10.3390/plants13050576

Academic Editors: Violetta Katarzyna Macioszek, Iwona Ciereszko and Andrzej K. Kononowicz

Received: 23 January 2024
Revised: 8 February 2024
Accepted: 19 February 2024
Published: 20 February 2024

1. Introduction

In nature, plants face numerous abiotic and biotic stress factors simultaneously, which increase due to climate change, thus affecting their growth, yield and survival. Among the possible stress combinations, the combination of drought and pathogens is one of the most devastating [1,2]. Drought is a frequent environmental stress that exacerbates the damage caused by bacterial pathogens [3,4]. For example, drought worsens Pierce's disease caused by *Xylella fastidiosa* (*Xf*) in *Vitis vinifera* [5], the bacterial blight caused by *Xanthomonas oryzae* pv. *oryzae* in rice [6], and the common scab disease caused by *Streptomyces* spp. in potato [7]. However, other works have reported that drought acts positively, improving the plant defence response against pathogens, as reported in tomato and grapevine against the necrotrophic fungus *Botrytis cinerea* [8,9]. Several studies [10–13] have also shown that Leccino's resistance to *Xf* may be attributed to its vulnerability to water deficit. This susceptibility could activate alternative defence mechanisms that assist the plant in response to pathogens.

To date, the relationship between the mechanisms of host resistance and drought tolerance has not been determined. At the molecular level, the only certainty is that both are controlled by a complex network of events involving the activation/inactivation of the expression of a large number of genes [14]. In particular, drought stress induces changes in the activity of genes encoding stress response proteins, including dehydrins (DHNs) and late embryogenesis-abundant (LEA) proteins, which enable plants to resist drought [15]. On the other hand, phytopathogens primarily induce the expression of genes encoding pathogenesis-related (PR) proteins [16], which are the molecules recruited to defend plants against pathogen attack and are also involved in the crosstalk of abiotic and biotic stress signalling.

In recent years, transcription factors (TFs) have been established as the main regulators of changes in gene expression and, thus, the major factors that facilitate stress responses in plants. In fact, TFs control the transcription rate by binding to cis-regulatory promoter elements and play a significant role in signal transduction networks. This thus leads to an improvement in plant tolerance.

In plants, more than 80 TF families have been identified. Some of these proteins, such as basic region-leucine zipper (bZIP), myeloblastosis-related proteins (MYB), NAC, WRKY and Zn finger proteins, are directly implicated in stress responses and are associated with enhanced resistance. In addition, they have been reported to impart plant cross-tolerance to abiotic and biotic stresses [17]. For instance, *Os*bZIP23 overexpression was shown to confer abscisic acid (ABA) hypersensitivity and increased salinity and drought tolerance in rice [18], and NAC TFs were shown to directly induce the pathogenesis-related genes *PR1*, *PR2*, and *PR5* [19] and enhance drought resistance [20]. WRKY proteins are particularly associated with the regulation of plant pathogen responses. However, recent functional analyses have also implicated WRKY TFs in abiotic stress responses [21]. Several studies have focused on the role of MYB TFs as key factors in regulating abiotic and biotic stress responses. As reviewed by Fang et al. [22] MYB TFs are active in stress signalling because they regulate downstream genes in response to stresses. Finally, Zn Finger enhances plant drought resistance by increasing the levels of osmotic adjustment substances. In fact, overexpressing *OsMSR15* [23] and *ZFP3* in transgenic *Arabidopsis* [24] results in an increase in drought tolerance by maintaining a higher proline content, reducing electrolyte leakage, and increasing stress-responsive gene expression. Generally, these TFs play a crucial role in abiotic and biotic stress responses [25]. Their ability to control a set of genes to modulate their expression via different pathways in response to various stimuli empowers plant defence. TFs are considered excellent targets for increasing plant adaptation to stress.

Xylella fastidiosa (*Xf*) is one of the most devastating pathogens and is able to infect a wide range of host plants, causing diseases that can cause severe yield losses in highly economically important crops, such as Pierce's disease in grapevine, citrus variegated chlorosis, olive quick decline syndrome (OQDS) and almond leaf scorch disease (ALSD) [26]. The latter is a severe disease that threatens almond (*Prunus dulcis* Mill. D. A. Webb) in several areas worldwide [27–29]. Recently, the pathogen caused severe yield losses in almond crops and eradicated 1000 trees in Spain. ALSD has also affected more than 79% of almond trees in Majorca [30]. Since 2017, symptoms of ALSD have also been observed on 30-year-old almond trees in mainland Spain [31]. According to Amanifar et al., 2022 [32], the severity of the disease is also related to the sensitivity of the cultivars.

Since 2013, *Xf* subsp. *pauca* has caused OQDS in Apulia and destroyed millions of olive trees, leading not only to massive damage to the local economy but also to complete changes in the landscape. In addition to olive trees, which exhibit lower infection levels than infected olive trees, almond trees were also found to be infected and symptomatic [33]. Although these features have not been fully investigated, they make almond trees effective for carrying out crop renewal to create new production chains and increase the biodiversity in an area entirely ravaged by the bacterium [34,35].

The almond tree has adapted to temperate and Mediterranean regions, as it can grow in conditions of water shortage without requiring irrigation, owing to adaptive mechanisms such as osmotic adjustment, stomatal conductance decreases, and the water loss rate increases [36]. In addition, the almond tree is an icon for the agricultural landscape of Apulia and is of significant economic importance. "Filippo Ceo" is the most appreciated cultivar in Apulia due to its high productivity and fruit yield. It has also been shown to have resistance traits that are similar to those of resistant olive tree varieties [35]. Considering the projections of the increasing impact of climate change, this work focuses on the resistance of "Filippo Ceo" to *Xf*. The study also evaluates the individual and combined effects of pathogens and water deficit to contribute to plant management and protection in areas threatened by *Xf*.

We thus evaluated the response of the "Filippo Ceo" cultivar to *Xf* infection and drought under individual stress (drought or pathogen) and combined stress in field conditions in a two-year trial. We assessed the changes in physiological parameters and studied the expression profiles of genes encoding proteins (DHN, LEA, and PR) and TFs (bZIP, MYB, NAC, WRKY, and Zn finger) involved in the response to these stresses. The overall aim of the study is to provide information on the resistance mechanisms of the almond tree Filippo Ceo for use in the recovery of an area compromised by the devastating action of the pathogen.

2. Results

2.1. Plant Health and Physiological Characterisation

No trees in the orchard under investigation experienced symptoms associated with *X. fastidiosa* infection throughout the whole trial period (2021–2022). Regarding the *Xf*-positive plants, the PCR-based analyses showed values not exceeding 10^3 CFU mL^{-1} (Figure 1). Thus, the bacterial concentration detected was rather low, which probably explains the absence of symptoms.

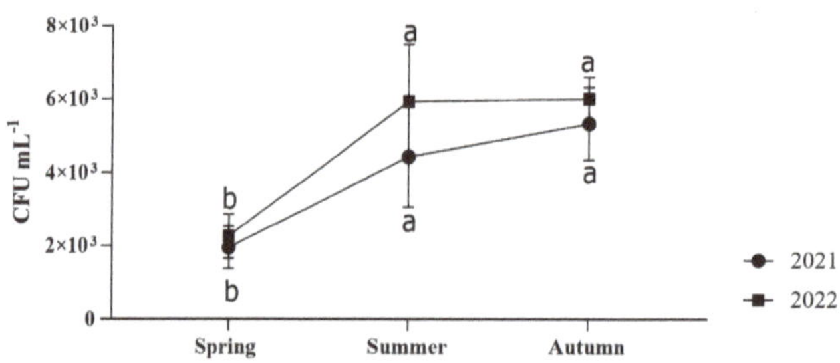

Figure 1. CFU mL^{-1} of *Xf*-positive almond plants sampled in this study. Statistical analysis was carried out by ANOVA followed by the Tukey HSD post hoc test. Different letters correspond to statistically different means.

The physiological characterisation of the cv. "Filippo Ceo" was performed by evaluating the relative water content (Figure 2). During the two seasons of the study, the control plants presented constant RWC values (RWC 95%), indicating an optimal and stable water status regardless of the sampling period. Under individual stress and in combination with drought/*Xf*, the RWC values slightly decreased compared to those of the control plants. This reduction did not exceed 10%, thus maintaining good water status, as shown by the high RWC values (82–89%). Notably, within each sampling period, the different stresses led to a substantially similar RWC, indicating that the water status of the leaves was essentially the same both when the plants were not irrigated and in the presence of the pathogen or when the two factors were combined.

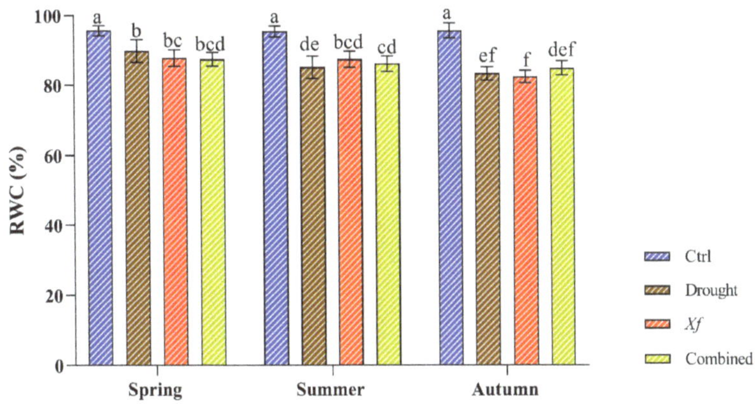

Figure 2. Relative water content (RWC) determined on cv. "Filippo Ceo" leaves subjected to drought, *Xylella fastidiosa* and combined over a two-year period of observation. Statistical analysis was carried out by ANOVA followed by the Tukey HSD post hoc test. Different letters correspond to statistically different means.

Compared with those of the control, the stress conditions significantly increased the content of free proline (Figure 3). Water deficit led to similar proline accumulation patterns regardless of the sampling period, with an average 3.46-fold increase compared to that of the control. Most strikingly, more proline accumulated in response to *Xf*. Our data showed an average increase of 8.14-fold compared to the control conditions. This trend was confirmed in all sampling seasons. The simultaneous action of drought and pathogen led to an average 6.43-fold increase compared to the non-stressed plants, higher than the water stress but lower than the pathogen alone.

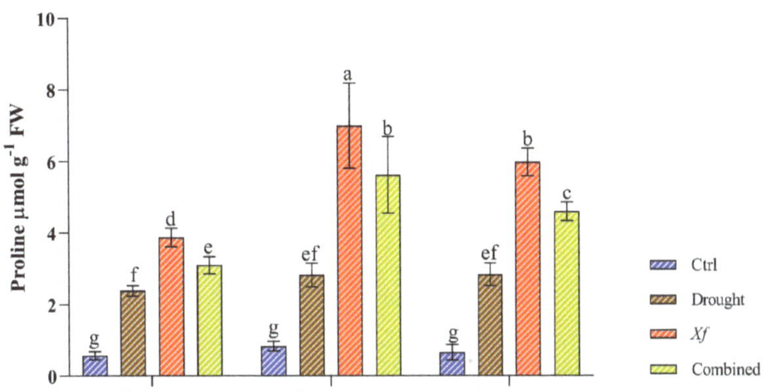

Figure 3. Proline content (μmol g^{-1} FW) determined in leaves of the "Filippo Ceo" cultivar subjected to drought, *Xylella fastidiosa* and combined stresses via two-year observation. Statistical analysis was carried out by ANOVA followed by the Tukey HSD post hoc test. Different letters correspond to significantly different means.

2.2. Gene Expression Analysis

To gain insights into the molecular mechanisms potentially involved in the resistance of "Filippo Ceo" to individual and combined stresses, we assessed the expression patterns of several marker genes encoding proteins and transcription factors known to be involved in drought and pathogen stress responses. The plant's response to different stress factors did not lead to specific gene modulation; however, the response was less clear.

Xf induced the expression of genes associated with abiotic stress the most (DHN and LEA), just as water stress better stimulated the expression of genes related to biotic stress (PR) (Figure 4). In fact, *Xf* significantly increased the expression of the LEA and DHN genes in spring and summer (e.g., up to 2.48 $\log_2$ FC for LEA and up to 2.71 $\log_2$ FC for DHN). Although these genes play a crucial role in plant adaptation to drought stress, the pathogen leads to their expression, suggesting their involvement in plant defence strategies. Similarly, the PR gene, which is usually known to be involved in the pathogen defence mechanism, exhibited a greater response to drought, regardless of the season. This trend was most evident in the spring when the relative expression level was 2.16 $\log_2$ FC. In contrast, the combination of abiotic and biotic stress did not result in differential expression of the three genes across the seasons considered, with similar or intermediate values to the single stress factors in spring or lower values in the rest of the year.

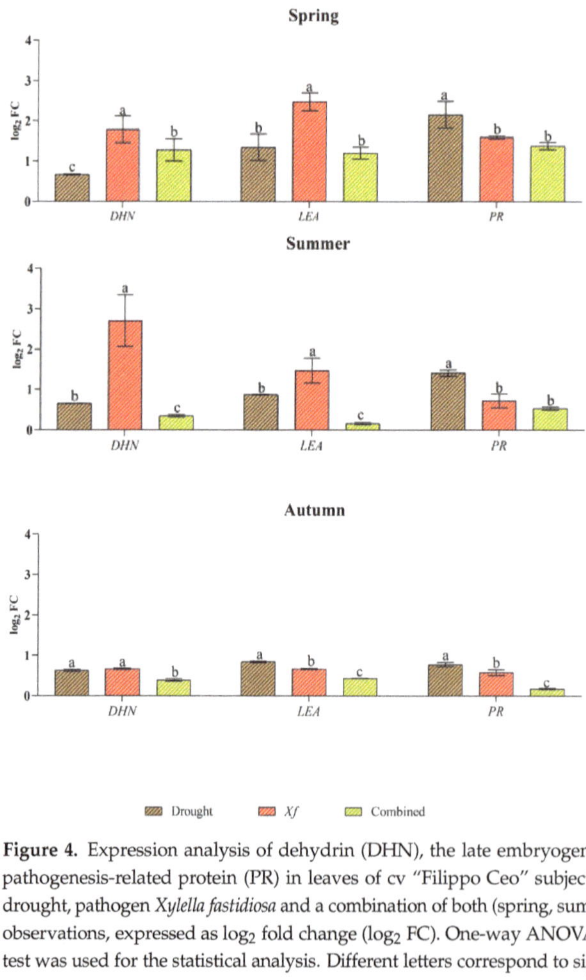

Figure 4. Expression analysis of dehydrin (DHN), the late embryogenesis-abundant protein (LEA) pathogenesis-related protein (PR) in leaves of cv "Filippo Ceo" subjected to different stress factors: drought, pathogen *Xylella fastidiosa* and a combination of both (spring, summer and autumn) in two-year observations, expressed as $\log_2$ fold change ($\log_2$ FC). One-way ANOVA with Tukey's HSD post hoc test was used for the statistical analysis. Different letters correspond to significantly different means.

Concerning the expression profiles of the genes encoding TFs (Figure 5), all the genes were induced under drought, although with generally lower expression than that of *Xf* or the combined stress factors. The only exceptions were observed in spring for bZIPs and, to a lesser extent, for Zn fingers because a greater response of bZIPs (4.63 $\log_2$ FC) was found together with a significant accumulation of transcripts for Zn fingers. The genes *MYB*, *NAC* and *WRKY* were not particularly affected by drought, and a very low expression level was observed for all the TFs analysed in the summer and autumn. The expression profile of the pathogen stress factor was completely different from that under the other stress conditions. With the exception of the aforementioned divergence for *bZIP* and *Zn fingers* in spring, *Xf* seemed to stimulate greater expression of TFs than water stress alone in all the seasons. This trend was most evident in the spring when the pathogen induced the expression of *MYB*, for which the value was 3.69 $\log_2$ FC, while the values of *NAC*, *WRKY*, *bZIP* and *Zn finger* were between 2.49 and 1.84 $\log_2$ FC. In the summer, the expression levels of all the genes considered were similar (range between 1.47 and 1.87 $\log_2$ FC), while in autumn, the expression level was lower than 1 $\log_2$ FC. *MYB* (2.26 $\log_2$ FC), *WRKY* (2.42 $\log_2$ FC), and *Zn finger* (2.22 $\log_2$ FC) genes were notably upregulated in response to the combined stress factors in the spring. A notable decrease in gene expression was observed in the summer and autumn, with values below 1 $\log_2$ FC, except for *MYB*, which showed an expression level equal to 1.60 $\log_2$ FC only in the summer. As with gene expression, combined stress factor-related stress sometimes led to similar or intermediate TF expression levels compared to those resulting from single stress.

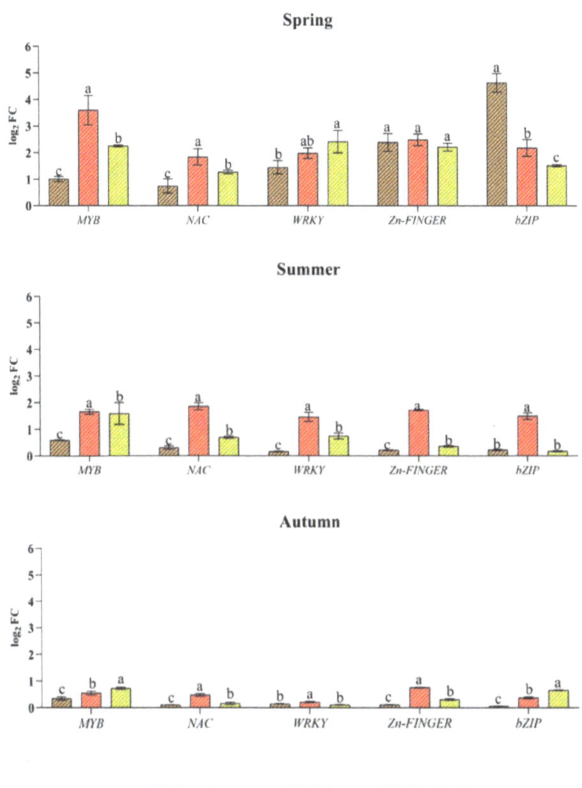

Figure 5. Expression analysis of transcription factors (TFs) involved in the regulation of defence and responses to different stress factors in plants: dehydration-responsive element binding myeloblastosis

(MYB); NAC; WRKY; Zn finger; basic leucine zipper (bZIP); in leaves of cv "Filippo Ceo" subjected to stresses: drought, pathogen *Xylella fastidiosa* combination of both (spring, summer and autumn) in two-year observations, expressed as $\log_2$ fold change ($\log_2$ FC). One-way ANOVA with Tukey's HSD post hoc test was used for the statistical analysis. Different letters correspond to statistically different means.

3. Discussion

In Salento, the spread of *Xf* has significantly damaged both the environment and the economy. This has led to the development of strategies aimed at restoring not only the landscape but also bolstering the local economy that has been severely affected by OQDS. One of the most effective strategies could be the varietal renewal of destroyed olive groves, not only through the use of cultivars that are resistant to bacteria but also through the introduction of new crops. Due to some of its particular characteristics, the "Filippo Ceo" almond tree could be an excellent candidate because it does not require frequent irrigation, and no appreciable symptoms attributable to *Xf* were observed during the trial, despite the high inoculum pressure present in the area. This finding suggested that "Filippo Ceo" is a cv resistant to the *Xf* subsp. *pauca* strain "De Donno".

We thus investigated the mechanisms underlying this ability. The RWC values measured were high both for the control plants and for the stressed plants, confirming that this almond cultivar is a hardy plant that is capable of resisting adverse factors. In fact, plants that maintain an excellent physiological balance under stress conditions have higher RWC values. In contrast, plants with lower RWC values are believed to be more sensitive to water deficit, making the RWC an excellent indicator of the plant's water status and resistance to water stress. In this study, the analysed "Filippo Ceo" almond plants exhibited consistent relative water content (RWC) values and health statuses, despite the different biotic, abiotic, or combinations of stressors. This difference may be due to distinct resistance mechanisms, possibly implicating proline accumulation as a contributing strategy. Proline stabilises the cellular structure, proteins, and enzymes, acts as an antioxidant and provides ROS defence [37]. It thus works as an osmoprotectant, enabling plants to tolerate stress [38]. By accumulating proline, plants lower their osmotic potential and delay drought-responsive stomatal closure through turgor maintenance and sustain normal photosynthesis and assimilation, thus maintaining plant growth and development [39]. Our findings showed elevated proline levels in almond trees subjected to stress conditions. In particular, the presence of the pathogen led to a more significant increase in the production of proline than that in the control, drought, or combined stress conditions, despite the similar RWC values. This finding suggested the active role of proline in the resistance of "Filippo Ceo" to *Xf*. However, further studies are required to fully understand the relationship between solute accumulation and stress adaptation.

At the molecular level, stress triggers cascading events that culminate in gene expression changes. It is widely reported that dehydrins and LEA proteins play a primary role in the response of plants to abiotic stress. They perform specific protective functions in plant cells, such as maintaining the integrity of crucial cell structures, and alleviate oxidative damage in stressed plants. In this work, the *DHN* and *LEA* genes were induced by all the stressors considered, confirming their involvement in stress responses. However, unexpectedly, *Xf* presented the greatest accumulation of transcripts compared to those under drought and combined stress. The roles of DHNs and LEA proteins in response to abiotic stress are well established, but their involvement in biotic stress responses is relatively unknown. However, some studies have shown that *DHNs* can be induced in response to attack by filamentous pathogens, such as *Erysiphe necator* [40], in grapevine plants or in combination with drought stress, as observed in oak plants infected with *Phytophthora cinnamomi* [41]. In olive, *DHNs* have also been shown to be induced by *Xf* in combination with drought [11]. Our results for almond suggest putative roles for DHNs and LEA proteins in modulating defence responses to vascular pathogens. These findings suggest that stress-related proteins may play a fundamental role in protecting plants against biotic stress.

In contrast, the *PR* genes are commonly induced by phytopathogens as well as defence-related signalling molecules, leading to increased resistance to pathogens [11,42,43]. Recent studies have reported that *PR* genes are also significantly induced by abiotic stressors, which makes them highly promising candidates for developing crop varieties that can tolerate multiple stresses [44–46]. In fact, our data showed that the transcript levels of *PR-1-like* were greater in response to drought stress than in response to *Xf* infection, confirming that these genes are activated not only in response to pathogen attack. Taken together, these data suggest significant crosstalk and trade-offs between almond tree responses to water deficit and *Xf*. In fact, the plant's response to different stress factors did not lead to specific gene modulation; however, the response appeared to be more unclear. This finding thus confirms that stressor responses share protective mechanisms (cross-tolerance) or signalling/regulatory pathways that activate independent protective mechanisms (cross-talk) [47].

TFs are being extensively studied because, similar to switches, they monitor the activity of stress responses in many genes in a coordinated manner and represent tools for enhancing abiotic or biotic stress tolerance in plants. TFs play a significant role in signal transduction networks, from the perception of a stress signal to the regulation and expression of almost any gene [48]. This work analysed the expression profiles of genes encoding transcription factors such as MYB, NAC, WRKY, Zn finger and bZIP. In particular, MYB proteins support a wide range of signalling cascades between abiotic and biotic stress signals [49,50]. The *MYB* gene also controls the production of dehydrins and LEA proteins, along with a greater accumulation of sugars and proline, and in *Vitis vinifera*, upregulation of the MYB transcription factor was observed in response to *Xf* [51]. In our work, the *MYB* gene was induced by all the stress factors considered; however, compared with drought and combined stress factors, *Xf* led to the greatest accumulation of transcripts during the period of greatest vegetative development. The same trend was observed for *DHN* and *LEA* gene expression, suggesting that MYB may be involved in their activation.

However, in the spring, the expression of the *ZIP1* gene increased in response to a water deficit compared with that in response to *Xf* and the combined stress factors. ZIP proteins control the signal transduction networks that mediate the response to drought [52].

In response to pathogen infection, numerous *NAC* genes are induced [53], and the overexpression or silencing of these genes results in enhanced or reduced resistance to pathogens [54]. In line with findings reported in the literature, the *NAC* gene is mainly induced by the pathogen, confirming that NAC TFs link signalling pathways to regulate resistance against pathogens. However, upregulation of the transcription factor NAC was recently not detected in the *Xf*-infected almond cv. Avijor [55], suggesting a different response of the cultivars to ALSD symptoms [26].

In the spring, the *WRKY* transcript level increased significantly in response to the combination of water deficit and pathogen stresses. Several WRKYs are active at the crossroads of plant responses to biotic and abiotic stresses [56]. In particular, a study conducted by Lee et al. in 2018 [57] demonstrated that OsWRKY11 serves as a positive regulator of plant defence responses against pathogen infection and drought stress.

Zn finger motifs (ZFPs) are crucial for plant growth and development, stress tolerance, transcriptional regulation, RNA binding and protein–protein interactions [58]. Several studies have reported that ZFPs play a significant role in the abiotic stress response in plants [59]. In particular, several ZFPs have been shown to play significant roles in enhancing drought [60]. In our work, however, drought stress led to a significant increase in the transcription of genes only in the spring, probably due to concomitant vegetative development and high water requirements. According to the gene expression profile results, the ZFP TF seems to be more involved in the pathogen response. Several studies have shown the involvement of ZFPs in plant–pathogen interactions. For example, the overexpression of these genes in transgenic tobacco plants has been found to enhance immunity against pathogens and induce the expression of defence-related genes [61]. ZFPs may thus also be involved in resistance to *Xf*.

Generally, the transcription factors analysed presented different expression profiles and were induced by all the stress factors in the spring. However, in the summer and autumn, the presence of the pathogen alone led to a significant accumulation of transcripts. These findings suggest the putative importance of these genes in the resistance of almond trees to *Xf*.

4. Materials and Methods

4.1. Plant Materials

The study was carried out in an 18-year-old commercial almond orchard located in Veglie (Lecce, Italy) during the 2021 and 2022 seasons. The experiments were conducted on the Apulian variety "Filippo Ceo" grafted onto GF-677 rootstock.

The planting layout consisted of trees distributed in 16 horizontal rows and a planting distance of 6 m × 4 m on sandy soil (average soil texture parameters: 78% sand, 15.4% silt, 5.1% clay, and 1.5% organic matter).

The orchard is located in an area where *Xf* has been present since 2015 [61]. One municipality within the area was declared to be infected in 2015 and overwhelmed by the pathogen. The orchard is thus subjected to the continuous pressure of the natural inoculum of *Xf*, as it is surrounded by olive groves that have been seriously affected by OQDS.

The experimental design followed a randomised block plan, and each experimental set consisted of three trees (in total, n = 12 trees per treatment). Sampling was carried out at three different climatic stages (spring, summer and autumn). The experimental design included four plant conditions: *Xf*-positive trees naturally infected and irrigated ("*X. fastidiosa*", three plants/cultivar); *Xf*-negative trees subjected to water deficit ("drought", three plants/cultivar); *Xf*-positive trees subjected to water deficit ("combined", three plants/cultivar); and *Xf*-negative trees and irrigated ("control", three plants/cultivar).

The selected almond trees had previously received the same agronomic treatments. The insect control and phytosanitary treatments, according to EU Decision 2015/789 [62], were carried out by the farmers. In addition, the trees were monitored for symptoms caused by natural infection with *Pseudococcus viburni*, *Pseudomonas siringae*, *Xanthomonas arboricola* pv. *pruni*, and *Candidatus Phytoplasma phoenicium* during sampling. Plants showing symptoms related to potential co-infections were excluded from the trial. Diagnostic tests (real-time PCR) for detecting Plum pox virus, according to Olmos et al. [63], were also carried out.

The *Xf*-positive or *Xf*-negative plants were assessed for the presence of symptoms using the severity scale of 1 to 3, proposed by Luvisi et al. [64] and the qPCR assay according to Harper et al. [65]; plants were tested in 2021 and in 2022 during the three sampling periods (spring, summer, autumn). The plants were considered *Xf*-negative when the twig samples were negative according to the *Xf* assay in each sampling period and positive when the twig samples of the Filippo Ceo trees were positive according to the *Xf* assay, with Cq values ≤ 32 in each sampling period. The *Xf* concentration, expressed as bacterial CFU mL^{-1}, was inferred from Cq values using a standard curve with dilutions ranging from 10^2 to 10^7 CFU mL^{-1}, as described by D'Attoma et al. [66].

The trees were subjected to two irrigation regimens. The "*X. fastidiosa*" and "Control" plants were watered weekly following local practices, maintaining at least 90% of the soil water capacity (SWC). For the "Drought" and "Combined" plants, the regime soil moisture was maintained at approximately 40% of the SWC. The distance between blocks of irrigated and non-irrigated plants was at least 10 m.

4.2. Relative Water Content (RWC)

To determine the leaf water status, ten leaves per tree were placed in tubes, and the tubes were closed on site. In the laboratory, several parameters were analysed: fresh weight (FW), turgid weight at full turgor (TW) (measured after the leaf petioles were immersed for

24 h in deionised water at 4 °C), and dry weight (DW) (measured after drying at 80 °C). The RWC was calculated as follows: RWC (%) = (FW − DW)/(TW − DW) × 100. The RWC measurements were performed in spring (May), summer (July), and autumn (September) in 2021 and 2022.

4.3. Free Proline Determination

A total of 0.5 g of almond leaves from control and stressed plants was homogenised in 10 mL of 3% aqueous sulfosalicylic acid to determine the free proline content. The proline concentration was calculated according to Bates et al. [67].

4.4. Total RNA Isolation, cDNA Synthesis, and Real-Time PCR Analysis

Total RNA was extracted from leaves via the CTAB-based procedure according to the methods of Gambino et al. [68]. RNA samples were treated with DNase I (Promega, Madison, WI, USA), after which the absorbance was read at 260 and 280 nm to determine the RNA concentration and purity. According to the manufacturer's instructions, cDNA synthesis was performed using TaqMan® Reverse Transcription Reagents (Applied Biosystems, Waltham, MA, USA) with oligo (dT)18 as a primer. RT–qPCR was carried out using SYBR Green fluorescent detection in a real-time PCR thermal cycler (QuantStudio™ 3 Real-Time PCR System, Applied Biosystems, Waltham, MA, USA). The PCR program was as follows: 2 min at 50 °C and 10 min at 95 °C, followed by 45 cycles of 95 °C for 15 s and 60 °C for 1 min [12]. The primers used (Table 1) were retrieved from the literature. The primers used were designed for genes related to drought responses, such as DHN and LEA; for genes involved in the pathogen stress response, namely, pathogenesis-related protein 1-like (PR); and finally, for genes encoding TFs such as bZIP, NAC, MYB, WRKY, and Zn Finger.

Table 1. Sequences of primers used in the RT–qPCR analysis.

Target Gene	Forward (5′ to 3′ Sequence)	Reverse (5′ to 3′ Sequence)	Primer Reference
Dehydrin	GTACTCTCATGACACCCACAAAACTAC	CCCGGCCCCACCGTAAGCTCCAGTT	[69]
LEA protein	GCAAAAGGTAGGGCAAACAG	TGGCTTTGCTTCTTTGGTCT	[69]
Zn Finger	ACACAGGCTTCCTCTACTCCATCTTT	GAACCCTCATTCCGAGACATTTATCAG	[69]
WRKY	GCCGAGAAATCACCGACTTC	GTTGTCTGAGGCTTGGGTTG	[70]
PR	GGAGATGCCTTTGATGTGGGA	AGCTTGAACTCGCCTTCTGG	[71]
NAC	GATAACCCAACTACCACTACCAC	GACAACTCCCAGATACCACG	[72]
b-ZIP	GGGTTGAAACACCCAAAAGA	GCGATTCGACAACATCCTCT	[73]
Actin	CAGATCATGTTTGAGACCTTCAATGT	CATCACCAGAGTCCAGCACAAT	[73]

To standardise the results, the relative abundance of the actin gene (Actin) was used as the internal standard (Table 1). Relative gene expression levels were calculated with the $\log_2 2^{-\Delta\Delta Ct}$ method [74,75]. The efficiency of the target amplification was evaluated for each primer pair, and the corresponding values were used to calculate the fold changes (FCs) with the following formula: FC = (1 + E) − ΔΔCt, where $\Delta\Delta Ct = (Ct_{target} - Ct_{UBQ})_{Treatment} - (Ct_{target} - Ct_{UBQ})_{Control}$.

4.5. Statistical Analysis

The means of the quantitative data related to the RWC, proline content and gene expression levels were determined for each season (spring, summer and autumn) and subjected to one-way ANOVA, followed by the Tukey HSD (honestly significant difference) post hoc test ($p < 0.05$). Analyses were carried out using GraphPad software, version 8.02.

5. Conclusions

Xf induces physiological conditions similar to those caused by a water deficit in plants. In this work, the "Filippo Ceo" almond plants maintained good water and health status regardless of abiotic, biotic or combined stress conditions, thus confirming their resistance to both water stress and the pathogen and the combination of both. However, these similar health statuses were achieved differently because of differences in proline accumulation and differential gene expression profiles during the pathogen response. These data suggest that almond tree resistance to *Xf* could be due to transcriptional reprogramming that results in the modulation of genes directly or indirectly involved in defence strategies.

We believe that this is the first study to explore the mechanisms underlying resistance to *Xfp* in the "Filippo Ceo" almond cultivar. This study provides a foundation for future research on identifying valuable traits to combat this pathogen in affected areas. Additionally, our work sheds light on the complex interactions among plant responses to multiple stress conditions, which is especially important for future climate change scenarios.

Author Contributions: Conceptualisation, M.D.P. and M.V.; methodology, M.D.P., D.G., M.V. and G.C.; validation, M.D.P. and M.V.; formal analysis, M.D.P.; data curation, M.V.; writing—original draft, M.D.P. and M.V.; writing—review and editing, L.D.B. and A.L.; supervision, L.D.B. and A.L.; funding acquisition, L.D.B. and A.L. All authors have read and agreed to the published version of the manuscript.

Funding: This research was partially funded by the Ministero dell'agricoltura, della sovranità alimentare e delle foreste research project "Approcci integrati per il miglioramento genetico, la selezione e l'ottenimento di materiali vegetali resistenti a *Xylella fastidiosa*-RIGENERA" (decreto direttoriale n. 665024 del 29/12/2022, CUP H93C22000750001).

Data Availability Statement: The data presented in this study are available on request from the corresponding author.

Acknowledgments: We would like to thank Luigi Costantini for kindly making his almond orchard available.

Conflicts of Interest: The authors declare no conflicts of interest.

References

1. Zandalinas, S.I.; Balfagón, D.; Gómez-Cadenas, A.; Mittler, R. Plant responses to climate change: Metabolic changes under combined abiotic stresses. *J. Exp. Bot.* **2022**, *73*, 3339–3354. [CrossRef] [PubMed]
2. Rivero, R.M.; Mittler, R.; Blumwald, E.; Zandalinas, S.I. Developing climate-resilient crops: Improving plant tolerance to stress combination. *Plant J.* **2022**, *109*, 373–389. [CrossRef]
3. Aung, K.; Jiang, Y.; He, S.Y. The role of water in plant–microbe interactions. *Plant J.* **2018**, *93*, 771–780. [CrossRef]
4. Cheng, Y.T.; Zhang, L.; He, S.Y. Plant-microbe interactions facing environmental challenge. *Cell Host Microbe* **2019**, *26*, 183–192. [CrossRef] [PubMed]
5. Choi, H.K.; Iandolino, A.; Da Silva, F.G.; Cook, D.R. Water deficit modulates the response of vitis vinifera to the Pierce's disease pathogen xylella fastidiosa. *Mol. Plant-Microbe Interact.* **2013**, *26*, 643–657. [CrossRef] [PubMed]
6. Dossa, G.S.; Torres, R.; Henry, A.; Oliva, R.; Maiss, E.; Cruz, C.V.; Wydra, K. Rice response to simultaneous bacterial blight and drought stress during compatible and incompatible interactions. *Eur. J. Plant Pathol.* **2017**, *147*, 115–127. [CrossRef]
7. Johansen, T.J.; Dees, M.W.; Hermansen, A. High soil moisture reduces common scab caused by Streptomyces turgidiscabies and Streptomyces europaeiscabiei in potato. *Acta Agric. Scand. Sect. B Soil. Plant Sci.* **2015**, *65*, 193–198.
8. Achuo, E.A.; Prinsen, E.; Höfte, M. Influence of drought, salt stress and abscisic acid on the resistance of tomato to Botrytis cinerea and Oidium neolycopersici. *Plant Pathol.* **2006**, *55*, 178–186. [CrossRef]
9. Hatmi, S.; Gruau, C.; Trotel-Aziz, P.; Villaume, S.; Rabenoelina, F.; Baillieul, F.; Eullaffroy, P.; Clément, C.; Ferchichi, A.; Aziz, A. Drought stress tolerance in grapevine involves activation of polyamine oxidation contributing to improved immune response and low susceptibility to Botrytis cinerea. *J. Exp. Bot.* **2015**, *66*, 775–787. [CrossRef]
10. Carluccio, G.; Greco, D.; Sabella, E.; Vergine, M.; De Bellis, L.; Luvisi, A. Xylem Embolism and Pathogens: Can the Vessel Anatomy of Woody Plants Contribute to *X. fastidiosa* Resistance? *Pathogens* **2023**, *12*, 825. [CrossRef]
11. De Pascali, M.; Vergine, M.; Sabella, E.; Aprile, A.; Nutricati, E.; Nicol, F.; Buja, I.; Negro, C.; Miceli, A.; Rampino, P.; et al. Molecular Effects of *Xylella fastidiosa* and Drought Combined Stress in Olive Trees. *Plants* **2019**, *8*, 437. [CrossRef]
12. De Pascali, M.; Vergine, M.; Negro, C.; Greco, D.; Vita, F.; Sabella, E.; De Bellis, L.; Luvisi, A. *Xylella fastidiosa* and Drought Stress in Olive Trees: A Complex Relationship Mediated by Soluble Sugars. *Biology* **2022**, *11*, 112. [CrossRef] [PubMed]

13. Sabella, E.; Aprile, A.; Genga, A.; Siciliano, T.; Nutricati, E.; Nicolì, F.; Vergine, M.; Negro, C.; De Bellis, L.; Luvisi, A. Xylem cavitation susceptibility and refilling mechanisms in olive trees infected by *Xylella fastidiosa*. *Sci. Rep.* **2019**, *9*, 9602. [CrossRef] [PubMed]
14. Rampino, P.; De Pascali, M.; De Caroli, M.; Luvisi, A.; De Bellis, L.; Piro, G.; Perrotta, C. Td4IN2: A drought-responsive durum wheat (*Triticum durum* Desf.) gene coding for a resistance like protein with serine/threonine protein kinase, nucleotide binding site and leucine rich domains. *Plant Physiol. Biochem.* **2017**, *120*, 223–231. [CrossRef]
15. Singh, B.; Bohra, A.; Mishra, S.; Joshi, R.; Pandey, S. Embracing new-generation 'omics' tools to improve drought tolerance in cereal and food-legume crops. *Biol. Plant* **2015**, *59*, 413–428. [CrossRef]
16. Jain, D.; Khurana, J.P. Role of Pathogenesis-Related (PR) Proteins in Plant Defense Mechanism. In *Molecular Aspects of Plant-Pathogen Interaction*; Singh, A., Singh, I.K., Eds.; Springer: Singapore, 2018; pp. 265–281.
17. Baillo, E.H.; Kimotho, R.N.; Zhang, Z.; Xu, P. Transcription factors associated with abiotic and biotic stress tolerance and their potential for crops improvement. *Genes* **2019**, *10*, 771. [CrossRef]
18. Xiang, Y.; Tang, N.; Du, H.; Ye, H.; Xiong, L. Characterization of OsbZIP23 as a key player of the basic leucine zipper transcription factor family for conferring abscisic acid sensitivity and salinity and drought tolerance in rice. *Plant Physiol.* **2008**, *148*, 1938–1952. [CrossRef]
19. Seo, P.J.; Park, C.M. A membrane-bound NAC transcription factor as an integrator of biotic and abiotic stress signals. *Plant Signal Behav.* **2010**, *5*, 481–483. [CrossRef]
20. Nakashima, K.; Ito, Y.; Yamaguchi-Shinozaki, K. Transcriptional Regulatory Networks in Response to Abiotic Stresses in Arabidopsis and Grasses. *Plant Physiol.* **2009**, *149*, 88–95. [CrossRef]
21. Li, B.; Feng, Y.; Zong, Y.; Zhang, D.; Hao, X.; Li, P. Elevated CO_2-induced changes in photosynthesis, antioxidant enzymes and signal transduction enzyme of soybean under drought stress. *Plant Physiol. Biochem.* **2020**, *154*, 105–114. [CrossRef]
22. Fang, Y.; Liao, K.; Du, H.; Xu, Y.; Song, H.; Li, X.; Xiong, L. A stress-responsive NAC transcription factor SNAC3 confers heat and drought tolerance through modulation of reactive oxygen species in rice. *J. Exp. Bot.* **2015**, *66*, 6803–6817. [CrossRef]
23. Zhang, X.; Zhang, B.; Li, M.J.; Yin, X.M.; Huang, L.F.; Cui, Y.C.; Wang, M.L.; Xia, X. OsMSR15 encoding a rice C2H2-type zinc finger protein confers enhanced drought tolerance in transgenic *Arabidopsis*. *J. Plant Biol.* **2016**, *59*, 271–281. [CrossRef]
24. Zhang, A.; Liu, D.; Hua, C.; Yan, A.; Liu, B.; Wu, M.; Liu, Y.; Huang, L.; Ali, I.; Gan, Y. The arabidopsis gene zinc finger protein 3(ZFP3) is involved in salt stress and osmotic stress response. *PLoS ONE* **2016**, *11*, e0168367. [CrossRef] [PubMed]
25. Erpen, L.; Devi, H.S.; Grosser, J.W.; Dutt, M. Potential use of the DREB/ERF, MYB, NAC and WRKY transcription factors to improve abiotic and biotic stress in transgenic plants. *Plant Cell Tissue Organ. Cult.* **2018**, *132*, 1–25. [CrossRef]
26. Greco, D.; Aprile, A.; De Bellis, L.; Luvisi, A. Diseases Caused by *Xylella fastidiosa* in Prunus Genus: An Overview of the Research on an Increasingly Widespread Pathogen. *Front. Plant Sci.* **2021**, *12*, 712452. [CrossRef] [PubMed]
27. Cao, T.; Connell, J.H.; Wilhelm, M.; Kirkpatrick, B.C. Influence of inoculation date on the colonization of *Xylella fastidiosa* and the persistence of almond leaf scorch disease among almond cultivars. *Plant Dis.* **2011**, *95*, 158–165. [CrossRef] [PubMed]
28. Sisterson, M.S.; Chen, J.; Viveros, M.A.; Civerolo, E.L.; Ledbetter, C.; Groves, R.L. Effects of almond leaf scorch disease on almond yield: Implications for management. *Plant Dis.* **2008**, *92*, 409–414. [CrossRef] [PubMed]
29. Amanifar, N.; Taghavi, M.; Salehi, M. *Xylella fastidiosa* from almond in Iran: Overwinter recovery and effects of antibiotics. *Phytopathol. Mediterr.* **2016**, *55*, 337–345. [CrossRef]
30. Moralejo, E.; Gomila, M.; Montesinos, M.; Borràs, D.; Pascual, A.; Nieto, A.; Adrover, F.; Gost, P.A.; Seguí, G.; Busquets, A.; et al. Phylogenetic inference enables reconstruction of a long-overlooked outbreak of almond leaf scorch disease (Xylella fastidiosa) in Europe. *Commun. Biol.* **2020**, *3*, 560. [CrossRef]
31. Marco-Noales, E.; Barbe, S.; Monterde, A.; Navarro-Herrero, I.; Ferrer, A.; Dalmau, V.; Aure, C.M.; Domingo-Calap, M.L.; Landa, B.B.; Rosello, M. Evidence that *Xylella fastidiosa* is the Causal Agent of Almond Leaf Scorch Disease in Alicante, Mainland Spain (Iberian Peninsula). *Plant Dis.* **2021**, *105*, 3349–3352. [CrossRef]
32. Amanifar, N.; Luvisi, A. Resistance of almond (*Prunus dulcis*) to *Xylella fastidiosa*: A comparative study on cultivars. *Plant Dis.* **2022**, *106*, 2625–2630. [CrossRef]
33. Saponari, M.; Boscia, D.; Altamura, G.; Loconsole, G.; Zicca, S.; D'Attoma, G.; Morelli, M.; Palmisano, F.; Saponari, A.; Tavano, D.; et al. Isolation and pathogenicity of *Xylella fastidiosa* associated to the olive quick decline syndrome in southern Italy. *Sci. Rep.* **2017**, *7*, 17723. [CrossRef]
34. Semeraro, T.; Buccolieri, R.; Vergine, M.; De Bellis, L.; Luvisi, A.; Emmanuel, R.; Marwan, N. Analysis of olive grove destruction by *Xylella fastidiosa* bacterium on the land surface temperature in Salento detected using satellite images. *Forests* **2021**, *12*, 1266. [CrossRef]
35. Commissione Europea. Regolamento di esecuzione (UE) 2020/1201 della commissione del 14 agosto 2020 relativo alle misure per prevenire l'introduzione e la diffusione nell'Unione della Xylella fastidiosa (Wells et al.). 2020; 17.
36. Torrecillas, A.; Alarcón, J.J.; Domingo, R.; Planes, J.; Sánchez-Blanco, M.J. Strategies for drought resistance in leaves of two almond cultivars. *Plant Sci.* **1996**, *118*, 135–143. [CrossRef]
37. Abobatta, W.F. Drought adaptive mechanisms of plants—A review. *Adv. Agric. Environ. Sci.* **2019**, *2*, 62–65. [CrossRef]
38. Alzahrani, Y.; Kuşvuran, A.; Alharby, H.F.; Kuşvuran, S.; Rady, M.M. The defensive role of silicon in wheat against stress conditions induced by drought, salinity or cadmium. *Ecotoxicol. Environ. Saf.* **2018**, *154*, 187–196. [CrossRef] [PubMed]

39. Mahmood, T.; Abdullah, M.; Ahmar, S.; Yasir, M.; Iqbal, M.S.; Yasir, M.; Rehman, S.U.; Ahmed, S.; Rana, R.M.; Ghafoor, A.; et al. Incredible role of osmotic adjustment in grain yield sustainability under water scarcity conditions in wheat (*Triticum aestivum* L.). *Plants* **2020**, *9*, 1208. [CrossRef] [PubMed]
40. Yang, Y.; He, M.; Zhu, Z.; Li, S.; Xu, Y.; Zhang, C.; Singer, D.S.; Wang, Y. Identification of the dehydrin gene family from grapevine species and analysis of their responsiveness to various forms of abiotic and biotic stress. *BMC Plant Biol.* **2012**, *12*, 140. [CrossRef] [PubMed]
41. Turco, E.; Close, T.J.; Fenton, R.D.; Ragazzi, A. Synthesis of dehydrin-like proteins in *Quercus ilex* L. and *Quercus cerris* L. seedlings subjected to water stress and infection with *Phytophthora cinnamomi*. *Physiol. Mol. Plant Pathol.* **2004**, *65*, 137–144. [CrossRef]
42. Kaur, J.; Fellers, J.; Adholeya, A.; Velivelli, S.L.S.; El-Mounadi, K.; Nersesian, N.; Clemente, T.; Shah, D. Expression of apoplast-targeted plant defensin MtDef4.2 confers resistance to leaf rust pathogen *Puccinia triticina* but does not affect mycorrhizal symbiosis in transgenic wheat. *Transgenic Res.* **2017**, *26*, 37–49. [CrossRef] [PubMed]
43. Mackintosh, C.A.; Lewis, J.; Radmer, L.E.; Shin, S.; Heinen, S.J.; Smith, L.A.; Wyckoff, M.N.; Dill-Macky, R.; Evans, C.K.; Kravchenko, S.; et al. Overexpression of defense response genes in transgenic wheat enhances resistance to Fusarium head blight. *Plant Cell Rep.* **2007**, *26*, 479–488. [CrossRef]
44. Ali, S.; Mir, Z.A.; Bhat, J.A.; Tyagi, A.; Chandrashekar, N.; Yadav, P.; Rawat, S.; Sultana, M.; Grover, A. Isolation and characterization of systemic acquired resistance marker gene PR1 and its promoter from Brassica juncea. *3 Biotech.* **2018**, *8*, 10. [CrossRef] [PubMed]
45. Gupta, P.; Ravi, I.; Sharma, V. Induction of β-1,3-glucanase and chitinase activity in the defense response of Eruca sativa plants against the fungal pathogen Alternaria brassicicola. *J. Plant Interact.* **2013**, *8*, 155–161. [CrossRef]
46. Jiang, L.; Wu, J.; Fan, S.; Li, W.; Dong, L.; Cheng, Q.; Xu, P.; Zhang, S. Isolation and characterization of a novel pathogenesis-related protein gene (*GmPRP*) with induced expression in soybean (*Glycine max*) during infection with *Phytophthora sojae*. *PLoS ONE* **2015**, *10*, e0129932. [CrossRef] [PubMed]
47. Sinclair, B.J.; Ferguson, L.V.; Salehipour-Shirazi, G.; Macmillan, H.A. Cross-tolerance and cross-talk in the cold: Relating low temperatures to desiccation and immune stress in insects. *Integr. Comp. Biol.* **2013**, *53*, 545–556. [CrossRef] [PubMed]
48. Khan, S.A.; Li, M.Z.; Wang, S.M.; Yin, H.J. Revisiting the role of plant transcription factors in the battle against abiotic stress. *Int. J. Mol. Sci.* **2018**, *19*, 1634. [CrossRef] [PubMed]
49. Grant, J.J.; Chini, A.; Basu, D.; Loake, G.J. Targeted activation tagging of the Arabidopsis NBS-LRR gene, ADR1, conveys resistance to virulent pathogens. *Mol. Plant-Microbe Interact.* **2003**, *16*, 669–680. [CrossRef] [PubMed]
50. Chini, A.; Grant, J.J.; Seki, M.; Shinozaki, K.; Loake, G.J. Drought tolerance established by enhanced expression of the CC-NBS-LRR gene, ADR1, requires salicylic acid, EDS1 and ABI1. *Plant J.* **2004**, *38*, 810–822. [CrossRef] [PubMed]
51. Zaini, P.A.; Nascimento, R.; Gouran, H.; Cantu, D.; Chakraborty, S.; Phu, M.; Goulart, L.R.; Dandekar, A.M. Molecular profiling of pierce's disease outlines the response circuitry of vitis vinifera to xylella fastidiosa infection. *Front. Plant Sci.* **2018**, *9*, 771. [CrossRef]
52. Hrmova, M.; Hussain, S.S. Plant transcription factors involved in drought and associated stresses. *Int. J. Mol. Sci.* **2021**, *22*, 5662. [CrossRef]
53. Puranik, S.; Sahu, P.P.; Srivastava, P.S.; Prasad, M. NAC proteins: Regulation and role in stress tolerance. *Trends Plant Sci.* **2012**, *17*, 369–381. [CrossRef]
54. Bian, Z.; Gao, H.; Wang, C. NAC transcription factors as positive or negative regulators during ongoing battle between pathogens and our food crops. *Int. J. Mol. Sci.* **2021**, *22*, 81. [CrossRef]
55. Moll, L.; Baró, A.; Montesinos, L.; Badosa, E.; Bonaterra, A.; Montesinos, E. Induction of Defense Responses and Protection of Almond Plants Against Xylella fastidiosa by Endotherapy with a Bifunctional Peptide. *Phytopathology* **2022**, *112*, 1907–1916. [CrossRef]
56. Bai, Y.; Sunarti, S.; Kissoudis, C.; Visser, R.G.F.; van der Linden, C.G. The role of tomato WRKY genes in plant responses to combined abiotic and biotic stresses. *Front. Plant Sci.* **2018**, *9*, 801. [CrossRef]
57. Lee, H.; Cha, J.; Choi, C.; Choi, N.; Ji, H.S.; Park, S.R.; Lee, S.; Hwang, D.J. Rice wrky11 plays a role in pathogen defense and drought tolerance. *Rice* **2018**, *11*, 5. [CrossRef] [PubMed]
58. Noman, A.; Aqeel, M.; Khalid, N.; Islam, W.; Sanaullah, T.; Anwar, M.; Khan, S.; Ye, W.; Lou, Y. Zinc finger protein transcription factors: Integrated line of action for plant antimicrobial activity. *Microb. Pathog.* **2019**, *132*, 141–149. [CrossRef] [PubMed]
59. Moulick, D.; Bhutia, K.L.; Sarkar, S.; Roy, A.; Mishra, U.N.; Pramanick, B.; Maitra, S.; Shankar, T.; Hazra, S.; Skalicky, M.; et al. The intertwining of Zn-finger motifs and abiotic stress tolerance in plants: Current status and future prospects. *Front. Plant Sci.* **2023**, *13*, 1083960. [CrossRef] [PubMed]
60. Huang, J.; Sun, S.J.; Xu, D.Q.; Yang, X.; Bao, Y.M.; Wang, Z.F.; Tang, H.J.; Zhang, H. Increased tolerance of rice to cold, drought and oxidative stresses mediated by the overexpression of a gene that encodes the zinc finger protein ZFP245. *Biochem. Biophys. Res. Commun.* **2009**, *389*, 556–561. [CrossRef] [PubMed]
61. Gupta, S.K.; Rai, A.K.; Kanwar, S.S.; Sharma, T.R. Comparative analysis of zinc finger proteins involved in plant disease resistance. *PLoS ONE* **2012**, *7*, e42578. [CrossRef] [PubMed]
62. European Commission. Commission Implementing Decision (EU) 2015/789 of 18 May 2015 as Regards Measures to Prevent the Introduction into and the Spread within the Union of *Xylella fastidiosa* (Wells et al.). Official Journal of the European Union. 2015. L 125 February 2014. pp. 36–53. Available online: http://eur-lex.europa.eu/legal-content/EN/TXT/?uri=uriserv:OJ.L_.2015.125.01.0036.01.ENG (accessed on 22 January 2024).

63. Olmos, A.; Bertolini, E.; Gil, M.; Cambra, M. Real-time assay for quantitative detection of non-persistently transmitted Plum pox virus RNA targets in single aphids. *J. Virol. Methods* **2005**, *128*, 151–155. [CrossRef] [PubMed]
64. Luvisi, A.; Aprile, A.; Sabella, E.; Vergine, M.; Nutricati, E.; Miceli, A.; Negro, C.; De Bellis, L. *Xylella fastidiosa* subsp. *pauca* (CoDiRO strain) infection in four olive (*Olea europaea* L.) cultivars: Profile of phenolic compounds in leaves and progression of leaf scorch symptoms. *Phytopathol. Mediterr.* **2017**, *56*, 259–273.
65. Harper, S.J.; Ward, L.I.; Clover, G.R.G. Development of LAMP and Real-Time PCR Methods for the Rapid Detection of *Xylella fastidiosa* for Quarantine and Field Applications. *Phytopathology* **2010**, *100*, 1282–1288. [CrossRef]
66. D'attoma, G.; Morelli, M.; Saldarelli, P.; Saponari, M.; Giampetruzzi, A.; Boscia, D.; Savino, V.N.; De La Fuente, L.; Cobine, P.A. Ionomic differences between susceptible and resistant olive cultivars infected by *Xylella fastidiosa* in the outbreak area of salento, Italy. *Pathogens* **2019**, *8*, 272. [CrossRef] [PubMed]
67. Bates, L.S.; Waldre, R.P.; Teare, I.D. Rapid determination of free proline for water-stress studies. *Plant Soil.* **1973**, *39*, 205. [CrossRef]
68. Gambino, G.; Perrone, I.; Gribaudo, I. A rapid and effective method for RNA extraction from different tissues of grapevine and other woody plants. *Phytochem. Anal.* **2008**, *19*, 520–525. [CrossRef]
69. Leida, C.; Conesa, A.; Llácer, G.; Badenes, M.L.; Ríos, G. Histone modifications and expression of *DAM6* gene in peach are modulated during bud dormancy release in a cultivar-dependent manner. *New Phytol.* **2012**, *193*, 67–80. [CrossRef] [PubMed]
70. Yu, Z.; Zhang, D.; Zeng, B.; Liu, X.; Yang, J.; Gao, W.; Ma, X. Characterization of the *WRKY* gene family reveals its contribution to the adaptability of almond (*Prunus dulcis*). *Peer J.* **2022**, *10*, e13491. [CrossRef] [PubMed]
71. Rubio, M.; Martinez-Garcia, P.J.; Nikbakht-Dehkordi, A.; Prudencio, A.S.; Gómez, E.; Rodamilans, B.; Dicenta, F.; García, J.A.; Martínez-Gómez, P. Gene Expression Analysis of Induced *Plum pox virus* (Sharka) Resistance in Peach (*Prunus persica*) by Almond (*P. dulcis*) Grafting. *Int. J. Mol. Sci.* **2021**, *22*, 3585. [CrossRef] [PubMed]
72. Zhuo, X.; Zheng, T.; Zhang, Z.; Zhang, Y.; Jiang, L.; Ahmad, S.; Sun, L.; Wang, J.; Cheng, T.; Zhang, Q. Genome-Wide Analysis of the NAC Transcription Factor Gene Family Reveals Differential Expression Patterns and Cold-Stress Responses in the Woody Plant *Prunus mume*. *Genes* **2018**, *9*, 494. [CrossRef]
73. Bielsa, B.; Leida, C.; Rubio-Cabetas, M.J. Physiological characterization of drought stress response and expression of two transcription factors and two LEA genes in three *Prunus* genotypes. *Sci. Hortic.* **2016**, *213*, 260–269. [CrossRef]
74. Livak, K.J.; Schmittgen, T.D. Analysis of relative gene expression data using real-time quantitative PCR and the 2-ΔΔCT method. *Methods* **2001**, *25*, 402–408. [CrossRef] [PubMed]
75. Quackenbush, J. Microarray data normalization and transformation. *Nat. Genet.* **2002**, *32*, 496–501. [CrossRef] [PubMed]

plants

Article

Phenological, Physiological, and Ultrastructural Analyses of 'Green Islands' on Senescent Leaves of Norway Maple (*Acer platanoides* L.)

Violetta Katarzyna Macioszek [1,*], Kamila Chalamońska [2], Jakub Oliwa [3], Aleksandra Maria Staszak [1] and Mirosław Sobczak [4,*]

1 Laboratory of Plant Physiology, Department of Biology and Plant Ecology, Faculty of Biology, University of Bialystok, 15-245 Bialystok, Poland
2 Faculty of Agriculture and Ecology, Warsaw University of Life Sciences (SGGW), 02-787 Warsaw, Poland
3 Institute of Biology and Earth Sciences, University of the National Education Commission, 31-054 Krakow, Poland; jakub.oliwa@gmail.com
4 Department of Botany, Institute of Biology, Warsaw University of Life Sciences (SGGW), 02-787 Warsaw, Poland
* Correspondence: v.macioszek@uwb.edu.pl (V.K.M.); miroslaw_sobczak@sggw.edu.pl (M.S.)

Citation: Macioszek, V.K.; Chalamońska, K.; Oliwa, J.; Staszak, A.M.; Sobczak, M. Phenological, Physiological, and Ultrastructural Analyses of 'Green Islands' on Senescent Leaves of Norway Maple (*Acer platanoides* L.). *Plants* **2025**, *14*, 909. https://doi.org/10.3390/plants14060909

Academic Editor: Vladimir V. Kuznetsov

Received: 5 February 2025
Revised: 10 March 2025
Accepted: 12 March 2025
Published: 14 March 2025

Citation: Macioszek, V.K.; Chalamońska, K.; Oliwa, J.; Staszak, A.M.; Sobczak, M. Phenological, Physiological, and Ultrastructural Analyses of 'Green Islands' on Senescent Leaves of Norway Maple (*Acer platanoides* L.). *Plants* **2025**, *14*, 909. https://doi.org/10.3390/plants14060909

Abstract: 'Green island' symptoms in the form of vivid green, round spots visible on the senescent leaves of many plants and trees are mostly the results of pathogenic colonization by fungi, and the greenish tissue is often dead. Therefore, this study investigates whether green spots observed on senescent Norway maple (*Acer platanoides* L.) leaves were still alive and photosynthetically active. The appearance of 'green islands' on the leaves of young Norway maple trees was observed from the autumn of 2019 to 2022 in an urban forest (Bialystok, eastern Poland). However, in the late summer (September) of 2023 and 2024, mostly tar spots caused by the fungus *Rhytisma* spp. on maple leaves could be observed, with only a few leaves having 'green island' symptoms. The percentage of 'green island' areas on senescent leaves observed during the 4 years (2019–2022) was influenced by a year of sampling ($p < 0.001$). A non-destructive physiological analysis of chlorophyll, flavonoids, and nitrogen balance index (NBI) in leaves revealed that these parameters were significantly lower in 'green islands' than in the summer leaves, but higher than in the senescent yellow area of the autumn leaves. In the case of anthocyanins, their level was significantly higher in 'green islands' than in yellow areas, although, in the summer leaves, anthocyanins were undetectable. The amount of chlorophyll and most photosynthetic parameters were significantly ($p < 0.05$) reduced in the 'green islands' of the senescent leaves compared to the mature green leaves. However, these parameters were significantly higher in the 'green islands' than in senescent yellow leaves. Carotenoid content in the 'green island' and yellow areas of senescent leaves were at the same level, twice as higher than in summer leaves. Green mature leaves and the 'green islands' on senescent leaves had the same structure and anatomy. The main differences concerned the chloroplasts, which were smaller and had less grana and starch grains, but had more plastoglobuli in 'green island' cells. The cells building the mesophyll in the yellow area of the leaf deteriorated and their chloroplasts collapsed. Epiphytes were present on the adaxial epidermis surface in all types of samples.

Keywords: chlorophyll; 'green islands'; Norway maple; phenolic compounds; photosynthesis; *Rhytisma* spp. senescence

1. Introduction

Norway maple (*Acer platanoides* L.) is a tree species from the soapberry family (*Sapindaceae* Juss.). It occurs naturally in Central and Eastern Europe and in the Balkans and Caucasus [1,2]. However, it is considered an invasive species in North America [3,4]. It is the most common of the maple species found in Poland and is characterized by a spreading, umbrella-like crown. *A. platanoides* grows to approx. 18–27 m in height, and the trunk width can reach up to 1.5 m. The trunk is covered with thin bark, red-brown in a young twig, and it darkens and becomes furrowed on old branches [5,6]. Maple stems are straight and short, with numerous perpendicular shoots, covered with opposite leaves. The leaves have five lobes with long, pointed teeth and smooth edges. Their size depends on the age and condition of the plant [7].

Norway maple is widely used as an ornamental, shade, and street tree due to its attractiveness and tolerance to specific urban conditions. It can regenerate vigorously after pruning, so it can be used as a living fence [8]. Due to the production of fine adventitious roots, maple can be planted in mountainous areas, thus protecting the soil against excessive erosion [9] and stabilizes slopes reducing rockfalls [10]. In natural habitats, Norway maple is almost completely free from serious diseases. However, in highly urbanized areas, it may suffer from various diseases caused by a combination of stresses resulting from excessive air and soil pollution [8].

In general, morphological changes in the form of 'green islands', i.e., fragments of the surface of leaf lamina retaining a green color on senescent and yellowing tree leaves, are most often associated with the local presence of endophytic microorganisms, mainly fungi and bacteria [11]. Endophytes are microorganisms usually having a beneficial effect on plants or benign parasites and pathogens that do not cause disease symptoms, inhabiting the interior of plant organs for most of their life cycle [12,13]. However, a term "endophyte" rather refers only to the place of their occurrence, and not to their impact on the metabolism and the physiological or health condition of plants, also concerning pathogenic microbes [14]. The employment of genome and metagenome sequencing enabled the identification of some endophyte species and the preliminary determination of the mechanisms of their interaction with various species of infected plants in 'green islands' [15]. The phenomenon of 'green island' is most often described and observed on plants infected with pathogenic microorganisms [11]. The area around the site of pathogen infection remains green while the surrounding leaf tissues degrade and naturally age.

However, 'green islands' formed on senescent and yellowing leaves of trees, leaves of herbaceous plants, and crops infected by various pathogens do not have the same structure. Thus, it enables easy identification of 'green islands' with living host cells and pathogens, as well as those in which pathogen cells remain alive, while the cells of the host plant are dead or dying [11,16]. The phenomenon of 'green islands' is related to interactions between the plant and biotrophic fungal pathogens, such as powdery mildew or brown rust [17], as well as with some hemibiotrophic fungi, e.g., *Plasmodiophora brassicae* and *Colletotrichum graminicola* [18], or with parasitism of some insects [19]. It has also been observed that 'green islands' may appear in places of infection with necrotrophic fungi such as *Pyrenophora teres* or *Alternaria brassicicola* [20,21]. Considering the number of pathosystems in which green islands are induced, it seems unlikely that there is one universal and common mechanism responsible for the formation or retention of photosynthetically active cells in degrading leaves. The appearance of 'green islands' on infected and degrading leaves is a beneficial phenomenon for fungal pathogens, as it significantly extends the life of host cells. However, the etiology, origin, structure, functions, and environmental significance of these structures have not yet been fully explained, although the scale of this phenomenon is extensively expanding [11]. Undoubtedly, 'green islands' that maintain photosynthesis function in infected areas have less impact on plant health in trees as they lose leaves in autumn than crops. However, this phenomenon can be a visible marker in crop plants' practical disease diagnosis and management.

'Green islands' in *A. platanoides* are visible in autumn when leaves begin to yellow. It has been found that this phenomenon in Norway maples in Europe is most probably induced by a pathogenic fungus *Sawadaea bicornis*. It has at least been recognized as a dominant species in *A. platanoides* 'green islands', together with other endophytic fungi and bacteria [15].

The aim of this study was to show the variability of dominant disease symptoms on leaves of Norway maple growing in the urban forest of Bialystok during the past 6 years, with an emphasis on 'green island' development. Moreover, investigating naturally occurring green islands might be a valuable model for investigating how pathogens manipulate the host plant metabolism. Therefore, physiological and ultrastructural changes within the 'green islands' of senescent leaves of *A. platanoides* were compared to alterations in the yellow areas of senescent leaves as well as mature summer leaves.

2. Results

All analyzed leaves with 'green island' symptoms were collected from young Norway maple trees in the last week of October/the first decade of November in the years 2019–2022 (Figure 1). The summer leaves were collected from the same trees in June 2020 and 2021.

Figure 1. Representative images of *A. platanoides* senescent leaves with 'green islands' symptoms harvested in October/November in the years 2019–2022. (**d**) bar = 10 mm.

2.1. 'Green Island' Symptoms

'Green islands' on the leaves of young Norway maples were first observed in late October 2019, when leaves began to yellow (Figure 1a). 'Green islands' are vividly green and mostly round, with various sizes from approximately 0.5 to 4.5 cm in diameter. Between one (relatively rarely) and ten 'green islands' per leaf were formed. These features differed 'green islands' from irregular patches of green tissue in senescent leaves. Most young and old Norway maple trees within the urban forest surrounding the University of Bialystok campus displayed 'green island' symptoms (Figure A1b).

Moreover, 'green islands' could also be observed on the leaves of another maple species, field maple (*Acer campestre* L.) (Figure A1c). Maples showing 'green island' symptoms could not be found in other locations of the urban forest in Bialystok. In the second half of November 2019, the fallen leaves around trees had still-evident 'green islands'. In 2020–2022, 'green islands' were major symptoms of biotic stress in maples (Figure 1), apart from the abundant leaf-biting areas caused by insects.

The mean percentage of 'green island' areas covering a leaf lamina changed within research years, showing significantly higher values in 2019 and 2021 compared to 2020 and 2022 (Table 1). A one-way analysis of variance revealed that the percentage of the 'green island' area was significantly influenced by a year of collection (F = 609.158, $p < 0.001$).

Table 1. Mean percentages of 'green island' area in *A. platanoides* leaves (n = 9–30) evaluated each year in November from 2019 to 2022. Data were gained using the WinDIAS system. Statistical differences between the means were labeled with different letters according to a post hoc Duncan's test ($p < 0.05$).

	2019	2020	2021	2022
Percentage of 'green' island area	23.99 a	15.60 b	26.17 a	19.49 ab
SD	9.62	4.61	6.43	4.42

From 2022, in the first decade of September, the tar spot, caused by a pathogenic fungus of *Rhytisma* spp., occurred on only a few Norway maple leaves (Figure A1d). This disease started in the form of yellowing leaf spots in late August 2023, and the round black tar spots were surrounded by yellow rings in the middle of September (Figure A2a,d). Tar spot was a major disease observed in Norway maples in 2023. Only a few young trees showed 'green island' symptoms. However, tar spots together with 'green islands' could also be found on the same leaves. A similar correlation of 'green islands' and tar spot appearance was also observed in 2024 (Figure A2).

2.2. Non-Destructive Assessment of Physiological Parameters

Three variants of leaves were assessed in the non-destructive physiological analysis of chlorophyll, flavonoids, anthocyanins, and nitrogen balance index (NBI): green mature summer leaves; and two areas of autumn leaves, being 'green islands' and senescent yellow areas. The analyses were performed every year on leaves collected from 2019 to 2021.

The content of chlorophyll and flavonoids and indices of nitrogen balance (NBIs) in 'green islands' were significantly lower than in the summer leaves, but higher compared to the senescent yellow area of the autumn leaves (Table 2). These parameters were similar in the 'green islands' of senescent leaves harvested in 2019 and 2021 and differed significantly compared to 2020. In the case of the summer leaves and the yellow areas of senescent leaves, chlorophyll, flavonoids, and NBIs were at the same level in each examined year. However, anthocyanin content was significantly higher in yellow areas than in 'green islands', although, in the summer leaves, anthocyanins were undetectable (Table 2).

Table 2. Non-destructive measurements of chlorophyll (Chl), flavonoids (Flav), anthocyanins (Anth), and nitrogen balance index (NBI) in *A. platanoides* leaves. Data were gained using a Dualex sensor. The means (n = 25–65) ± SD were obtained from senescent leaves harvested each year in November from 2019 to 2021 and from summer leaves harvested between June 2020 and 2021. Different letters within each year indicate significant differences between leaf variants according to a post hoc Duncan's test ($p < 0.05$).

Year	Leaf Variant	Chl (μg/cm^2)	Flav Index	Anth Index	NBI
2019	'Green island'	11.34 ± 5.40 a	1.06 ± 0.22 a	0.27 ± 0.07 a	11.09 ± 5.60 a
	Yellow area	3.35 ± 1.96 b	1.25 ± 0.20 b	0.38 ± 0.05 b	2.74 ± 1.68 b
2020	Summer leaves	35.91 ± 2.63 c	0.90 ± 0.06 c	0.00	40.10 ± 3.77 c
	'Green island'	15.85 ± 5.25 d	1.02 ± 0.22 a	0.24 ± 0.05 a	16.50 ± 7.37 d
	Yellow area	2.47 ± 1.33 b	1.24 ± 0.16 b	0.38 ± 0.05 b	2.02 ± 1.10 b
2021	Summer leaves	34.15 ± 3.62 c	0.80 ± 0.14 c	0.00	43.99 ± 9.71 e
	'Green island'	12.75 ± 2.29 a	0.96 ± 0.16 a	0.63 ± 0.05 c	13.38 ± 2.27 a
	Yellow area	1.99 ± 0.97 b	1.06 ± 0.21 a	0.88 ± 0.06 d	1.89 ± 0.78 b

A two-way analysis of variance revealed that the content of chlorophyll, flavonoids, and anthocyanins were significantly influenced by a year of sampling (at least $p < 0.005$) and leaf variant (at least $p < 0.005$) (Supplementary Materials Table S1). Only the NBIs were not influenced by a year of sampling (F = 0.07, p = 0.8), although the NBIs showed significant dependence on a leaf variant (F = 5429.69, p < 0.001).

2.3. Chlorophyll and Carotenoid Contents

An analysis of chlorophyll and carotenoid content was performed in three variants of leaves: green mature summer leaves, and two areas of the autumn leaves: 'green islands' and senescent yellow areas. The leaves were collected in the summer and autumn 2020.

The content of chlorophyll *a*, *b*, and total chlorophyll differed significantly between leaf variants, showing the highest values for the summer leaves and 4–7 times lower values in the yellow areas of the autumn leaves (Table 3). In the case of 'green islands', the amounts of chlorophyll were only 1.3–1.4 times lower than in the summer leaves and approximately 3–4.5 higher than in the yellow areas. Surprisingly, the ratio of chlorophyll *a*:*b* was the same in the summer leaves and 'green islands', regardless of the lower content of chlorophylls in 'green islands'. This parameter was significantly lower in the yellow areas of the autumn leaves than in the other two leaf variants ($p < 0.001$; Table 3).

Table 3. Chlorophyll (Chl) and carotenoid (Car) content in *A. platanoides* leaves determined spectrophotometrically. The means (n = 8) ± SD were obtained from senescent leaves harvested in November 2020 and summer leaves harvested in June 2020. Different letters within each parameter indicate significant differences between leaf variants according to a post hoc Duncan's test ($p < 0.05$).

Parameters	Summer Leaves	Senescent Leaves	
(μg mg^{-1} F.W.)		'Green Islands'	Yellow Area
Chl*a*	1.933 ± 0.10 a	1.395 ± 0.27 b	0.268 ± 0.09 c
Chl*b*	1.156 ± 0.19 a	0.814 ± 0.14 b	0.272 ± 0.08 c
Chl*a*:*b*	1.709 ± 0.27 a	1.709 ± 0.08 a	1.00 ± 0.22 b
Total Chl	3.09 ± 0.24 a	2.209 ± 0.42 b	0.540 ± 0.15 c
Car	0.190 ± 0.66 a	0.400 ± 0.06 b	0.431 ± 0.07 b
Total Chl:Car	18.627 ± 8.63 a	5.636 ± 1.40 b	1.201 ± 0.37 b

Carotenoid content had almost the same value in the 'green island' and yellow areas of the autumn senescent leaves. In the summer leaves, there were significantly fewer carotenoids than in the autumn leaves ($p < 0.001$). The ratio of total chlorophyll:carotenoids was extremely high in the summer leaves (18.627 ± 8.63) and over three times lower in 'green islands'. In the yellow areas of the autumn leaves, this parameter was significantly lower than in the other examined leaf variants (Table 3). One-way analysis of variance revealed that all calculated parameters were significantly influenced by a leaf variant (at least $p < 0.001$).

2.4. Analysis of Photosynthetic Parameters

A non-destructive analysis of chlorophyll fluorescence was performed on green mature summer leaves and two areas of the autumn leaves: the 'green islands' and senescent yellow areas collected in the summer and autumn of 2020.

In the 'green island' areas, the JIP test parameters took values intermediate between those for the green summer leaves and the yellow areas of the senescent leaves (Table 4). The maximum photochemical efficiency of the PSII (Fv/Fm) of 'green islands' on senescent leaves was slightly reduced compared to the summer leaves (10%) and more than two times higher than in the yellow areas. Despite this, a significant decrease in the PI abs value was revealed. A significant decrease in Fv/F0 was also observed, with a simultaneous increase in the values of parameters describing energy absorption by active reaction centers (ABS/RC, TR0/RC) and energy dissipation in non-photochemical form (DI0/RC) in green islands compared to summer leaves. However, in senescent leaves, the fluctuations of the above parameters were significantly greater than in the 'green island' areas (Table 4). The probability (t0) that a trapped exciton moved an electron into the electron transport chain beyond QA (ΨEo) was at a similar level in the 'green island' and yellow parts of the senescent leaf (more than half as low as in summer leaves). The efficiency with which an electron from the intersystem carriers moves to reduce end electron acceptors at the PSI acceptor side (at t0) increased in senescent leaves and reached the highest values in yellow areas; however, the quantum yield for reduction in end electron acceptors at the PSI acceptor side (ϕRo) was similar in all of the analyzed leaf tissues (Table 4).

Table 4. Parameters of chlorophyll fluorescence in *A. platanoides* leaves. The means (n = 21–58) ± SD were obtained from summer leaves in June 2020 and autumn senescent leaves in November 2020. Data were gained using a PocketPEA fluorimeter. Different letters indicate significant differences between leaf variants according to a post hoc Duncan's test ($p < 0.05$).

Parameters	Summer Leaves	Senescent Leaves	
		'Green Islands'	Yellow Area
Measured parameters and basic JIP-test parameters			
Fo	866.7 ± 143.78 a	397.1 ± 125.22 b	561.74 ± 226.7 c
Fm	34,039 ± 1447 a	27,345 ± 5723 b	4452 ± 2818 c
Fv	26,032 ± 1232 a	18,997 ± 4913 b	1702 ± 1453 c
Fv/Fm	0.765 ± 0.012 a	0.688 ± 0.07 b	0.340 ± 0.128 c
Fv/Fo	3.261 ± 0.215 a	2.349 ± 0.641 b	0.581 ± 0.374 c
Vj	0.380 ± 0.032 a	0.733 ± 0.032 b	0.730 ± 0.102 b
Vi	0.929 ± 0.013 a	0.919 ± 0.023 a	0.827 ± 0.053 b
PI abs	4.268 ± 1.065 a	0.283 ± 0.151 b	0.029 ± 0.031 c
Specific energy fluxes expressed per active RC of PSII			
ABS/RC	1.294 ± 0.125 a	3.500 ± 0.875 b	10.392 ± 5.032 c
DIo/RC	0.306 ± 0.043 a	1.151 ± 0.623 b	7.389 ± 5.038 c
TRo/RC	0.988 ± 0.084 a	2.349 ± 0.291 b	3.003 ± 0.286 c
ETo/RC	0.611 ± 0.039 a	0.621 ± 0.067 a	0.813 ± 0.315 b
REo/RC	0.070 ± 0.012 a	0.194 ± 0.075 b	0.523 ± 0.167 c
Quantum yields parameters			
ϕ(Po)	0.765 ± 0.012 a	0.688 ± 0.074 b	0.340 ± 0.128 c
ψ(Eo)	0.620 ± 0.032 a	0.267 ± 0.032 b	0.270 ± 0.102 b
ϕ(Eo)	0.474 ± 0.029 a	0.185 ± 0.036 b	0.088 ± 0.041 c
Δ(Ro)	0.114 ± 0.019 a	0.308 ± 0.107 b	0.668 ± 0.181 c
ϕ(Ro)	0.054 ± 0.010 a	0.054 ± 0.010 a	0.056 ± 0.02 a

2.5. Anatomy and Ultrastructure of Leaves

The anatomy of leaf blades was clearly recognizable in all collected samples (Figure 2). The leaves were covered by single-cell layers of adaxial and abaxial epidermis surrounding the mesophyll, which was composed of a single tier of relatively short palisade mesophyll cells and 3–4 tiers of loosely arranged spongy mesophyll cells. The extensive intercellular spaces were present in the latter. The uncommon feature of epidermal cells was the strong thickening of inner cell walls which were infiltrated with mucilage, and thus were strongly stainable with Toluidine Blue (Figures 2 and 3f,i). The mucilaginous cell walls were present in most of the adaxial epidermis cells, except the cells above the vascular bundles, which retained non-modified call walls. In the abaxial epidermis, their development was more random, but most of the epidermises had mucilaginous cell walls (Figure 2). Cross-sections of the 'green island' regions show a slightly less compact arrangement of palisade and spongy parenchyma cells in comparison to the mature summer leaves (Figure 2b). The outlines of mesophyll cells were strongly wavy. Green-stained chloroplasts were still clearly recognizable and located along the cell walls. Large intercellular spaces were formed even between palisade mesophyll cells. Additionally, groups of strongly hypertrophied cells were often found next to the vascular bundles. The anatomical organization was still well-preserved in the yellow areas of the leaf blade, but mesophyll cells seemed to be almost empty or contained single patches of strongly stained protoplast remnants (Figure 2c). The vascular bundles in 'green island' regions contained well-preserved phloem and xylem parenchyma cells, whereas these cells were degraded in yellow areas.

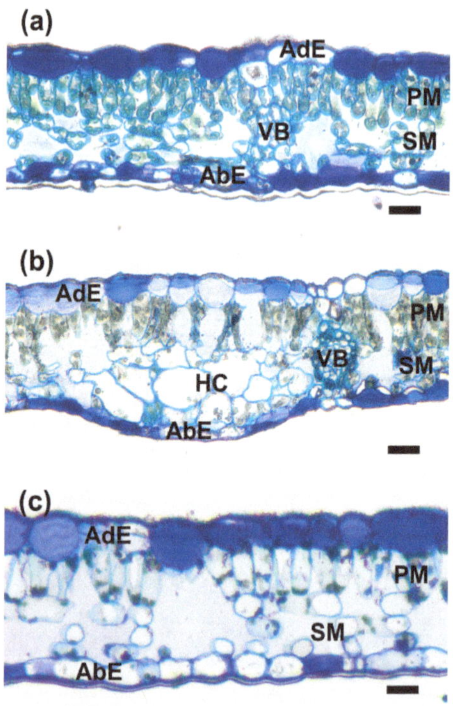

Figure 2. Light microscopy images of Toluidine Blue-stained sections taken from the following: (**a**) mature summer leaf; (**b**) 'green island' induced on autumn senescent leaf; (**c**) the yellow area of a senescent leaf. Abbreviations: AbE, abaxial epidermis; AdE, adaxial epidermis; HC, hypertrophied cells; PM, palisade mesophyll; SM, spongy mesophyll; VB, vascular bundle. Scale bars: 20 μm.

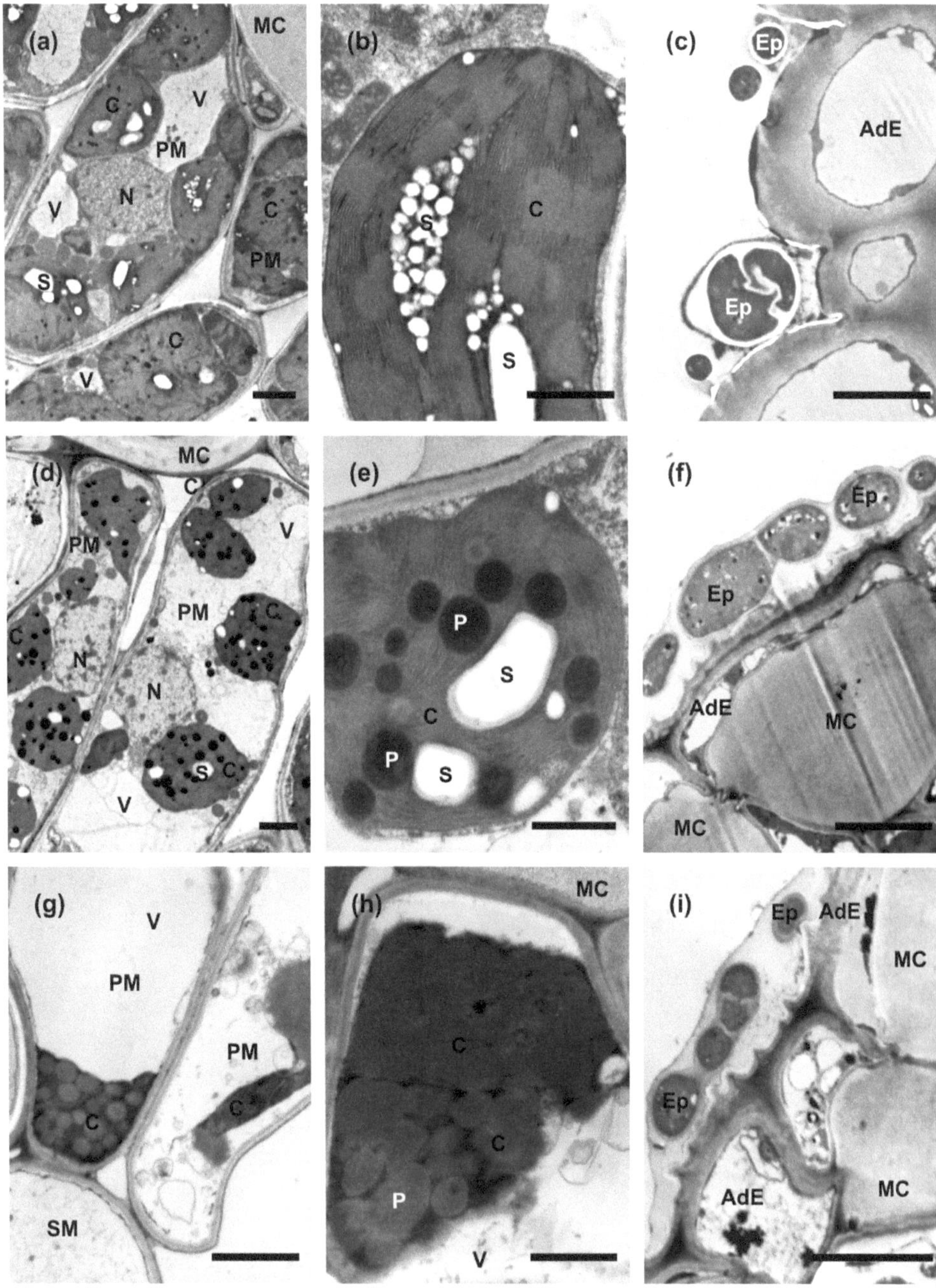

Figure 3. Transmission electron microscopy images of sections taken from summer green leaf (**a–c**), 'green island' induced on autumn senescent leaf (**d–f**), and the yellow area of a senescent leaf (**g–i**). (**a**,**d**,**g**) Overview of palisade mesophyll cell ultrastructure. (**b**,**e**,**h**) Overview of chloroplast's ultrastructure. (**c**,**f**,**i**) Epiphytes located on the surface of adaxial epidermis. Abbreviations: AdE, adaxial epidermis; C, chloroplast; Ep, epiphyte; MC, mucilaginous cell wall; N, nucleus; P, plastoglobuli; PM, palisade mesophyll cell; S, starch grain; SM, spongy mesophyll cell; V, vacuole. Scale bars: 5 μm (**c**,**f**,**i**); 2 μm (**a**,**d**,**g**), and 1 μm (**b**,**e**,**h**).

Ultrastructural examinations of control summer leaves indicated that palisade mesophyll cells protoplasts contain numerous large chloroplasts, nuclei with an electron-dense nucleoplasm, and relatively small vacuoles (Figure 3a). Spongy mesophyll cells contained fewer and smaller chloroplasts and large central vacuoles. Chloroplasts were generally round in outlines on sections. Their stroma was strongly electron dense; the thylakoid system was arranged in extensive grana, and large starch grains were formed. Almost no plastoglobuli were observed (Figure 3b). In contrast, palisade mesophyll cells in 'green island' samples contained large central vacuoles, and chloroplasts were located in cytoplasm along the cell walls (Figure 3d). Additionally, nucleoplasm was also more electron-translucent (Figure 3d). The chloroplasts were also smaller and had more irregular outlines than chloroplasts in mature summer leaves. Moreover, chloroplasts had electron-dense stroma and a well-developed system of thylakoids, but the grana were hardly discernible (Figure 3e). In samples collected from the yellow areas of autumn leaves, mesophyll cells contained only remnants of degraded protoplasts (Figure 3g). Chloroplasts could be recognized only as strongly osmiophilic spots with numerous plastoglobuli inside (Figure 3h), and they contained smaller and fewer starch grains in contrast to mature green leaves.

Spores or hyphae of epiphytes were attached to the surface of adaxial epidermis in all types of samples. In mature summer leaves, they were relatively few, and only sections of single cells were observed (Figure 3c). However, in 'green island' and yellow areas, they appeared frequently as sections of multicellular hyphae embedded in a mucilaginous sheath (Figure 3f,i). No section showing the formation of invasion hyphae or hyphae inside or between mesophyll cells was found. Thus, the nature and function of these epiphytes could not be established clearly.

3. Discussion

Green spots on senescent leaves, called 'green islands', are a seasonal phenomenon associated with autumn and may be caused by pathogens attacking a weakened plant and senescent tissues enhance their visibility. They are most often related to the colonization of plant tissues with pathogenic fungi that are obligate biotrophs or hemibiotrophs at their biotrophic stage [11,22]. In the case of biotrophic pathogens, the inhibition of naturally occurring or induced senescence is beneficial for these pathogens, as they rely on living host cells. However, it is more challenging to explain the formation of 'green islands' in the case of necrotrophs, as they require dying or dead cells to feed on [21].

'Green islands' are leaf lamina fragments in which there is an increased concentration of cytokinins and organic nutrients [23]. As plant hormones, cytokinins are involved in various biological processes, including the inhibition of plant cell senescence. They are also responsible for regulation and maintaining an appropriate level of chlorophyll, thus increasing the mobilization of nutrients [24]. Environmental factors, such as temperature or humidity, may also affect 'green island' development, increasing the intensity of plant–pathogen interactions [25]. 'Green islands' on Norway and field maple leaves could be observed in the following years across the temperate climate zone of Europe, which has four clear seasons [15]. Moreover, they were also described on apple leaves, induced by leaf-mining insects (*Phyllonorycter blancardella*), and the presence of the endosymbiotic bacteria *Wolbachia* enhanced their appearance by the modulation of host physiology [26].

This work studied the appearance of 'green islands' on Norway maple trees in eastern Poland during 6 years, 2019–2024 (Figures A1 and 1). However, very dry winters almost without snow or rain and warm summers (hydrological drought) probably led to the disappearance of 'green islands' on Norway maples in 2023 and 2024 [27,28]. Instead, tar spot, another fungus-induced disease of maples, occurred (Figures A1 and A2). However, how drought and elevated temperature may influence 'green island' formation needs to be investigated.

Although, 'green islands' were vividly greenish even on fallen senescent leaves of Norway maples in the late autumn (November), we decided to investigate whether these green spots were still alive and photosynthetically active. Photosynthesis is the primary process of the plant metabolism, which, in addition to biotic and abiotic environmental factors, is influenced by the structure of assimilation organs (mainly leaves). Significant differences in the efficiency of photochemical processes in individual leaf parts has already been well understood, as well as influence of the efficiency of the photosynthesis light phase on other ontogenetic processes occurring in the leaf, e.g., those related to generative development [29]. Thus, non-invasive estimations of chlorophyll, defense-related metabolites (flavonoids), and the NBI index of plant health status indicated that 'green island' tissues were still alive. However, the mean values of those parameters were significantly lower than in summer healthy leaves (Tables 2 and 3). Chlorophyll content is very often negatively affected by both biotrophic and necrotrophic pathogens and depends on water status, and enhanced hydrological drought (which fluctuated between years [27,28]) may affect 'green island' formation and chlorophyll maintenance [21,30]. In this study, the highest amounts of chlorophylls (total Chl, Chl *a* and *b*) were observed in summer leaves, significantly lower in 'green islands', and the lowest chlorophyll contents were noticed in yellow area of senescent leaves (Table 3). Such a pattern of chlorophyll contents could be observed, for, e.g., in yellow areas and dark 'green islands' (DGIs) induced by Cucumber Mosaic Virus in *Nicotiana tabacum* [31]. A negative correlation between the 'green island' area and chlorophyll content over the years could be related to the fluctuation in carotenoid content, which was relatively high in 'green islands' (similar to yellow areas), but we investigated it only in one year, or else unbalanced nitrogen content as NBI also showed negative correlation to green island areas (and chlorophyll contains nitrogen). One more possible explanation for these negative correlations could be a fluctuating microbiome composition over the years, but it needs to be investigated. In general, the biosynthesis of flavonoids in many cases increases during a pathogen infection as a part of the activity of a host defense system [32,33]. So, it is unsurprising that flavonoid contents were significantly higher in senescent leaves both in 'green islands' and yellow areas. Moreover, the mean values of flavonoids were the least fluctuating parameters over the years in all investigated leaf variants. As expected, anthocyanins were undetectable in summer leaves compared to the high amounts of anthocyanins in the yellow areas of senescent leaves. Also, significantly increased amounts of anthocyanins were detected in 'green islands' (Table 2). All leaf variants had similar mean anthocyanin contents in 2019 and 2020. However, they significantly increased (twice as much) in 2021 both in the 'green islands' and yellow areas of senescent leaves, which is difficult to explain based on the other obtained results. Anthocyanins are often associated with low chlorophyll levels; thus, they appear in abundance in senescent tissues [34]. Moreover, they also play a photoprotective role during senescence to cope with excess light similarly as carotenoids [35,36]. It has to be emphasized that such physiological analyses are rarely performed during 'green island' appearances, so it is difficult to compare our results to other research concerning 'green island' development.

Moreover, a detailed analysis of photosynthetic parameters has been performed (Table 4). The description of light energy transfer within PSII is possible by using the analysis of chlorophyll *a* fluorescence kinetics parameters. The light-harvesting complex of photosystem II absorbs light reaching the leaf. Then the electrons are transported to the central part of the antenna and chlorophyll molecules constituting the reaction centers (RC). This results in charge separation across the membrane and splitting of water molecules into molecular oxygen protons and electrons on the donor side of PSII. Furthermore, electron transport from PSII involves their transfer to QA and QB via b6f, and plastocyanin to PSI, where a second charge separation and further reduction in ferredoxin and NADP+ occur [37]. Thanks to the advanced methodology, the efficiency of the above processes can be effectively analyzed using the JIP test, which allows for the early detection of disturbances in the light phase of photosynthesis [38,39].

The maximum photosynthetic quantum efficiency (Fv/Fm) in 'green islands' was slightly reduced (10%) compared to summer leaves (Table 4). Related to 'green islands' resulting from the impact of biotrophic fungal pathogens, this decrease was small (in barley leaves during *Blumeria graminis* infection, a decrease in the quantum efficiency was 47% in the 'green island' tissues) [11]. At the same time, Fv/Fm remained significantly higher in the 'green islands' compared to the yellow areas of the senescing leaf, which is consistent with previous observations made on plants of other species [40]. Despite this, we observed a significant decrease in Fv/F0 (compared to summer leaves), which indicated the disturbance of activity of the water-splitting complex on the donor side of PSII or the damage of thylakoid structure [38,41]. However, the literature provides conflicting information about whether chloroplasts in 'green islands' remain intact and photosynthetically active. The decrease in the PI abs value, with a simultaneous increase in TR0/RC and DI0/RC, suggests a decrease in the number of active reaction centers and decreased efficiency of electron transport beyond QA, with a simultaneous high degree of electron trapping in the PSII antenna (increase in the size of the PSII antenna) and dissipation of a significant amount of absorbed energy as heat [42]. This phenomenon is probably related to the delayed distribution of antennal chlorophyll in the 'green island' tissues (Table 4).

The only research on 'green islands' on maples revealed that Norway maple leaves were mainly infected by endophytic bacteria from the *Gramproteobacteria* family, and the most numerous endophytic fungi belonged to *Dothideomycetes* and *Leotiomycetes* [15]. In this study, only the spores and hyphae of epiphytic fungi were detected on all leaf variants during microscopic analyses. Moreover, they were present in abundance on both the yellow areas and 'green islands' of senescent leaves (Figure 3). However, no fungal cells were observed inside mesophyll.

An ultrastructural analysis of the leaves showed that cells located in the unevenly distributed 'green islands' on senescent leaves differed from those building a healthy summer leaf. 'Green islands' have more developed mesophylls than in the yellow area of the senescent leaf. So, 'green island' development leads to an extension of the photosynthesis process, which is beneficial for biotrophic pathogenic and non-pathogenic epiphytes and endophytes to be active longer in autumn season [43,44]. However, extended photosynthesis can also lead to a deficiency of nutrients, such as nitrogen, resulting in a weakening of the plant, which may have difficulties to survive the winter, the consequences of which will be visible in spring when the trees re-green [45,46]. The most conspicuous features of cells in 'green island' mesophylls were smaller chloroplasts with numerous plastoglobuli compared to summer leaves (Figure 3). This phenomenon was also observed in *Triticum aestivum* 'green island' cells during infection with *Puccinia striiformis* [40] and in *Brassica juncea* 'green islands' infected with *A. brassicicola* [21]. The abundance of plastoglobuli could be a response to extensive lipid peroxidation that occurred in 'green island' cells, which was detected as a significant increase in malondialdehyde (MDA) content in comparison to their amounts in summer leaves and the yellow areas of senescent leaves (Figure S1). Elevated MDA content is a marker of oxidative stress, and it is often observed in host cells during pathogen infection and abiotic stress [47].

Further research on 'green island' origin and functioning will allow for a more detailed understanding of the causes of this phenomenon and its role in plant–pathogen interactions.

4. Materials and Methods

4.1. Plant Material

All experiments were performed on Norway maple (*Acer platanoides* L.) leaves plucked from young 6–8-year-old trees (Figure A1). The trees were localized in the urban forest surrounding the University of Bialystok campus (53°06′56″ N 23°09′54″ E) in Bialystok, a city in eastern Poland. The young summer leaves were collected from Norway maples on 27 June 2020 and 7 June 2021, whereas the senescent leaves showing 'green island' symptoms were collected in the first week of November in the years 2019–2022. The leaves were torn from young maples (fallen leaves were not collected) and carefully selected to avoid collecting leaves with blurred, irregular green spots, which could result from a delayed aging process. In each experiment, three variants of leaf areas were analyzed: summer leaves and two areas within senescent leaves, yellow and 'green island' areas. The Norway maple trees were also observed for 'green island' symptoms in the following years, 2023 and 2024. However, due to the small number of leaves showing 'green islands', the leaves were not subjected to analysis.

4.2. Non-Destructive Evaluation of Physiological Parameters

Each year, beginning from the autumn of 2019 until November 2022, the 'green island' areas were measured in senescent leaves using a WinDIAS Leaf Image Analysis System (Delta-T Devices, Cambridge, UK) and are expressed as a percentage of the whole leaf surface. Each year, between 9 and 30 leaves per variant were analyzed.

The same leaves plus the additional number of those harvested on the other day of November were subjected to non-destructive measurements of physiological parameters. Chlorophyll content in the mesophyll, anthocyanin and flavonoid content in the leaf epidermis, and the Nitrogen Balance Index (NBI; the ratio of chlorophylls and flavonoids) were measured with a Dualex Optical Leaf-Clip Sensor (ForceA, Orsay, France). This method is based on leaf transmittance and chlorophyll fluorescence [48]. The content of chlorophylls is expressed in $\mu g\ cm^{-2}$, whereas anthocyanins, flavonoids, and NBI have no units [49]. The measurements were also performed on 25–65 summer leaves. A detailed statistical analysis of these parameters is available in Supplementary Materials Table S1.

4.3. Chlorophyll and Carotenoid Content

The 150 mg samples from the summer and autumn leaves were collected on 27 June and 10 November 2020, respectively, and stored at −80 °C until used. Chlorophylls and carotenoids were extracted in 100% methanol (HPLC grade, Avantor Performance Materials, Gliwice, Poland), as described by Macioszek et al. [32], analyzed using a Hitachi U-5100 spectrophotometer (Hitachi Ltd., Tokyo, Japan), and calculated according to Wellburn [50]. Eight summer and eight autumn leaf samples were used for analyses, with four technical replicates.

4.4. Analysis of Photosynthesis

For the chlorophyll fluorescence analyses, from 21 to 58 detached summer and senescent leaves were immediately dark-adapted for 20 min using leaf clips. Then, dark-adapted areas were subjected to measurements conducted with a Pocket PEA chlorophyll fluorimeter (Hansatech Instruments Ltd., Norfolk, UK) equipped with a high-intensity focus LED of 3500 $\mu mol\ m^{-2}\ s^{-1}$ photon flux with a peak wavelength of 627 nm at the sample surface according to the manufacturer's built-in protocol. Leaves for analyzes were collected in June and November 2020.

4.5. Microscopic Analysis

The leaves for light and transmission electron microscopy investigation were collected on 7 June and 10 November 2021. The samples were dissected from green summer leaves and two areas of autumn leaves: yellow and the center of the 'green islands'. They were collected from at least three different trees. The dissected specimens were immediately transferred into a modified Karnovsky fixative and processed for light and transmission electron microscopy and sectioning, as described by Crespo et al. [51].

4.6. Statistical Analysis

Each parameter's means and standard deviations (SD) were calculated using MS Office Excel 2021 software. An analysis of variance (ANOVA) and post hoc Duncan's test ($p < 0.05$) of all the data obtained in this work were performed using STATISTICA v.13.3 (Tibco Software Inc., StatSoft, Krakow, Poland).

The figures were composed using Adobe Photoshop v. 6.0 and Corel Software v. 11.

5. Conclusions

'Green islands' have been described in limited but variable pathosystems, mainly affecting crop plants such as maize, barley, and tobacco [16,18,20]. However, naturally occurring 'green islands' were observed only in a few tree species, such as maples and apples [15,26]. Pathogens and insects inducing 'green islands' also have different life cycles and feeding strategies. Therefore, the structural, functional, and environmental factors of 'green islands' require further research, as do the biotic and abiotic factors influencing their appearance. The physiological, metabolic, and genetic aspects of 'green islands' also should be investigated.

The potential usefulness of 'green islands' cannot be overestimated. 'Green islands' can indicate specific pathogen infection; virus-induced 'green islands' look different from those on maple leaves. Moreover, they can be an interesting model for studying how pathogens manipulate host metabolism (e.g., this study) or for observing localized immune responses similarly to hypersensitive responses. A more far-reaching use of 'green islands' will be the identification of the genetic traits associated with resistance to pathogens that induce 'green islands' to support breeding programs for disease-resistant varieties. 'Green islands' constitute a research challenge due to their low impact and economic significance on tree leaves and their visibility being mostly in autumn.

Supplementary Materials: The following supporting information can be downloaded at: https://www.mdpi.com/article/10.3390/plants14060909/s1, Table S1: Two-way ANOVA of non-destructive parameters: chlorophyll, flavonoids, anthocyanins, and nitrogen balance index (NBI) in summer and autumn Norway maple leaves; Figure S1: Content of malondialdehyde (MDA) in summer and senescent leaves of Norway maple collected in June and November 2020.

Author Contributions: Conceptualization, V.K.M.; methodology, V.K.M. and M.S.; formal analysis, V.K.M., K.C., J.O. and M.S.; investigation, V.K.M. (analyses of physiological parameters and photosynthesis), K.C. (microscopic analysis), A.M.S. (analysis of lipid peroxidation, MDA); writing—original draft preparation, V.K.M., K.C., J.O. and M.S.; writing—review and editing, V.K.M. and M.S.; visualization, V.K.M. and M.S.; supervision, V.K.M. and M.S.; funding acquisition, V.K.M. and M.S. All authors have read and agreed to the published version of the manuscript.

Funding: This research was supported by the Faculty of Biology, University of Bialystok, Poland (statutory funds), and Warsaw University of Life Sciences, Poland (WULS-SGGW).

Data Availability Statement: Data are contained within the article or Supplementary Materials.

Acknowledgments: The authors would like to thank Weronika Mierzejewska (an undergraduate student of University of Bialystok, Bialystok, Poland) for taking photos of maple leaves with 'green island' and tar spot symptoms in the autumn of 2022 using a WinDIAS system, and Justyna Frankowska-Łukawska (WULS-SGGW) for their excellent ultramicrotomy assistance.

Conflicts of Interest: The authors declare no conflicts of interest. The funders had no role in the design of the study; in the collection, analyses, or interpretation of data; in the writing of the manuscript; or in the decision to publish the results.

Appendix A

An abundant appearance of 'green islands' on Norway maple leaves was observed in the small area of the urban forest surrounding the University of Bialystok campus in Bialystok, eastern Poland (Figure A1). Tar spot, a fungal disease, could be observed on Norway maple leaves since 2022, and it became the major disease of Norway maple since 2023. Last year, only a few young trees showed symptoms of 'green islands', sometimes in combination with tar spots (Figure A2).

Figure A1. Young maple trees from the urban forest on the University of Bialystok campus (Bialystok, Poland). (**a**) Green leaves of Norway maple on 27 June 2021; (**b**) Norway maple senescent leaves with 'green islands' symptoms on 5 November 2022; (**c**) young field maple trees with 'green islands' symptoms on 9 November 2020; (**d**) Norway maple leaves with tar spot symptoms on 27 September 2024.

Figure A2. Tar spot symptoms on leaves of young Norway maple trees in the urban forest on the University of Bialystok campus (Bialystok, Poland). (**a**–**c**) Tar spots with yellow rings in late August 2023; (**d**) tar spots in September 2024; (**e**) 'green island' symptoms with a single tar spot in November 2024.

References

1. Akhmetov, A.; Ianbaev, R.; Boronnikova, S.; Yanbaev, Y.; Gabitova, A.; Kulagin, A. Norway maple (*Acer platanoides*) and pedunculate oak (*Quercus robur*) demonstrate different patterns of genetic variation within and among populations on the eastern border of distribution ranges. *J. For. Sci.* **2021**, *67*, 522–532. [CrossRef]
2. Caudullo, G.; de Rigo, D. Acer platanoides in Europe: Distribution, habitat, usage and threats. In *European Atlas of Forest Tree Species*; San-Miguel-Ayanz, J., De Rigo, D., Caudullo, G., Houston Durrant, T., Mauri, A., Eds.; Joint Research Centre, Publications Office of the European Union, European Commission: Luxembourg, 2016; pp. 54–55. Available online: https://data.europa.eu/doi/10.2788/4251 (accessed on 23 November 2024).
3. Fang, W.; Wang, X.Z. Impact of invasion of *Acer platanoides* on canopy structure and understory seedling growth in a hardwood forest in North America. *Trees-Struct. Func.* **2011**, *25*, 455–464. [CrossRef]
4. Fang, W.; Wang, X.Z. A field experimental study on the impact of *Acer platanoides*, an urban tree invader, on forest ecosystem processes in North America. *Ecol. Process.* **2020**, *9*, 9. [CrossRef]
5. Mitchell, A.F. *A Field Guide to the Trees of Britain and Northern Europe*, 1st ed.; Collins: London, UK, 1974.
6. Kerr, G.; Niles, J. Growth and provenance of Norway maple (*Acer platanoides*) in lowland Britain. *Forestry* **1998**, *71*, 219–224. [CrossRef]
7. Gilman, E.F.; Watson, D.G. *Acer platanoides—Norway Maple. Fact Sheet ST-28*; University of Florida: Gainesville, FL, USA, 1993; Available online: https://hort.ifas.ufl.edu/trees/ACEPLAA.pdf (accessed on 12 December 2024).
8. Praciak, A.; Pasiecznik, N.; Sheil, D.; Van Heist, M.; Sassen, M.; Correia, C.S.; Dixon, C.; Fyson, G.; Rushford, K.; Teeling, C. *The CABI Encyclopedia of Forest Trees*; CABI: Oxfordshire, UK, 2013.
9. Bosco, C.; de Rigo, D.; Dewitte, O.; Poesen, J.; Panagos, P. Modelling soil erosion at European scale: Towards harmonization and reproducibility. *Nat. Hazards Earth Syst. Sci.* **2015**, *15*, 225–245. [CrossRef]
10. Stokes, A.; Norris, J.E.; van Beek, L.P.H.; Bogaard, T.; Cammeraat, E.; Mickovski, S.B.; Jenner, A.; Di Iorio, A.; Fourcaud, T. How Vegetation Reinforces Soil on Slopes. In *Slope Stability and Erosion Control: Ecotechnological Solutions*; Norris, J.E., Stokes, A., Mickovski, S.B., Cammeraat, E., van Beek, L.P.H., Achim, A., Eds.; Springer: Dordrecht, The Netherlands, 2008. [CrossRef]

11. Walters, D.R.; McRoberts, N.; Fitt, B.D. Are green islands red herrings? Significance of green islands in plant interactions with pathogens and pests. *Biol. Rev.* **2008**, *83*, 79–102. [CrossRef]
12. Le Cocq, K.; Gurr, S.J.; Hirsch, P.R.; Mauchline, T.H. Exploitation of endophytes for sustainable agricultural intensification. *Mol. Plant Pathol.* **2016**, *18*, 469–473. [CrossRef]
13. van Overbeek, L.S.; Saikkonen, K. Impact of bacterial–fungal interactions on the colonization of the endosphere. *Trends Plant Sci.* **2016**, *21*, 230–242. [CrossRef]
14. Hardoim, P.R.; van Overbeek, L.S.; Berg, G.; Pirttila, A.M.; Compant, S.; Campisano, A.; Doring, M.; Sessitsch, A. The hidden world within plants: Ecological and evolutionary considerations for defining functioning of microbial endophytes. *Microbiol. Mol. Biol. Rev.* **2015**, *79*, 293–320. [CrossRef]
15. Wemheuer, F.; Wemheuer, B.; Daniel, R.; Vidal, S. Deciphering bacterial and fungal endophyte communities in leaves of two maple trees with green islands. *Sci. Rep.* **2019**, *9*, 14183. [CrossRef]
16. Moore, C.J.; Sutherland, P.W.; Forster, R.L.S.; Gardner, R.C.; MacDiarmid, R.M. Dark green islands in plant virus infection are the result of posttranscriptional gene silencing. *Mol. Plant-Microbe Interact.* **2001**, *14*, 939–946. [CrossRef] [PubMed]
17. Walters, D.R.; McRoberts, N. Plants and biotrophs: A pivotal role for cytokinins? *Trends Plant Sci.* **2006**, *11*, 581–586. [CrossRef] [PubMed]
18. Behr, M.; Humbeck, K.; Hause, G.; Deising, H.B.; Wirsel, S.G.R. The hemibiotroph *Colletotrichum graminicola* locally induces photosynthetically active green islands but globally accelerates senescence on aging maize leaves. *Mol. Plant-Microbe Interact.* **2009**, *23*, 879–892. [CrossRef] [PubMed]
19. Gutzwiller, F.; Dedeine, F.; Kaiser, W.; Giron, D.; Lopez-Vaamonde, C. Correlation between the green-island phenotype and *Wolbachia* infections during the evolutionary diversification of *Gracillariidae* leaf-mining moths. *Ecol. Evol.* **2015**, *5*, 4049–4062. [CrossRef]
20. Angra-Sharma, R.; Sharma, D. Cytokinins in pathogenesis and disease resistance of *Pyrenophora teres*-barley and *Dreschslera maydis*-maize interactions during early stages of infection. *Mycopathologia* **2000**, *148*, 87–95. [CrossRef]
21. Macioszek, V.K.; Sobczak, M.; Skoczowski, A.; Oliwa, J.; Michlewska, S.; Gapinska, M.; Ciereszko, I.; Kononowicz, A.K. The Effect of Photoperiod on Necrosis Development, Photosynthetic Efficiency and 'Green Islands' Formation in *Brassica juncea* Infected with *Alternaria brassicicola*. *Int. J. Mol. Sci.* **2021**, *22*, 8435. [CrossRef]
22. Ashby, A.M. Biotrophy and the cytokinin conundrum. *Physiol. Mol. Plant Pathol.* **2000**, *57*, 147–158. [CrossRef]
23. Giron, D.; Frago, E.; Glevarec, G.; Pieterse, C.M.J.; Dicke, M. Cytokinins as key regulators in plant–microbe–insect interactions: Connecting plant growth and defence. *Funct. Ecol.* **2013**, *27*, 599–609. [CrossRef]
24. Naseem, M.; Wölfling, M.; Dandekar, T. Cytokinins for immunity beyond growth, galls and green islands. *Trends Plant Sci.* **2014**, *19*, 481–484. [CrossRef]
25. Bushnell, W.R. Symptom development in mildewed and rusted tissue. In *The Dynamic Role of Molecular Constituents in Plant-Parasite Interactions*; Mirocha, C.J., Uritani, I., Eds.; Bruce Publishing Company: St. Paul, MN, USA, 1967; pp. 21–39.
26. Kaiser, W.; Huguet, E.; Casas, J.; Commin, C.; Giron, D. Plant green-island phenotype induced by leaf-miners is mediated by bacterial symbionts. *Proc. R. Soc. B Biol. Sci.* **2010**, *277*, 2311–2319. [CrossRef]
27. Pińskwar, I.; Choryński, A.; Kundzewicz, Z.W. Severe Drought in the Spring of 2020 in Poland—More of the Same? *Agronomy* **2020**, *10*, 1646. [CrossRef]
28. Hejduk, L.; Kaznowska, E.; Wasilewicz, M.; Hejduk, A. Hydrological Droughts in the Białowieża Primeval Forest, Poland, in the Years 1951–2020. *Forests* **2021**, *12*, 1744. [CrossRef]
29. Skoczowski, A.; Rut, G.; Oliwa, J.; Kornaś, A. Sporulation modifies the photosynthetic activity of sporotrophophyll leaves of *Platycerium bifurcatum*. *Photosynthetica* **2020**, *58*, 488–496. [CrossRef]
30. Lu, Y.; Yao, J. Chloroplasts at the Crossroad of Photosynthesis, Pathogen Infection and Plant Defense. *Int. J. Mol. Sci.* **2018**, *19*, 3900. [CrossRef]
31. Shang, J.; Xi, D.-H.; Yuan, S.; Xu, F.; Xu, M.-Y.; Qi, H.-L.; Huang, Q.-R.; Wen, L.; Lin, H.-H.; Wang, S.-D. Difference of Physiological Characters in Dark Green Islands and Yellow Leaf Tissue of Cucumber mosaic Virus (CMV)-Infected *Nicotiana tabacum* Leaves. *Z. Naturforsch. C* **2010**, *65*, 73–78. [CrossRef]
32. Macioszek, V.K.; Wielanek, M.; Morkunas, I.; Ciereszko, I.; Kononowicz, A.K. Leaf position-dependent effect of *Alternaria brassicicola* development on host cell death, photosynthesis and secondary metabolites in *Brassica juncea*. *Physiol. Plant.* **2020**, *168*, 601–616. [CrossRef]
33. Mierziak, J.; Kostyn, K.; Kulma, A. Flavonoids as Important Molecules of Plant Interactions with the Environment. *Molecules* **2014**, *19*, 16240–16265. [CrossRef]
34. Lee, D.W.; Gould, K.S. Anthocyanins in leaves and other vegetative organs: An introduction. *Adv. Bot. Res.* **2002**, *3*, 1–16. [CrossRef]
35. Verhoeven, A.; Southwick, C.; Miller, E.; Blood, M.; Thibodeau, A. Do red and yellow autumn leaves make use of different photoprotective strategies during autumn senescence? *Physiol. Plant.* **2024**, *176*, e14327. [CrossRef]

36. Gómez-Sagasti, M.T.; López-Pozo, M.; Artetxe, U.; Becerril, J.M.; Hernández, A.; García-Plazaola, J.I.; Esteban, R. Carotenoids and their derivatives: A "Swiss Army knife-like" multifunctional tool for fine-tuning plant-environment interactions. *Environ. Exp. Bot.* **2023**, *207*, 105229. [CrossRef]
37. Rochaix, J.-D. Regulation of photosynthetic electron transport. *Biochim. Biophys. Acta—Bioenerg.* **2011**, *1807*, 878–886. [CrossRef] [PubMed]
38. Strasser, R.J.; Tsimilli-Michael, M.; Srivastava, A. Analysis of the chlorophyll *a* fluorescence transient. In *Chlorophyll a Fluorescence: A Signature of Photosynthesis*; Advances in Photosynthesis and Respiration; Papageorgiou, G.C., Govindjee, Eds.; Springer: Dordrecht, Germany, 2004; pp. 321–362. [CrossRef]
39. Kalaji, H.M.; Bąba, W.; Gediga, K.; Goltsev, V.; Samborska, I.A.; Cetner, M.D.; Dimitrova, S.; Piszcz, U.; Bielecki, K.; Karmowska, K.; et al. Chlorophyll fluorescence as a tool for nutrient status identification in rapeseed plants. *Photosynth. Res.* **2018**, *136*, 329–343. [CrossRef]
40. Aldesuquy, H.S.; Abdel-Fattah, G.M.; Baka, Z.A. Changes in chlorophyll, polyamines and chloroplast ultrastructure of *Puccinia striiformis* induced 'green islands' on detached leaves of *Triticum aestivum*. *Plant Physiol. Biochem.* **2000**, *38*, 613–620. [CrossRef]
41. Pereira, W.E.; de Siqueira, D.L.; Martínez, C.A.; Puiatti, M. Gas exchange and chlorophyll fluorescence in four citrus rootstocks under aluminium stress. *J. Plant Physiol.* **2000**, *157*, 513–520. [CrossRef]
42. Gilmore, A.M.; Itoh, S.; Govindjee. Global spectral–kinetic analysis of room temperature chlorophyll *a* fluorescence from light-harvesting antenna mutants of barley. *Philos. Trans. R. Soc. B* **2000**, *355*, 1371–1384. [CrossRef]
43. Rozpądek, P.; Wężowicz, K.; Nosek, M.; Ważny, R.; Tokarz, K.; Lembicz, M.; Miszalski, Z.; Turnau, K. The fungal endophyte *Epichloë typhina* improves photosynthesis efficiency of its host orchard grass (*Dactylis glomerata*). *Planta* **2015**, *242*, 1025–1035. [CrossRef] [PubMed]
44. Gomes, T.; Pereira, J.A.; Benhadi, J.; Lino-Neto, T.; Baptista, P. Endophytic and Epiphytic Phyllosphere Fungal Communities Are Shaped by Different Environmental Factors in a Mediterranean Ecosystem. *Microb. Ecol.* **2018**, *76*, 668–679. [CrossRef]
45. Estiarte, M.; Peñuelas, J. Alteration of the phenology of leaf senescence and fall in winter deciduous species by climate change: Effects on nutrient proficiency. *Glob. Change Biol.* **2015**, *21*, 1005–1017. [CrossRef]
46. Shangguan, Z.; Shao, M.; Dyckmans, J. Effects of Nitrogen Nutrition and Water Deficit on Net Photosynthetic Rate and Chlorophyll Fluorescence in Winter Wheat. *J. Plant Physiol.* **2000**, *156*, 46–51. [CrossRef]
47. Heath, R.L.; Packer, L. Photoperoxidation in isolated chloroplasts: I. Kinetics and stoichiometry of fatty acid peroxidation. *Arch. Biochem. Biophys.* **1968**, *125*, 189–198. [CrossRef]
48. Goulas, Y.; Cerovic, Z.G.; Cartelat, A.; Moya, I. Dualex: A new instrument for field measurements of epidermal ultraviolet absorbance by chlorophyll fluorescence. *Appl. Opt.* **2004**, *43*, 4488–4496. [CrossRef] [PubMed]
49. Cerovic, Z.G.; Masdoumier, G.; Ghozlen, N.B.; Latouche, G. A new optical leaf-clip meter for simultaneous non-destructive assessment of leaf chlorophyll and epidermal flavonoids. *Physiol. Plant.* **2012**, *146*, 251–260. [CrossRef] [PubMed]
50. Wellburn, A.R. The Spectral determination of chlorophylls a and b, as well as total carotenoids, using various solvents with spectrophotometers of different resolution. *J. Plant Physiol.* **1994**, *144*, 307–313. [CrossRef]
51. Crespo-Martínez, S.; Sobczak, M.; Różańska, E.; Forneck, A.; Griesser, M. The role of the secondary phloem during the development of the grapevine Berry Shrivel ripening disorder. *Micron* **2019**, *116*, 36–45. [CrossRef]

plants

MDPI

Article

Exogenous Melatonin Enhances Cold Tolerance by Regulating the Expression of Photosynthetic Performance, Antioxidant System, and Related Genes in Cotton

Jincheng Zhu [1,2,†], Hui Lou [1,†], Chen Yan [1], Wei Zhang [1,*] and Zhibo Li [1,*]

1 The Key Laboratory of Oasis Eco-Agriculture, Agriculture College, Shihezi University, Shihezi 832000, China; zhujincheng@stu.shzu.edu.cn (J.Z.); louhui@stu.shzu.edu.cn (H.L.); yanchen@stu.shzu.edu.cn (C.Y.)

2 Biotechnology Research Institute, Xinjiang Academy of Agricultural and Reclamation Science, Shihezi 832000, China

* Correspondence: zhw_agr@shzu.edu.cn (W.Z.); lzb_oea@shzu.edu.cn (Z.L.)

† These authors contributed equally to this work.

Abstract: In China, cotton is a significant cash crop, and cold stress negatively impacts the crop's development, production, and quality formation. Recent studies have shown that melatonin (MT) can alleviate the damage to plants under cold stress and promote good growth and development. In this study, the morphological and physiological changes induced by exogenous melatonin pretreatment on 'Xinluzao 33' cotton seedlings under cold stress were examined to investigate its defensive effects. The results showed that 100 μM MT pretreatment improved the cold resistance of cotton most significantly. It also improved the wilting state of cotton under cold stress, greatly increased the photosynthetic rate (Pn), stomatal conductance (Gs), maximum photochemical efficiency (Fv/Fm), and photosynthetic performance index (PIabs) by 116.92%, 47.16%, 32.30%, and 50.22%, respectively, and mitigated the adverse effects of low-temperature. In addition, MT supplementation substantially reduced the accumulation of superoxide anion ($O_2^{\bullet-}$) and hydrogen peroxide (H_2O_2) by 14.5% and 45.49%, respectively, in cold-stressed cotton leaves by modulating the antioxidant system, thereby mitigating oxidative damage. Furthermore, MT pretreatment increased the endogenous melatonin content (23.80%) and flavonoid content (21.44%) and considerably induced the expression of biosynthesis enzyme-related genes. The above results indicate that exogenous melatonin improves the low-temperature resistance of cotton seedlings by regulating photosynthetic performance, antioxidant enzyme activity, antioxidant content, endogenous melatonin and flavonoid content, and the expression levels of genes related to their synthesis.

Keywords: melatonin; antioxidant defense; cold stress; photosynthetic performance

Citation: Zhu, J.; Lou, H.; Yan, C.; Zhang, W.; Li, Z. Exogenous Melatonin Enhances Cold Tolerance by Regulating the Expression of Photosynthetic Performance, Antioxidant System, and Related Genes in Cotton. *Plants* **2024**, *13*, 2010. https://doi.org/10.3390/plants13152010

Academic Editors: Violetta Katarzyna Macioszek, Iwona Ciereszko and Andrzej K. Kononowicz

Received: 27 May 2024
Revised: 15 July 2024
Accepted: 16 July 2024
Published: 23 July 2024

1. Introduction

Plant growth and development are severely hampered by cold stress, which also restricts the range of locations where plants can be found in nature [1]. Reduced photosynthesis, altered gene expression levels, disruption of intracellular redox equilibrium, damage to membrane systems, and perturbation of basal metabolism are the principal impacts of low-temperature stress on plants [2]. Low temperature induces the production of reactive oxygen species (ROS) such as superoxide ($O_2^{\bullet-}$), hydrogen peroxide (H_2O_2), and hydroxyl radicals (-OH). High concentrations of ROS can induce oxidative damage in plants; however, spatiotemporal transient accumulation of ROS, such as H_2O_2, is often used as a signal [3]. Plants have evolved a range of defense mechanisms, including enzymatic and non-enzymatic antioxidant systems, and ROS are partially scavenged by enzymes such as superoxide dismutase (SOD), peroxidase (POD), catalase (CAT), ascorbate peroxidase (APX), and glutathione peroxidase (GPX) [4]. Malondialdehyde (MDA), relative conductivity (REC), photosynthetic rate (Pn), stomatal conductance (Gs), maximum photochemical efficiency (Fv/Fm), and photosynthetic performance index (PIabs) can be used to assess the degree of damage and photosynthetic performance of plants under adversity [5]. An increase in MDA indicates damage to cell membranes, an increase in REC reflects an increase in cell membrane permeability, a decrease in Pn indicates a decrease in photosynthetic capacity, a decrease in GS affects photosynthesis, a decrease in Fv/Fm indicates a decrease in the efficiency of light energy conversion, and a decrease in PIabs reflects an overall impaired photosynthetic performance of plants [6]. The decrease in PIabs reflects the overall impaired photosynthetic performance of the plant. These indicators provide key information for understanding the response mechanism of plants to adversity. Previous studies have identified chalcone synthase (*CHS*), flavanone 3′-hydroxylase (*F3′H*), and flavanone 3-hydroxylase (*F3H*) as key enzymes in flavonoid biosynthesis. Chalconeisomerase (*CHI*) is a key enzyme in flavonoid metabolism. Dihydroflavonol-4-reductase (*DFR*) catalyzes the production of anthocyanin precursors from dihydroflavonols and flavanones [7]. Chlorophyllate a-oxygenase (*CAO*) is involved in the conversion of chlorophyll from a to b; Glutamic acid-1-semialdehyde transaminase (*HEML*) plays a key role in chlorophyll synthesis; Glutamyl t-RNA reductase (*HEMA*) catalyzes ALA synthesis and plays an important role in plant growth regulation; Pheide a oxygenase (*PAO*), 7-Hydroxymethyl Chlorophyll a reductase (*HCAR*), and Pheophytinase (*PPH*) are involved in chlorophyll degradation [8]. Under cold conditions, some cold-tolerant plants mainly exhibit elevated content of secondary metabolites, increased accumulation of osmoregulatory substances, enhanced antioxidant capacity, and up-regulation of cold resistance-related gene expression [9,10]. Most researchers have tried to enhance plant hardiness by screening and breeding new varieties, but in general, yield and quality are negatively correlated with adversity traits, and the long breeding period of new varieties makes it challenging to obtain desirable results in a short time. Therefore, there is an urgent need to find a method that is non-toxic, efficient, non-polluting, and effective in improving plant resistance.

Numerous scholars have successfully explored and applied various means, such as the exogenous application of phytohormones, osmoregulatory substances, gas molecules, and micronutrients, to significantly enhance the tolerance of plants to low-temperature stress [11–13]. Chang and colleagues discovered that the application of exogenous melatonin may mitigate the growth-inhibiting effects of low temperatures on barley seedlings. This was achieved by controlling the expression of the circadian clock genes *HvCCA1* and *HvTOC1* and the accumulation of several important physiological markers [14]. Zhang et al. found that melatonin improved cucumber cold tolerance by activating antioxidant enzymes and inducing key PSI, PSII-related genes, and carbon assimilation genes, and finally alleviated the damage to the photosynthetic apparatus of cucumber seedlings under cold stress [15]. Zhao et al. found that exogenous melatonin increased the expression of *CsZat12* and controlled the metabolism of pure amine (PA) and abscisic acid (ABA), thereby mitigating the damage caused by low-temperature stress in cucumbers [16].

Since it originated in the tropics and subtropics, cotton (*Gossypium* spp.) is susceptible to cold temperatures when it is growing and developing [17]. With the increasing demand, cotton cultivation areas have expanded to high latitudes and altitudes where the climate is unsuitable [18,19]. Therefore, the relevant planting areas are very prone to "reverse spring" weather, resulting in delayed growth and development, especially frequent cold damage during the budding and seedling stages, seriously affecting the final yield and quality of cotton [4,20], improving the cold tolerance of cotton seedlings is of great significance to increasing cotton yield.

There are currently studies available on the use of MT to increase resistance to verticillium wilt, heavy metal stress, drought, and salt [21–24], but little research has been performed on whether MT treatment regulates cold tolerance in cotton. Our focus was on studying growth situation, photosynthetic performance, the degree of cellular damage, and ROS homeostasis in cotton seedlings under cold stress by MT. Precisely, we also investigated the content of endogenous melatonin and flavonoids and their biosynthetic enzymes encoding the expression of related genes. This study attempted to provide a theoretical basis for MT to improve cold tolerance in cotton.

2. Results

2.1. 100 µM Exogenous Melatonin Alleviates Cold Stress Injury

Following low-temperature stress, cotton leaves were subjected to an analysis of the effects of varying doses of MT pretreatment on malondialdehyde (MDA), relative electrical conductivity, net photosynthetic rate, and stomatal conductance. The results showed that MDA content and relative conductivity (REC) in cotton leaves were drastically increased after cold stress by 148.28% and 84.52%, respectively (Figure 1A,B). However, Pn, Gs, Fv/Fm, and PIabs decreased markedly by 71.23%, 58.35%, 36.99%, and 72.42%, respectively (Figure 1C–F). Compared with spraying with 0 µM MT, spraying with 25–200 µM MT restored the Pn, Gs, Fv/Fm, and PIabs values to some extent, inhibited the accumulation of MDA, and eased the degree of cell damage. Among all MT treatments, 100 µM pretreatment had the best effect; Pn, Gs, Fv/Fm, and PIabs were greatly increased by 116.92%, 47.16%, 32.30%, and 50.22%, respectively, and MDA and REC were appreciably decreased by 34.09% and 33.79%, respectively, so this concentration was chosen for the follow-up test.

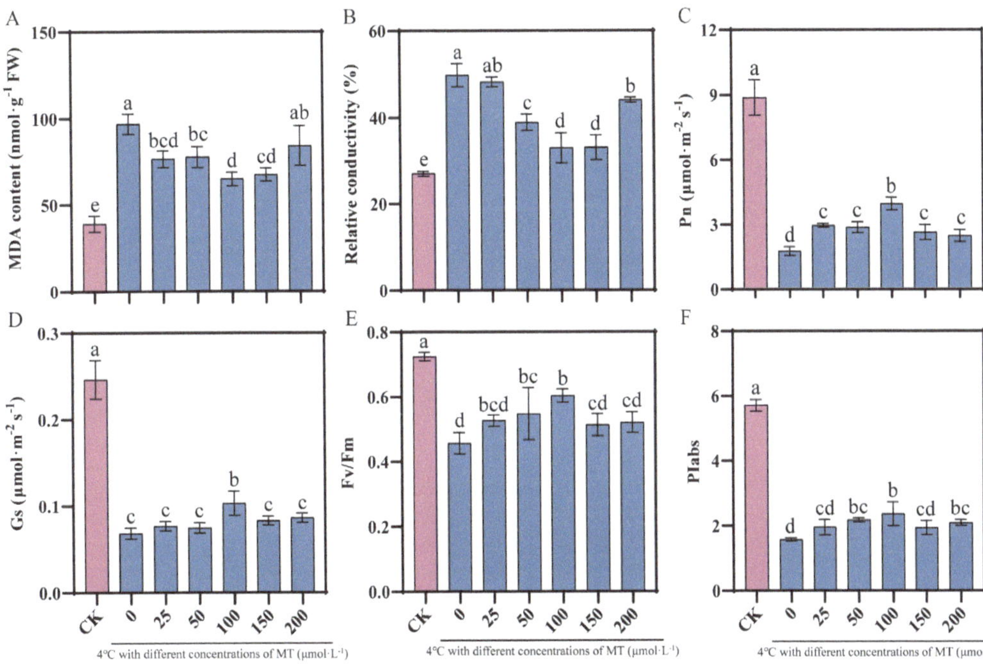

Figure 1. Effects of melatonin pretreatment at different concentrations on MDA. CK: under normal growth conditions after distilled water spray treatment (pink column). 0, 25, 50, 100, 150, 200: 4 °C with different concentrations of MT (0, 25, 50, 100, 150, 200 µmol·L^{-1}) (pink column). MDA (**A**), REC (**B**), Pn (**C**), GS (**D**), Fv/Fm (**E**), and PIabs (**F**) in cotton under cold stress. Vertical bars in each column represent ±SD of three replicates, and different letters represent significant differences ($p < 0.05$).

2.2. *Effect of Exogenous Melatonin on Cotton Growth*

At the end of the stress, both CS seedlings and MT + CS seedlings showed different degrees of leaf wilting, dehydration (Figure 2), and drooping, but the MT + CS group had markedly fewer symptoms than the CS group. After pretreatment with MT, growth limitations caused by CS stress were improved, and fewer reductions in fresh weight

(7.69%) and dry weight (9.30%) were observed (Figure 2A,B). These findings imply that pretreating cotton seedlings with exogenous MT can lessen the harm that cold stress causes.

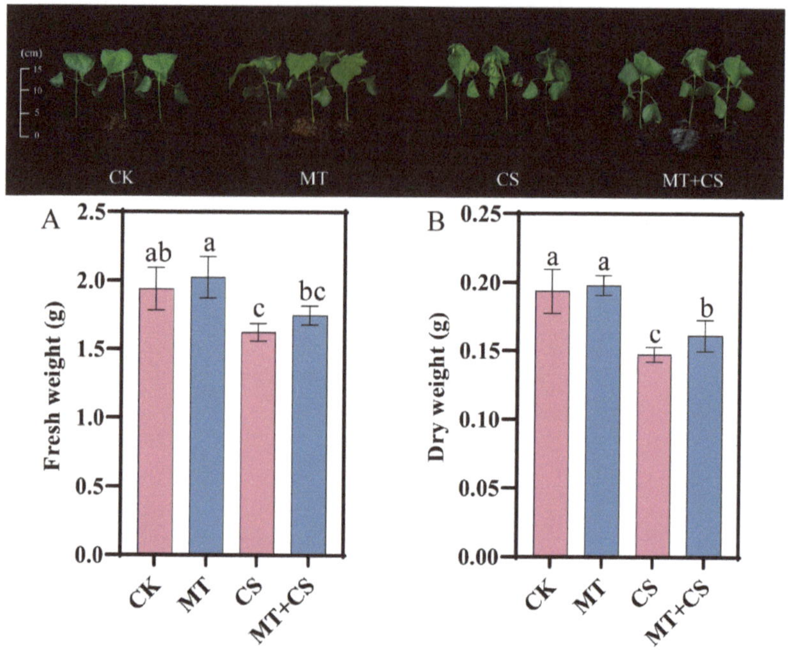

Figure 2. Effect of different treatments on the growth of cotton seedlings. CK: under normal growth conditions after distilled water spray treatment. MT: under normal growth conditions after 100 μM MT (the optimal concentration in the pre-screening) spray treatment. CS: spray distilled water treatment followed by 4 °C treatment for 24 h. MT + CS: spray 100 μM MT treatment followed by 4 °C treatment for 24 h. (**A**) Fresh weight, (**B**) Dry weight. Vertical bars in each column represent ±SD of three replicates, and different letters represent significant differences ($p < 0.05$).

2.3. *Effects of Exogenous Melatonin on the Antioxidant System*

Plants produce large amounts of ROS species as signaling molecules to deal with stress, but too much ROS can cause oxidative damage to themselves. Exogenous MT pretreatment under normal conditions did not change H_2O_2 and $O_2^{\bullet-}$ content in leaves (Figure 3A,B). The H_2O_2 range and $O_2^{\bullet-}$ in the CS group were substantially higher by 38.27% and 118.71%, respectively, as compared to the CK group. MT pretreatment may be able to lessen the harm produced by cold induction, as seen by the MT + CS group's much lower values, which decreased by 14.50% and 45.49%, respectively, after the CS group.

In addition, we measured antioxidant enzyme (POD, CAT, SOD, APX, and GPX) activities (Figure 3C–G) and antioxidant (AsA and GSH) content (Figure 3H,I) in the different treatments. Compared with CK plants, the activities of POD and GPX enzymes in CS plants increased by 97.43% and 17.34%, respectively, while the activity of the APX enzyme decreased by 18.20% (Figure 3C–F). Compared with the CS plants, the activities of these five antioxidant enzymes were all increased to different degrees in the MT + CS plants, with POD, APX, and GPX activities notably increased by 29.96%, 39.78%, and 25.03%, respectively. AsA and GSH contents in MT + CS seedlings increased by 17.43% and 15.87%, respectively, and were substantially higher than those in CS seedlings.

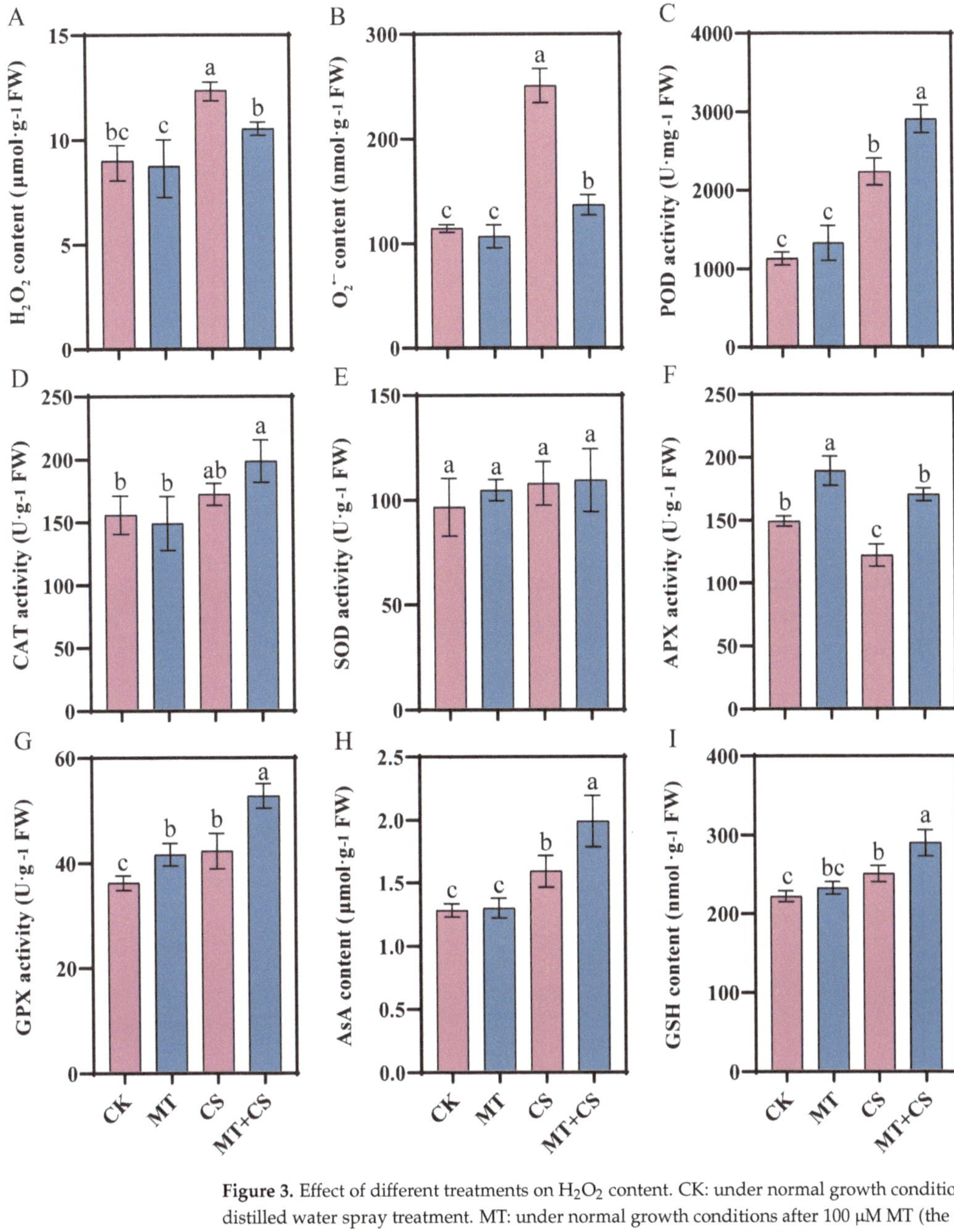

Figure 3. Effect of different treatments on H_2O_2 content. CK: under normal growth conditions after distilled water spray treatment. MT: under normal growth conditions after 100 μM MT (the optimal concentration in the pre-screening) spray treatment. CS: spray distilled water treatment followed by 4 °C treatment for 24 h. MT + CS: spray 100 μM MT treatment followed by 4 °C treatment for 24 h. H_2O_2 content (**A**), $O_2^{\bullet-}$ content (**B**), antioxidant enzyme (POD (**C**), CAT (**D**), SOD (**E**), APX (**F**), GPX (**G**)) activities, and antioxidant (AsA (**H**), GSH (**I**)) content of cotton seedlings. Vertical bars in each column represent ±SD of three replicates, and different letters represent significant differences ($p < 0.05$).

2.4. *Effects of Exogenous Melatonin on Photosynthetic Pigment Content*

Photosynthesis is carried out with the participation of photosynthetic pigments, trapping light radiation and converting it into CO_2^- assimilated energy. According to Figure 4, compared with the CK seedlings, chlorophyll a, chlorophyll b, and carotenoid in CS seedlings decreased markedly by 25.10%, 31.90%, and 42.50%, respectively. On the contrary, the chlorophyll a, b, and carotenoid contents of MT + CS seedlings were reduced by 11.8%, 16.6%, and 19.7%, respectively, compared with CK seedlings, which was a significantly smaller decrease.

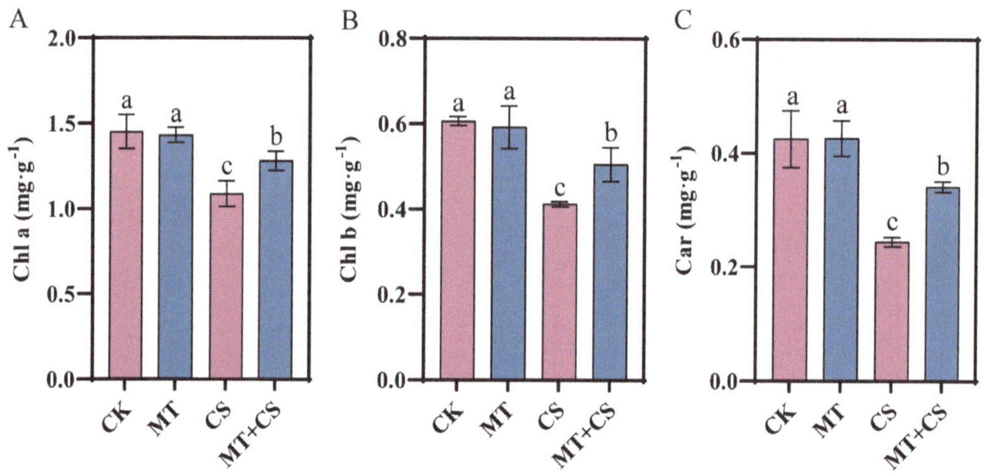

Figure 4. Effect of different treatments on chlorophyll a (**A**), chlorophyll b (**B**), and carotenoid (**C**) content of cotton seedlings. CK: under normal growth conditions after distilled water spray treatment. MT: under normal growth conditions after 100 µM MT (the optimal concentration in the pre-screening) spray treatment. CS: spray distilled water treatment followed by 4 °C treatment for 24 h. MT + CS: spray 100 µM MT treatment followed by 4 °C treatment for 24 h. Vertical bars in each column represent ±SD of three replicates, and different letters represent significant differences ($p < 0.05$).

2.5. *Effects of Exogenous Melatonin on Flavonoid Content and Biosynthesis Enzyme-Related Genes Expression*

One of the most significant and prevalent secondary metabolites in plants, flavonoids can improve the plant's ability to withstand stress and produce antioxidants naturally. In the present study, the flavonoid content in the MT group was slightly higher than that in the CK group, but the difference was not significant. Compared with the CK group, the flavonoid content in the CS and MT + CS groups increased by 10.20% and 23.80%, respectively (Figure 5).

The expression levels of flavonoid synthesis genes *GhDFR*, *GhF3H*, *GhCHI*, and *GhF3′H* were up-regulated to varying degrees in the MT plants compared to the CK plants, and only *GhCHS* was slightly down-regulated, but there was no significant difference. Compared with CK seedlings, the expression levels of *GhDFR*, *GhF3H*, *GhCHI*, and *GhF3′H* were up-regulated to varying degrees in MT + CS seedlings, among which *GhDFR*, *GhCHI*, and *GhF3′H* were considerably up-regulated, increasing by 671.58%, 22.65% and 241.29%, respectively.

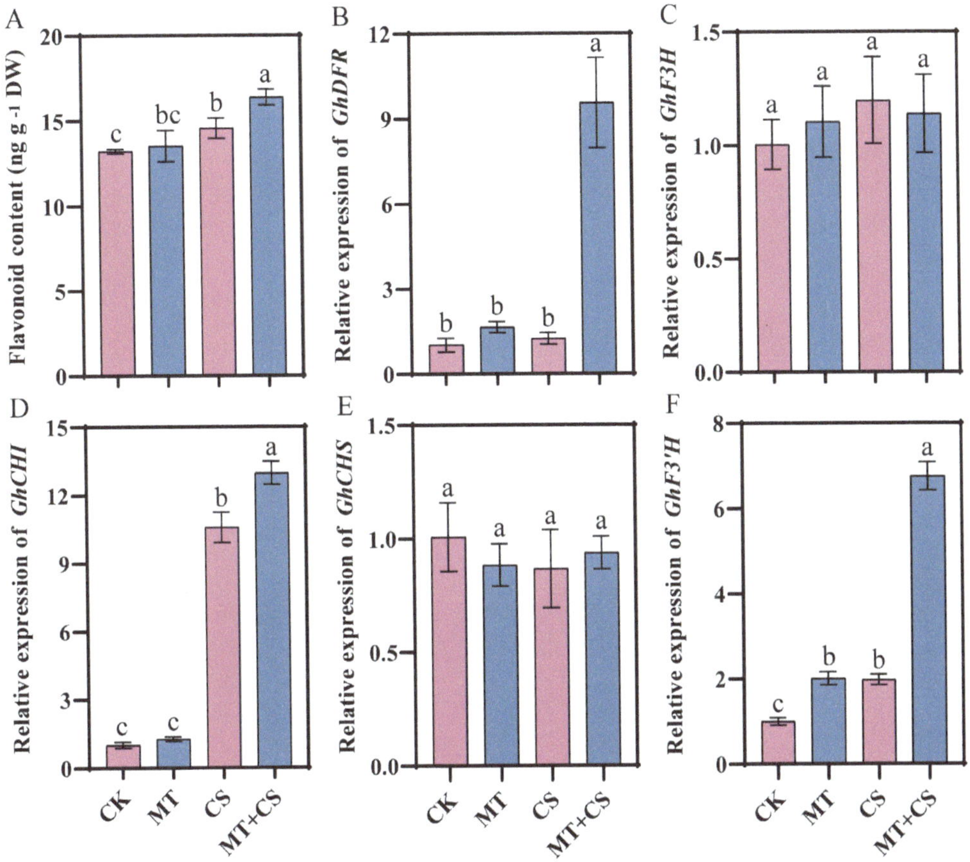

Figure 5. Effect of different treatments on flavonoid content (**A**) and flavonoid biosynthesis enzyme-related gene (*GhDFR* (**B**), *GhF3H* (**C**), *GhCHI* (**D**), *GhCHS* (**E**), *GhF3′H* (**F**)) expression in cotton seedlings. CK: under normal growth conditions after distilled water spray treatment. MT: under normal growth conditions after 100 μM MT (the optimal concentration in the pre-screening) spray treatment. CS: spray distilled water treatment followed by 4 °C treatment for 24 h. MT + CS: spray 100 μM MT treatment followed by 4 °C treatment for 24 h. Vertical bars in each column represent ±SD of three replicates, and different letters represent significant differences ($p < 0.05$).

2.6. Effects of Exogenous Melatonin on Endogenous Melatonin Content and Synthesis-Related Genes Expression

Endogenous melatonin content was increased by 21.44% in MT + CS plants compared to CK plants (Figure 6). The expression of *GhSNAT* and *GhCOMT* was up-regulated in the CS group compared to the CK group, increasing by 68.44% and 240.14%, respectively, while the expression of *GhTDC* was down-regulated compared to the CK group, decreasing by 30.19%. The gene expression levels of melatonin biosynthesis genes *GhSNAT*, *GhCOMT*, *GhTDC*, and *GhASMT* in the MT + CS group were up-regulated to varying degrees compared with those in the CK group, among which *GhCOMT* and *GhASMT* were greatly up-regulated by 109.61% and 595.61%, respectively.

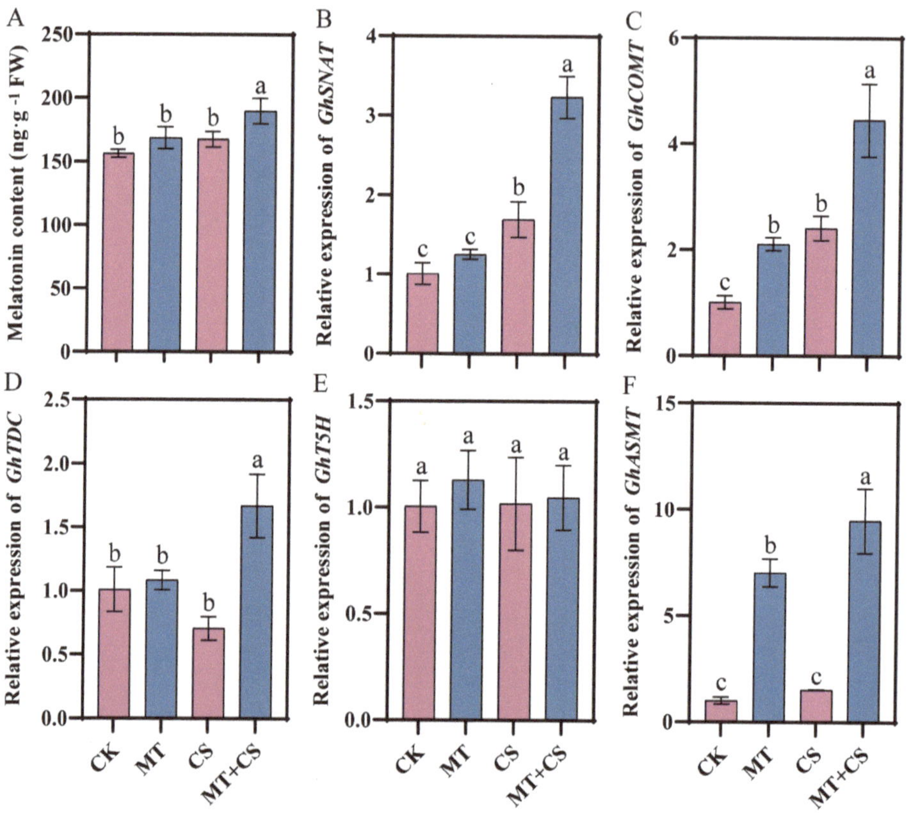

Figure 6. Effects of different treatments on endogenous melatonin content (**A**) and melatonin biosynthesis enzyme-related gene (*GhSNAT* (**B**), *GhCOMT* (**C**), *GhTDC* (**D**), *GhT5H* (**E**), *GhASMT* (**F**)) expression in cotton seedlings. CK: under normal growth conditions after distilled water spray treatment. MT: under normal growth conditions after 100 μM MT (the optimal concentration in the pre-screening) spray treatment. CS: spray distilled water treatment followed by 4 °C treatment for 24 h. MT + CS: spray 100 μM MT treatment followed by 4 °C treatment for 24 h. Vertical bars in each column represent ±SD of three replicates, and different letters represent significant differences ($p < 0.05$).

2.7. Effects of Exogenous Melatonin on the Expression of Genes Related to Chlorophyll Degradation and Synthesis in Cotton under Low-Temperature Stress

Chlorophylls are a class of green pigments contained in photosynthesizing organisms and are magnesium porphyrin compounds that belong to a family of lipid-containing pigments. According to Figure 7, the expression levels of chlorophyll synthesis genes *GhCAO*, *GhHEML*, and *GhHEMA* were apparently not dramatically different in the CS group compared with the CK group. However, the expression of *GhCAO*, *GhHEML*, and *GhHEMA* were remarkably up-regulated in the MT + CS group, which increased by 135.06%, 488.47%, and 131.02%, respectively. The expression levels of chlorophyll-degrading genes *GhPAO*, *GhHCAR*, and *GhPPH* were appreciably up-regulated in the CS group compared with the CK group, which increased by 78.67%, 18.54%, and 64.79%, respectively. Meanwhile, the expression of *GhPAO* was significantly down-regulated in the MT + CS group, with a reduction of 48.71%. Among them, the *GhPPH* gene expression was not different from that of the MT group.

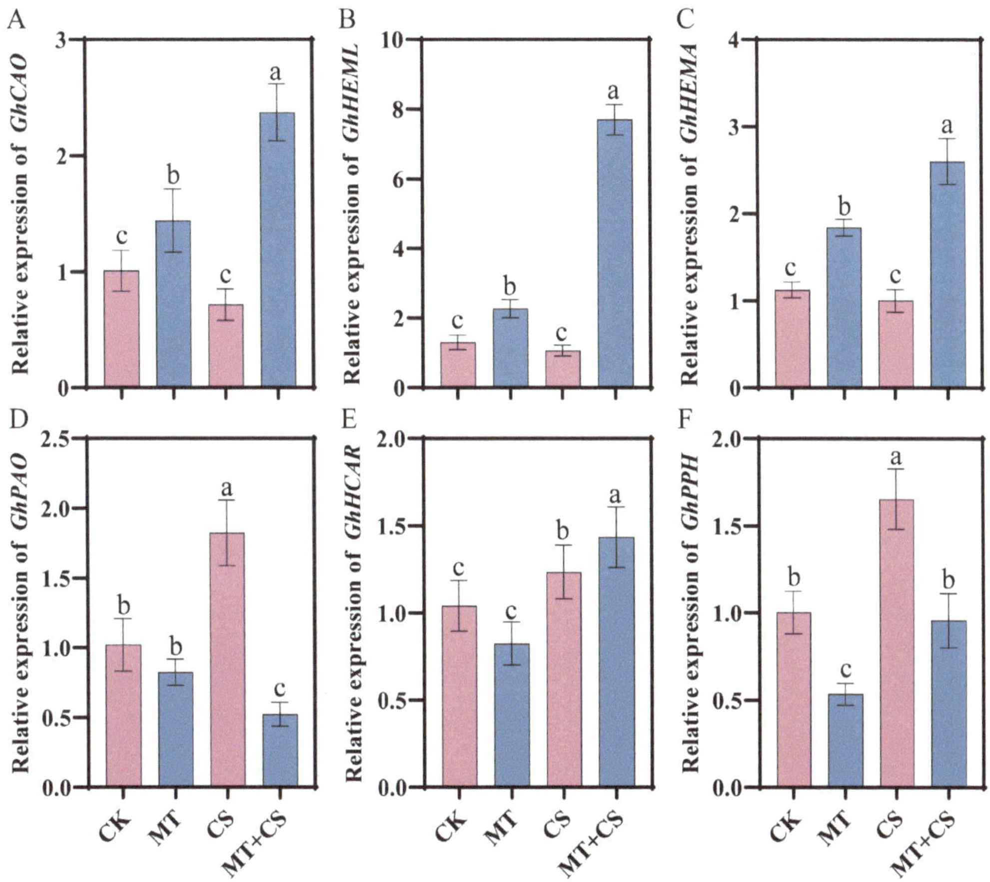

Figure 7. Effects of different treatments on the expression of genes related to chlorophyll degradation and synthesis in cotton (*GhCAO* (**A**), *GhHEML* (**B**), *GhHEMA* (**C**), *GhPAO* (**D**), *GhHCAR* (**E**), *GhPPH* (**F**)). CK: under normal growth conditions after distilled water spray treatment. MT: under normal growth conditions after 100 μM MT (the optimal concentration in the pre-screening) spray treatment. CS: spray distilled water treatment followed by 4 °C treatment for 24 h. MT + CS: spray 100 μM MT treatment followed by 4 °C treatment for 24 h. Vertical bars in each column represent ±SD of three replicates, and different letters represent significant differences ($p < 0.05$).

2.8. Correlation Analysis of Cotton Antioxidant System with Flavonoids and Melatonin

Through a comprehensive analysis of the correlation of antioxidant system parameters with flavonoids and melatonin in cotton, the antioxidant system parameter POD showed a highly significant positive correlation with flavonoid content and its synthesized genes, *GhDFR*, *GhCHI*, and *GhFNS*. Meanwhile, there was a highly significant positive correlation between melatonin content and its synthesizing genes, *GhSNAT* and *GhCOMT*, and a significant negative correlation with Chl a, Chl b, and Car (Figure 8).

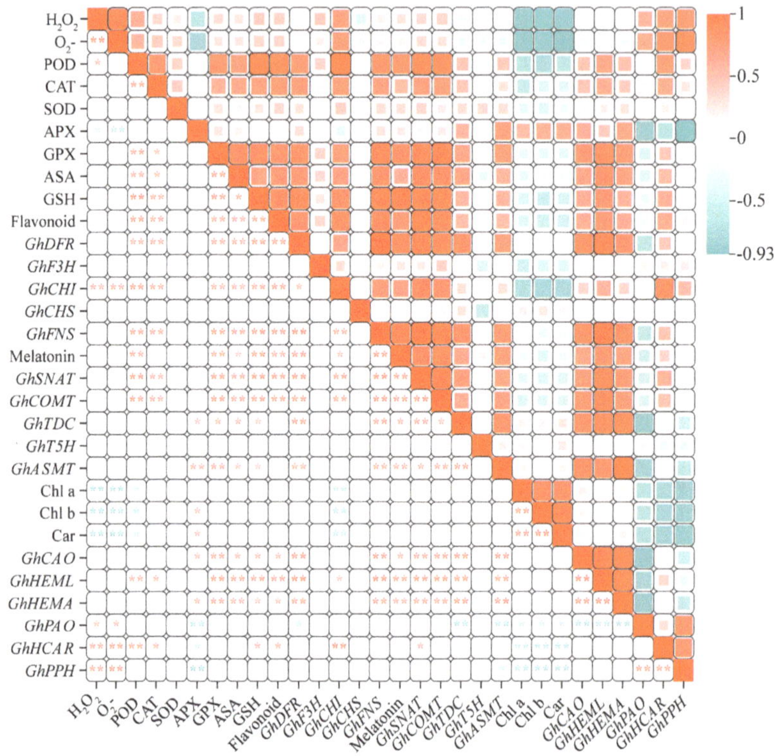

Figure 8. Correlation analysis of cotton antioxidant system with flavonoids and melatonin. (Different size of colored cells map the magnitude of correlation,* $p < 0.05$; ** $p < 0.01$).

3. Discussion

Cold stress greatly affects the yield of cotton by disrupting its physiological metabolism and its morphological structure and stunting its growth [25]. Previous studies have found that MT, a ubiquitous multifunctional signaling molecule, can improve plant tolerance to low-temperature stress by regulating physiological and biochemical processes [11]. This study showed that the leaves of cotton seedlings exhibited significant wilting and crinkling after cold stress treatment, and MT pretreatment could reduce the damage caused by the stress (Figure 2). Similar findings were also verified in banana seedlings, pepper seedlings, and strawberry seedlings under cold stress [26–28].

In plants, photosynthesis is a vital physiological mechanism that produces 95% of the dry matter accumulation in plants [29]. The chlorophyll content is crucial for the smooth progress of photosynthesis, and its level determines the rate of photosynthesis, which directly affects plant growth and yield [30–32]. In this study, we found that the net photosynthetic rate, stomatal conductance, and chlorophyll content of cotton seedlings were considerably decreased under cold stress, but MT pretreatment alleviated the degradation of photosynthetic pigments caused by cold stress, indicating that MT can effectively promote the synthesis of photosynthetic pigments under stress conditions (Figure 4). According to earlier research, exogenous MT dramatically up-regulates gene expression linked to chlorophyll production and down-regulates gene expression related to chlorophyll breakdown, hence maintaining chlorophyll concentration under stress [33–35]. Furthermore, chlorophyll fluorescence measurements are thought of as internal sensors for researching the connection between plant photosynthesis and the environment. To a certain extent, these metrics can accurately indicate a plant's ability to withstand cold temperatures [36,37].

Studies have shown that Fv/Fm is relatively stable when plants are not under stress, but it will be markedly reduced when plants are under pressure or mechanical damage [38]. The present study revealed that low-temperature stress significantly reduced the Fv/Fm values of cotton leaves, which is a clear indication of the inhibition of the photosynthetic system. However, it is noteworthy that the Fv/Fm values of the optimal melatonin pretreated group were significantly elevated compared with those of the untreated group, a finding that suggests that the application of exogenous melatonin effectively mitigated the inhibition of the photosynthetic capacity of cotton by low-temperature stress (Figure 1E). In addition, our study also found that the PIabs values of the melatonin-pretreated group were significantly higher than those of the untreated group under low-temperature conditions (Figure 1F), which coincided with the results of a study on bananas conducted by Liu et al. [26] and further confirmed the positive role of melatonin in enhancing plant stress tolerance.

Cold stress disrupts the plant's own ROS scavenging mechanism, causing excessive ROS accumulation and lipid peroxidation, which affects plant growth and development [39]. MT directly scavenges ROS to act as an antioxidant and lessen damage to plants [40]. It was shown that in order to keep the balance between ROS generation and scavenging, a single MT molecule may interact with eight or more ROS molecules continually [41]. The MT congener 5-methoxyindole and the downstream product of MT metabolism, N1-acetyl-N2-formyl-5-methoxykynura-mine, were also found to have different degrees of antioxidant activity [42–44]. According to earlier research, MT supplementation under stressful circumstances boosted the expression of genes encoding related synthetic enzymes and stimulated the manufacture of endogenous MT [45]. In this study, we observed that the endogenous melatonin (MT) content and the expression levels of its biosynthesis genes (*GhSNAT*, *GhCOMT*, *GhTDC*, *GhASMT*) were up-regulated to different degrees in cotton seedlings after cold stress. It is noteworthy that these up-regulation phenomena showed more significance when cotton seedlings were pretreated with MT (Figure 6). Accordingly, we hypothesized that cotton seedlings would actively increase the level of melatonin in response to cold stress, and MT pretreatment could accelerate the process of melatonin synthesis, thus scavenging reactive oxygen species (ROS) more efficiently and mitigating the damage caused by cold stress. This finding provides a new perspective for understanding how cotton seedlings respond and adapt to cold stress.

Important components of the antioxidant defense system include both non-enzymatic and enzyme-based antioxidants [46]. According to earlier research, MT pretreatment can lessen the harm that cold stress causes to oilseed rape and watermelon by increasing the activity of antioxidant enzymes and the expression of genes encoding them [47,48]. MT pretreatment significantly enhanced the activities of antioxidant enzymes, including SOD, CAT, POD, APX, and GPX, as well as increased the contents of antioxidants AsA and GSH in cotton under stress conditions (Figure 3). These results fully demonstrated that MT pretreatment had a positive effect on enhancing the performance of the cotton antioxidant system and could maintain the balance between ROS production and accumulation more effectively. These findings further validate the hypothesis that MT scavenges ROS directly or indirectly through the activation of antioxidant enzymes and fit with previous research results [49].

It has been shown that by taking part in the manufacture of secondary metabolites in plants, MT increases tolerance to a range of stressors. A significant portion of the phenylpropane metabolic system, flavonoid molecules play a major role in the growth, development, and stress tolerance of plants [50,51]. Previous studies have found that the content of flavonoids in Arabidopsis leaves is positively correlated with cold tolerance, proving that flavonoids are one of the key factors determining cold acclimation and freezing tolerance of Arabidopsis [52–54]. Sun et al. found that exogenous melatonin increased the expression of *CcF3′h-5* through the transcription factor *CcPCL1*, promoted the accumulation of luteolin and its derivatives, and speculated that flavonoids could be used as essential mediators in melatonin to enhance the tolerance of plants to various stresses [55]. In the present study, it was found that there was a significant increase in flavonoid content in cotton leaves under low temperature stress conditions. However, when cotton was pretreated with melatonin, the increase in flavonoid content was more significant. Meanwhile, we observed that the up-regulated expression of key genes in the flavonoid synthesis pathway (*GhDFR*, *GhCHI*, and *GhF3′H*) was more prominent after melatonin pretreatment (Figure 5), and this finding provides a new perspective for an in-depth understanding of flavonoid biosynthesis and regulatory mechanisms. Therefore, we believe that MT pretreatment enhanced the resistance of cotton to cold stress can be reflected by the increase in flavonoid content and biosynthetic gene expression levels to be regulated.

4. Materials and Methods

4.1. Effects of Exogenous Melatonin on Photosynthetic Pigment Content

The experimental material 'Xinluzao 33' was purified, preserved, and provided by the Cotton Molecular Breeding Laboratory of Oasis Ecology Laboratory of Shihezi University. After choosing the seeds with whole granules, they were steeped in 75% alcohol for 10 min and then rinsed 3–4 times in ultrapure water. The sterilized seeds were soaked in 50 °C–60 °C warm water for 20 min and then placed in a germination box with sterilized gauze. After being accelerated to sprout at 25 °C in the dark for 48 h (keeping the gauze moist all the time), the germinated seeds were planted in a nutrient bowl and cultured in an artificial intelligence climate box (ZDN-1000g-2, Ningbo, China). The culture substrate was peat: vermiculite = 3:1, and the climate box conditions were set as follows: the temperature was 28 °C/25 °C (day/night), relative humidity was 60%, light intensity was 32,000 Lux, and photoperiod was 16 h/8 h (day/night).

After cotton seedlings grew to 'three leaves in one heart,' different concentrations (0, 25, 50, 100, 150, and 200 μM) of MT (CAS: 73-31-4, Sigma, St. Louis, MO, USA) were applied. The solution required uniform spraying and spraying to the leaf surface as the standard, once in the morning and once in the evening (9 a.m. and 9 p.m.), with continuous spraying for three days while avoiding light after spraying. Each treatment was applied to 20 seedlings. After the last spraying treatment, the cotton was allowed to stand for 24 h in preparation for low-temperature stress treatment. The procedure was performed as follows: cotton was placed in a climate chamber with a temperature of 4 °C/4 °C (day/night), relative humidity of 60%, a light intensity of 32,000 Lux, and photoperiod of 16 h/8 h (day/night) for 24 h. After the stress, the gas exchange measurements, fast chlorophyll fluorescence parameters, malondialdehyde content, and electrical conductivity of the third genuine leaf of cotton seedlings were measured to screen the optimal melatonin concentration for improving the cold tolerance of cotton seedlings.

To cultivate new cotton seedlings, seed disinfection, germination promotion, and seedling cultivation were the same as above. When cotton seedlings reached the "three leaves and one heart" stage, they were divided into two groups. One group was treated with 100 μM MT, and the other group was treated with distilled water (CK), with the same treatment method as above. The seedlings of each group were placed in incubations at 28 °C/25 °C (day/night) and 4 °C/4 °C (day/night) for 24 h (CK and MT seedlings treated with 4 °C low-temperature were named CS and MT + CS, respectively). Thirty cotton seedlings were collected from each treatment. The third true leaves of cotton seedlings under different treatments were taken for the determination of physiological indicators and gene expression.

4.2. Measurement of Photosynthetic Parameters

Photosynthetic parameters, including Pn (net photosynthetic rate) and Gs (stomatal conductance), were assayed (9 to 11 a.m.) by LI-COR 6400 Photosynthesis System (Li-Cor, Lincoln, NE, USA). The light source was the instrument's red and blue built-in light source, and the light intensity was set at 1000 $\mu mol \cdot m^{-2} \cdot s^{-1}$. The maximum photochemical efficiency of PSII (Fv/Fm) and light energy absorption performance index (PIabs) were measured by a portable pulse adjustment fluorometer (Handy PEA, Hansatech, UK). The leaves were fully dark-adapted for 30 min before the assay and then induced using

3000 μmol·m^{-2}·s^{-1} red light for 2 s. Seedlings with the same growth rate were selected from the same leaf position for each treatment, and five biological replicates were set up. Acetone extraction was used to measure the amount of chlorophyll. The UV1800 Spectrophotometer was used to measure the absorbance at 645, 663, and 440 nm, respectively, to compute the amounts of chlorophyll a (Chl a), chlorophyll b (Chl b), and carotenoid (Caro) [56].

4.3. Measurement of Malondialdehyde Content and Relative Conductivity

In a test tube, 0.4 g of leaves were mixed with 10 mL of distilled water, and the vacuum was evacuated for 20 min. A conductivity meter (DDSJ-308A, Shanghai Kanglu Instrument Equipment Co., Ltd., Shanghai, China) was used to measure the mixture's initial conductivity (S1). After incubating in a boiling water bath for 10 min, the mixture was cooled and its conductivity was measured (S2). A conductivity test was also conducted on the distilled water (S0). Electrolyte leakage (%) = [(S1 − S0)/(S2 − S0)] × 100 [57]. Using the thiobarbituric acid method, malondialdehyde (MDA) content was calculated from absorption values at 450, 532, and 600 nm [58], and three independent replications were conducted.

4.4. Measurement of H_2O_2 and $O_2^{\bullet-}$ Content, Antioxidant Enzyme Activity, and Antioxidant Content

The contents of H_2O_2, $O_2^{\bullet-}$, ascorbic acid (AsA), reduced glutathione (GSH), and the activities of peroxidase (POD), catalase (CAT), superoxide dismutase (SOD), ascorbic acid peroxidase (APX), and glutathione peroxidase (GPX) were measured by using the kit on a UV-1800 spectrophotometer. APX) and glutathione peroxidase (GPX) activities. These kits (H_2O_2-2-Y, SA-2-G, ASA-2-W, GSH-2-W, POD-2-Y, CAT-2-Y, SOD-2-Y, APX-2-W, GPX-2-W, Suzhou, China) were produced by Suzhou Comin Biotechnology Co.

H_2O_2 forms a complex with titanium sulfate with characteristic absorption at 415 nm. $O_2^{\bullet-}$ reacts with hydroxylamine hydrochloride to form NO_2^-, and in the presence of p-aminobenzenesulfonic acid and α-naphthylamine, an azo compound is formed with a characteristic absorption peak at 530 nm. In acetic acid solution, ASA reacts with the solid blue salt B to form a yellow derivative of oxalylhydrazide-2-hydroxybutyrolactone, the absorbance of which is measured at a maximum absorption wavelength of 420 nm. GSH reacts with DTNB to form a complex with a characteristic absorption peak at 412 nm. POD catalyzes the oxidation of specific substrates by H_2O_2 with characteristic light absorption at 470 nm. CAT is able to decompose H_2O_2 so that the absorbance of the reaction solution at 240 nm decreases with reaction time, and CAT activity can be calculated from the rate of change of absorbance. The SOD activity was determined by the NBT method, i.e., the reduced riboflavin is highly susceptible to re-oxidation and generates superoxide anion radicals, and reduces NBT to methanone, with a characteristic absorption peak at 560 nm. APX catalyzed the reduction of H_2O_2 by AsA, and the activity of APX could be calculated by measuring the amount of AsA oxidized. GPX catalyzes the peroxide oxidation of GSH to GSSG, and glutathione reductase uses NADPH to reduce GSSG and regenerate GSH, which has a characteristic absorption peak at 340 nm, and the rate of reduction of light absorption in this band was measured to calculate GPX activity. The procedure is described in the instructions (Supplementary Materials S2), and three biological replicates were used for each indicator.

4.5. Measurement of Endogenous Melatonin Content and Flavonoid Content

Melatonin content was determined by plant melatonin enzyme-linked immunosorbent assay analytical kit (96T/48T, Shanghai, China). The 0.1 g sample was mashed with normal saline and centrifuged at 3000 rpm for 10 min. A Plant MT ELISA Kit was used to detect melatonin. Using the kit of Suzhou Comin Biotechnology Co., Ltd. (LHT-2-G, Suzhou, China), flavonoid content was determined by measuring absorbance values at 510 nm. A total of three biological replicates were tested on each sample.

4.6. Real-Time Polymerase Chain Reaction Analysis of Gene Expression

EasyPure Plant RNA Kit was used to extract total RNA, and EasyScript One-Step gDNA Removal and cDNA Synthesis SuperMix kit was used to reverse transcribe it into cDNA. SYBR Green was used in qRT-PCR reactions on a Light Cycler 480II sequence detection system, following the method of Xiong et al. [59]. *GhUBQ7* (DQ116441) was used as the internal control since it is expressed stably in cotton plants and is unaffected by treatment or genotype. Suitable primers were designed using Primer Premier 5.0 software (Table S1), and relative expression was calculated using the $2^{-\Delta\Delta Ct}$ method [60].

4.7. Statistical Analysis

Statistical analysis of all data was performed by SPSS 24.0 (IBM, Armonk, NY, USA) software with single factor variance (ANOVA). The least significant difference method (LSD) was used for multiple comparisons. Statistical significance was defined as $p < 0.05$. The error bars in all graphs represent the standard error of the mean.

5. Conclusions

In this research, we examined the notable decline in resistance of cotton seedlings to cold stress due to impaired leaf dehydration, inhibited photosynthetic performance, excessive accumulation of ROS, and increased degree of lipid peroxidation. In contrast, MT treatment increased the resistance of cotton seedlings to cold stress. MT pretreatment enhanced the antioxidant defense system of cotton seedlings and delayed leaf wilting under low-temperature stress, thus enhancing photosynthetic performance. In addition, MT pretreatment resisted cold stress by affecting endogenous melatonin content, flavonoid content, and biosynthesis-related genes' expression. In conclusion, exogenous melatonin supplementation effectively sustained photosynthesis, ROS scavenging, and redox mechanisms in cotton under cold stress, thereby mitigating damage to the seedlings.

Supplementary Materials: The following supporting information can be downloaded at https://www.mdpi.com/article/10.3390/plants13152010/s1, Table S1: List of primers. S2: Kit Instruction Manual.

Author Contributions: Conceptualization, W.Z. and Z.L.; methodology, J.Z. and H.L.; software, J.Z.; formal analysis, J.Z.; investigation, H.L. and C.Y.; data curation, J.Z.; writing—original draft preparation, J.Z. and H.L; writing—review and editing, W.Z. and Z.L; visualization, J.Z. and H.L.; supervision, W.Z. and Z.L.; project administration, W.Z. and Z.L. All authors have read and agreed to the published version of the manuscript.

Funding: This work was sponsored by the State Key Laboratory of Cotton Biology Open Fund (CB2021A09) and the National Natural Science Foundation of China (31560407, 31560074).

Data Availability Statement: Data is contained within the article and Supplementary Materials.

Conflicts of Interest: The authors declare no conflicts of interest.

References

1. Li, Q.; Cai, Y.M.; Gu, L.Q.; Yu, X.H.; Wang, Y.; Zhang, G.L.; Zhao, Y.Q.; Abdullah, S.; Li, P. Transcriptome reveals molecular mechanism of cabbage response to low temperature stress and functional study of *BoPYL8* gene. *Sci. Hortic.* **2024**, *323*, 112523. [CrossRef]
2. Meng, A.J.; Wen, D.X.; Zhang, C.Q. Maize Seed Germination Under Low-Temperature Stress Impacts Seedling Growth Under Normal Temperature by Modulating Photosynthesis and Antioxidant Metabolism. *Front. Plant Sci.* **2022**, *13*, 843033. [CrossRef] [PubMed]
3. Xiao, L.J.; Asseng, S.; Wang, X.T.; Xia, J.X.; Zhang, P.; Liu, L.L.; Tang, L.; Cao, W.X.; Zhu, Y.; Liu, B. Simulating the effects of low-temperature stress on wheat biomass growth and yield. *Agric. For. Meteorol.* **2022**, *326*, 109191. [CrossRef]
4. Malko, M.M.; Peng, X.Y.; Gao, X.; Cai, J.; Zhou, Q.; Wang, X.; Jiang, D. Effect of Exogenous Calcium on Tolerance of Winter Wheat to Cold Stress during Stem Elongation Stage. *Plants* **2023**, *12*, 3784. [CrossRef] [PubMed]
5. Ding, Y.L.; Shi, Y.T.; Yang, S.H. Molecular Regulation of Plant Responses to Environmental Temperatures. *Mol. Plant* **2020**, *13*, 544–564. [CrossRef] [PubMed]

6. Zheng, S.; Su, M.; Wang, L.; Zhang, T.G.; Wang, J.; Xie, H.C.; Wu, X.X.; Ul Haq, S.I.; Qiu, Q.S. Small signaling molecules in plant response to cold stress. *J. Plant Physiol.* **2021**, *266*, 153534. [CrossRef] [PubMed]
7. Wang, K.; Li, J.; Fan, Y.; Yang, J. Temperature Effect on Rhizome Development in Perennial rice. *Rice* **2024**, *17*, 32. [CrossRef] [PubMed]
8. Sun, Y.J.; He, Y.H.; Irfan, A.R.; Liu, X.M.; Yu, Q.Q.; Zhang, Q.; Yang, D.G. Exogenous Brassinolide Enhances the Growth and Cold Resistance of Maize (*Zea mays* L.) Seedlings under Chilling Stress. *Agronomy* **2020**, *10*, 488. [CrossRef]
9. Kosakivska, I.V.; Vedenicheva, N.P.; Babenko, L.M.; Voytenko, L.V.; Romanenko, K.O.; Vasyuk, V.A. Exogenous phytohormones in the regulation of growth and development of cereals under abiotic stresses. *Mol. Biol. Rep.* **2022**, *49*, 617–628. [CrossRef]
10. John, R.; Anjum, N.A.; Sopory, S.K.; Akram, N.A.; Ashraf, M. Some key physiological and molecular processes of cold acclimation. *Biol. Plant.* **2016**, *60*, 603–618. [CrossRef]
11. Qari, S.H.; Hassan, M.U.; Chattha, M.U.; Mahmood, A.; Naqve, M.; Nawaz, M.; Barbanti, L.; Alahdal, M.A.; Aljabri, M. Melatonin Induced Cold Tolerance in Plants: Physiological and Molecular Responses. *Front. Plant Sci.* **2022**, *13*, 843071. [CrossRef]
12. Zhao, D.K.; Wang, H.P.; Chen, S.Y.; Yu, D.Q.; Reiter, R.J. Phytomelatonin: An Emerging Regulator of Plant Biotic Stress Resistance. *Trends Plant Sci.* **2021**, *26*, 70–82. [CrossRef]
13. Li, J.; Song, Q.Q.; Zuo, Z.F.; Liu, L. MicroRNA398: A Master Regulator of Plant Development and Stress Responses. *Int. J. Mol. Sci.* **2022**, *23*, 10803. [CrossRef]
14. Chang, T.L.; Zhao, Y.; He, H.Y.; Xi, Q.Q.; Fu, J.Y.; Zhao, Y.W. Exogenous melatonin improves growth in hulless barley seedlings under cold stress by influencing the expression rhythms of circadian clock genes. *PeerJ* **2021**, *9*, e10740. [CrossRef] [PubMed]
15. Zhang, X.W.; Feng, Y.Q.; Jing, T.T.; Liu, X.T.; Ai, X.Z.; Bi, H.A. Melatonin Promotes the Chilling Tolerance of Cucumber Seedlings by Regulating Antioxidant System and Relieving Photoinhibition. *Front. Plant Sci.* **2021**, *12*, 789617. [CrossRef]
16. Zhao, H.L.; Zhang, K.; Zhou, X.T.; Xi, L.J.; Wang, Y.P.; Xu, H.J.; Pan, T.H.; Zou, Z.R. Melatonin alleviates chilling stress in cucumber seedlings by up-regulation of CsZat12 and modulation of polyamine and abscisic acid metabolism. *Sci. Rep.* **2017**, *7*, 4998. [CrossRef]
17. Wei, W.; Ju, J.S.; Zhang, X.L.; Ling, P.J.; Luo, J.; Li, Y.; Xu, W.J.; Su, J.J.; Zhang, X.L.; Wang, C.X. *GhBRX.1*, *GhBRX.2*, and *GhBRX4.3* improve resistance to salt and cold stress in upland cotton. *Front. Plant Sci.* **2024**, *15*, 1353365. [CrossRef]
18. Pham, H.M.; Kebede, H.; Ritchie, G.; Trolinder, N.; Wright, R.J. *Alternative oxidase* (*AOX*) over-expression improves cell expansion and elongation in cotton seedling exposed to cool temperatures. *Theor. Appl. Genet.* **2018**, *131*, 2287–2298. [CrossRef] [PubMed]
19. Hu, G.H.; Chao, M.N.; Zhou, X.R.; Fu, Y.Z. Comparative Transcriptome Analysis of Seed Germination of a Cotton Variety with High Tolerance to Low Temperature. *Phyton-Int. J. Exp. Bot.* **2023**, *92*, 2535–2554. [CrossRef]
20. Li, N.; Lin, H.X.; Wang, T.X.; Li, Y.; Liu, Y.; Chen, X.G.; Hu, X.T. Impact of climate change on cotton growth and yields in Xinjiang, China. *Field Crops Res.* **2020**, *247*, 107590. [CrossRef]
21. Li, C.; He, Q.L.; Zhang, F.; Yu, J.W.; Li, C.; Zhao, T.L.; Zhang, Y.; Xie, Q.W.; Su, B.R.; Mei, L.; et al. Melatonin enhances cotton immunity to *Verticillium* wilt via manipulating lignin and gossypol biosynthesis. *Plant J.* **2019**, *100*, 784–800. [CrossRef] [PubMed]
22. Jiang, D.; Lu, B.; Liu, L.T.; Duan, W.J.; Meng, Y.J.; Li, J.; Zhang, K.; Sun, H.C.; Zhang, Y.J.; Dong, H.Z.; et al. Exogenous melatonin improves the salt tolerance of cotton by removing active oxygen and protecting photosynthetic organs. *BMC Plant Biol.* **2021**, *21*, 331. [CrossRef] [PubMed]
23. Supriya, L.; Durgeshwar, P.; Muthamilarasan, M.; Padmaja, G. Melatonin Mediated Differential Regulation of Drought Tolerance in Sensitive and Tolerant Varieties of Upland Cotton (*Gossypium hirsutum* L.). *Front. Plant Sci.* **2022**, *13*, 821353. [CrossRef] [PubMed]
24. Li, L.; Yan, X.Y.; Li, J.; Wu, X.; Wang, X.K. Metabolome and transcriptome association analysis revealed key factors involved in melatonin mediated cadmium-stress tolerance in cotton. *Front. Plant Sci.* **2022**, *13*, 995205. [CrossRef] [PubMed]
25. Xia, J.; Kong, X.; Shi, X.; Hao, X.; Li, N.; Khan, A.; Luo, H. Physio-biochemical characteristics and correlation analysis of the seeds of some cotton (*Gossypium hirsutum* L.) genotypes under cold temperature stress. *Appl. Ecol. Environ. Res.* **2020**, *18*, 89–105. [CrossRef]
26. Liu, J.P.; Wu, H.; Wang, B.; Zhang, Y.Y.; Wang, J.S.; Cheng, C.Z.; Huang, Y.J. Exogenous Melatonin Enhances Cold Resistance by Improving Antioxidant Defense and Cold-Responsive Genes' Expression in Banana. *Horticulturae* **2022**, *8*, 260. [CrossRef]
27. Korkmaz, A.; Deger, Ö.; Szafranska, K.; Köklü, S.; Karaca, A.; Yakupoglu, G.; Kocaçinar, F. Melatonin effects in enhancing chilling stress tolerance of pepper. *Sci. Hortic.* **2021**, *289*, 110434. [CrossRef]
28. Hayat, F.; Sun, Z.X.; Ni, Z.J.; Iqbal, S.; Xu, W.Y.; Gao, Z.H.; Qiao, Y.S.; Tufail, M.A.; Jahan, M.S.; Khan, U.; et al. Exogenous Melatonin Improves Cold Tolerance of Strawberry (*Fragaria x ananassa* Duch.) through Modulation of DREB/CBF-COR Pathway and Antioxidant Defense System. *Horticulturae* **2022**, *8*, 194. [CrossRef]
29. Long, H.J.; Li, Z.Y.; Suo, H.; Ou, L.J.; Miao, W.; Deng, W.Q. Study on the Mechanism of Grafting to Improve the Tolerance of Pepper to Low Temperature. *Agronomy* **2023**, *13*, 1347. [CrossRef]
30. Gao, Y.; Stein, M.; Oshana, L.; Zhao, W.; Matsubara, S.; Stich, B. Exploring natural genetic variation in photosynthesis-related traits of barley in the field. *J. Exp. Bot.* **2024**. [CrossRef]
31. Liu, Z.; Wang, J.; Luo, S.; Yue, Z.; Li, Z.; Wang, J.; Hu, L.; Yu, J. Effects of Xanthomonas campestris pv. campestris on the photosynthesis of cabbage in the early stage of infection. *Sci. Hortic.* **2024**, *324*, 112620. [CrossRef]
32. Soualiou, S.; Duan, F.Y.; Li, X.; Zhou, W.B. Nitrogen supply alleviates cold stress by increasing photosynthesis and nitrogen assimilation in maize seedlings. *J. Exp. Bot.* **2023**, *74*, 3142–3162. [CrossRef]

33. Wu, C.H.; Cao, S.F.; Xie, K.Q.; Chi, Z.Y.; Wang, J.; Wang, H.F.; Wei, Y.Y.; Shao, X.F.; Zhang, C.D.; Xu, F.; et al. Melatonin delays yellowing of broccoli during storage by regulating chlorophyll catabolism and maintaining chloroplast ultrastructure. *Postharvest Biol. Technol.* **2021**, *172*, 111378. [CrossRef]
34. Jahan, M.S.; Guo, S.R.; Baloch, A.R.; Sun, J.; Shu, S.; Wang, Y.; Ahammed, G.J.; Kabir, K.; Roy, R. Melatonin alleviates nickel phytotoxicity by improving photosynthesis, secondary metabolism and oxidative stress tolerance in tomato seedlings. *Ecotoxicol. Environ. Saf.* **2020**, *197*, 110593. [CrossRef]
35. Khan, M.N.; Zhang, J.; Luo, T.; Liu, J.H.; Rizwan, M.; Fahad, S.; Xu, Z.H.; Hu, L.Y. Seed priming with melatonin coping drought stress in rapeseed by regulating reactive oxygen species detoxification: Antioxidant defense system, osmotic adjustment, stomatal traits and chloroplast ultrastructure perseveration. *Ind. Crops Prod.* **2019**, *140*, 111597. [CrossRef]
36. Ogunyemi, S.O.; Luo, J.; Abdallah, Y.; Yu, S.; Wang, X.; Alkhalifah, D.H.M.; Hozzein, W.N.; Wang, F.; Bi, J.; Yan, C.; et al. Copper oxide nanoparticles: An effective suppression tool against bacterial leaf blight of rice and its impacts on plants. *Pest. Manag. Sci.* **2024**, *80*, 1279–1288. [CrossRef] [PubMed]
37. Dogru, A.; Çakirlar, H. Effects of leaf age on chlorophyll fluorescence and antioxidant enzymes activity in winter rapeseed leaves under cold acclimation conditions. *Braz. J. Bot.* **2020**, *43*, 11–20. [CrossRef]
38. Tang, Y.W.; Li, J.H.; Song, Q.Q.; Cheng, Q.; Tan, Q.L.; Zhou, Q.G.; Nong, Z.M.; Lv, P. Transcriptome and WGCNA reveal hub genes in sugarcane tiller seedlings in response to drought stress. *Sci. Rep.* **2023**, *13*, 12823. [CrossRef]
39. Mittler, R.; Zandalinas, S.I.; Fichman, Y.; Van Breusegem, F. Reactive oxygen species signalling in plant stress responses. *Nat. Rev. Mol. Cell Biol.* **2022**, *23*, 663–679. [CrossRef]
40. Song, R.X.; Ritonga, F.N.; Yu, H.Y.; Ding, C.J.; Zhao, X.Y. Plant Melatonin: Regulatory and Protective Role. *Horticulturae* **2022**, *8*, 810. [CrossRef]
41. Ahmad, J.; Hayat, F.; Khan, U.; Ahmed, N.; Li, J.; Ercisli, S.; Iqbal, S.; Javed, H.U.; Alyas, T.; Tu, P.; et al. Melatonin: A promising approach to enhance abiotic stress tolerance in horticultural plants. *South Afr. J. Bot.* **2024**, *164*, 66–76. [CrossRef]
42. Khan, M.N.; Siddiqui, M.H.; AlSolami, M.A.; Siddiqui, Z.H. Melatonin-regulated heat shock proteins and mitochondrial ATP synthase induce drought tolerance through sustaining ROS homeostasis in H(2)S-dependent manner. *Plant Physiol. Biochem. PPB* **2024**, *206*, 108231. [CrossRef]
43. Tan, D.X.; Reiter, R.J. An evolutionary view of melatonin synthesis and metabolism related to its biological functions in plants. *J. Exp. Bot.* **2020**, *71*, 4677–4689. [CrossRef]
44. Kong, M.M.; Liang, J.; Ali, Q.B.; Wen, W.; Wu, H.J.; Gao, X.W.; Gu, Q. 5-Methoxyindole, a Chemical Homolog of Melatonin, Adversely Affects the Phytopathogenic Fungus *Fusarium graminearum*. *Int. J. Mol. Sci.* **2021**, *22*, 10991. [CrossRef]
45. Jahan, M.S.; Guo, S.R.; Sun, J.; Shu, S.; Wang, Y.; Abou El-Yazied, A.; Alabdallah, N.M.; Hikal, M.; Mohamed, M.H.M.; Ibrahim, M.F.M.; et al. Melatonin-mediated photosynthetic performance of tomato seedlings under high-temperature stress. *Plant Physiol. Biochem.* **2021**, *167*, 309–320. [CrossRef]
46. Bi, G.Z.; Hu, M.; Fu, L.; Zhang, X.J.; Zuo, J.R.; Li, J.Y.; Yang, J.; Zhou, J.M. The cytosolic thiol peroxidase PRXIIB is an intracellular sensor for H_2O_2 that regulates plant immunity through a redox relay. *Nat. Plants* **2022**, *8*, 1160–1175. [CrossRef]
47. Chang, J.J.; Guo, Y.L.; Zhang, Z.X.; Wei, C.H.; Zhang, Y.; Ma, J.X.; Yang, J.Q.; Zhang, X.; Li, H. CBF-responsive pathway and phytohormones are involved in melatonin-improved photosynthesis and redox homeostasis under aerial cold stress in watermelon. *Acta Physiol. Plant.* **2020**, *42*, 159. [CrossRef]
48. Lei, Y.; He, H.; Raza, A.; Liu, Z.; Ding, X.Y.; Wang, G.J.; Yan, L.; Yong, C.; Zou, X.L. Exogenous melatonin confers cold tolerance in rapeseed (*Brassica napus* L.) seedlings by improving amtioxidanis and genes expression. *Plant Signal. Behav.* **2022**, *17*, 2129289. [CrossRef]
49. Sharma, A.; Wang, J.F.; Xu, D.B.; Tao, S.C.; Chong, S.L.; Yan, D.L.; Li, Z.; Yuan, H.W.; Zheng, B.S. Melatonin regulates the functional components of photosynthesis, antioxidant system, gene expression, and metabolic pathways to induce drought resistance in grafted *Carya cathayensis* plants. *Sci. Total Environ.* **2020**, *713*, 136675. [CrossRef]
50. Xiao, Y.; Xie, L.; Li, Y.; Li, C.; Yu, Y.; Hu, J.; Li, G. Impact of low temperature on the chemical profile of sweet corn kernels during post-harvest storage. *Food Chem.* **2024**, *431*, 137079. [CrossRef]
51. Yang, X.; Han, Y.; Huo, G.; Ge, G.; He, S.; Yang, X.; Zhang, L.; Wei, S.; Luo, L. Integrated transcriptomic and metabolomic analysis provides insights into cold tolerance in lettuce (*Lactuca sativa* L.). *BMC Plant Biol.* **2024**, *24*, 442. [CrossRef]
52. Schulz, E.; Tohge, T.; Zuther, E.; Fernie, A.R.; Hincha, D.K. Natural variation in flavonol and anthocyanin metabolism during cold acclimation in *Arabidopsis thaliana* accessions. *Plant Cell Environ.* **2015**, *38*, 1658–1672. [CrossRef]
53. Schulz, E.; Tohge, T.; Zuther, E.; Fernie, A.R.; Hincha, D.K. Flavonoids are determinants of freezing tolerance and cold acclimation in *Arabidopsis thaliana*. *Sci. Rep.* **2016**, *6*, 34027. [CrossRef]
54. Schulz, E.; Tohge, T.; Winkler, J.B.; Albert, A.; Schäffner, A.R.; Fernie, A.R.; Zuther, E.; Hincha, D.K. Natural Variation among Arabidopsis Accessions in the Regulation of Flavonoid Metabolism and Stress Gene Expression by Combined UV Radiation and Cold. *Plant Cell Physiol.* **2021**, *62*, 502–514. [CrossRef]
55. Song, Z.H.; Yang, Q.; Dong, B.Y.; Li, N.; Wang, M.Y.; Du, T.T.; Liu, N.; Niu, L.L.; Jin, H.J.; Meng, D.; et al. Melatonin enhances stress tolerance in pigeon pea by promoting flavonoid enrichment, particularly luteolin in response to salt stress. *J. Exp. Bot.* **2022**, *73*, 5992–6008. [CrossRef]

56. Keya, S.S.; Mostofa, M.G.; Rahman, M.M.; Das, A.K.; Sultana, S.; Ghosh, P.K.; Anik, T.R.; Ahsan, S.M.; Rahman, M.A.; Jahan, N.; et al. Salicylic Acid Application Improves Photosynthetic Performance and Biochemical Responses to Mitigate Saline Stress in Cotton. *J. Plant Growth Regul.* **2023**, *42*, 5881–5894. [CrossRef]
57. Zhao, P.; Xu, Y.; Chen, W.; Sang, X.; Zhao, Y.; Wang, H. A bZIP transcription factor *GhVIP1* increased drought tolerance in upland cotton. *J. Cotton Res.* **2023**, *6*, 11. [CrossRef]
58. Song, S.Y.; Chen, Y.; Chen, J.; Dai, X.Y.; Zhang, W.H. Physiological mechanisms underlying OsNAC5-dependent tolerance of rice plants to abiotic stress. *Planta* **2011**, *234*, 331–345. [CrossRef]
59. Xiong, X.P.; Sun, S.C.; Zhang, X.Y.; Li, Y.J.; Liu, F.; Zhu, Q.H.; Xue, F.; Sun, J. *GhWRKY70D13* Regulates Resistance to *Verticillium dahliae* in Cotton Through the Ethylene and Jasmonic Acid Signaling Pathways. *Front. Plant Sci.* **2020**, *11*, 1045. [CrossRef]
60. Zhao, W.; Li, S.; Dong, L.; Wang, P.; Lu, X.; Zhang, X.; Su, Z.; Guo, Q.; Ma, P. Effects of different crop rotations on the incidence of cotton Verticillium wilt and structure and function of the rhizospheric microbial community. *Plant Soil* **2023**, *485*, 457–474. [CrossRef]

plants

MDPI

Communication

Diurnal High Temperatures Affect the Physiological Performance and Fruit Quality of Highbush Blueberry (*Vaccinium corymbosum* L.) cv. Legacy

Jorge González-Villagra [1,2,*], Kevin Ávila [1], Humberto A. Gajardo [3], León A. Bravo [3,4], Alejandra Ribera-Fonseca [4,5], Emilio Jorquera-Fontena [1], Gustavo Curaqueo [1,2], Cecilia Roldán [6], Priscilla Falquetto-Gomes [7], Adriano Nunes-Nesi [7] and Marjorie M. Reyes-Díaz [4,8]

1 Departamento de Ciencias Agropecuarias y Acuícolas, Facultad de Recursos Naturales, Universidad Católica de Temuco, Temuco P.O. Box 15-D, Chile; oavila@uct.cl (K.Á.); ejorquera@uct.cl (E.J.-F.); gcuraqueo@uct.cl (G.C.)
2 Núcleo de Investigación en Producción Alimentaria, Facultad de Recursos Naturales, Universidad Católica de Temuco, Temuco P.O. Box 15-D, Chile
3 Departamento de Ciencias Agronómicas y Recursos Naturales, Facultad de Ciencias Agropecuarias y Medioambiente, Universidad de La Frontera, Temuco P.O. Box 54-D, Chile; h.gajardo.balboa@gmail.com (H.A.G.); leon.bravo@ufrontera.cl (L.A.B.)
4 Center of Plant, Soil Interaction and Natural Resources Biotechnology, Scientific and Technological Bioresource Nucleus (BIOREN), Universidad de La Frontera, Temuco P.O. Box 54-D, Chile; alejandra.ribera@ufrontera.cl (A.R.-F.); marjorie.reyes@ufrontera.cl (M.M.R.-D.)
5 Centro de Fruticultura, Facultad de Ciencias Agropecuarias y Medioambiente, Universidad de La Frontera, Temuco P.O. Box 54-D, Chile
6 Área de Desarrollo Rural, Instituto Nacional de Tecnología Agropecuaria, EEA INTA Bariloche, San Carlos de Bariloche 8400, Argentina; roldan.cecilia@inta.gob.ar
7 National Institute of Science and Technology on Plant Physiology under Stress Conditions, Departamento de Biologia Vegetal, Universidade Federal de Viçosa, Viçosa 36570-900, Brazil; priscilla.gomes@ufv.br (P.F.-G.); nunesnesi@ufv.br (A.N.-N.)
8 Departamento de Ciencias Químicas y Recursos Naturales, Facultad de Ingeniería y Ciencias, Universidad de La Frontera, Temuco P.O. Box 54-D, Chile
* Correspondence: jorge.gonzalez@uct.cl

Citation: González-Villagra, J.; Ávila, K.; Gajardo, H.A.; Bravo, L.A.; Ribera-Fonseca, A.; Jorquera-Fontena, E.; Curaqueo, G.; Roldán, C.; Falquetto-Gomes, P.; Nunes-Nesi, A.; et al. Diurnal High Temperatures Affect the Physiological Performance and Fruit Quality of Highbush Blueberry (*Vaccinium corymbosum* L.) cv. Legacy. *Plants* **2024**, *13*, 1846. https://doi.org/10.3390/plants13131846

Academic Editors: Violetta Katarzyna Macioszek, Iwona Ciereszko and Andrzej K. Kononowicz

Received: 19 June 2024
Revised: 2 July 2024
Accepted: 3 July 2024
Published: 4 July 2024

Abstract: In this study, the physiological performance and fruit quality responses of the highbush blueberry (*Vaccinium corymbosum*) cultivar Legacy to high temperatures (HTs) were evaluated in a field experiment. Three-year-old *V. corymbosum* plants were exposed to two temperature treatments between fruit load set and harvest during the 2022/2023 season: (i) ambient temperature (AT) and (ii) high temperature (HT) (5 °C ± 1 °C above ambient temperature). A chamber covered with transparent polyethylene (100 µm thick) was used to apply the HT treatment. In our study, the diurnal temperature was maintained with a difference of 5.03 °C ± 0.12 °C between the AT and HT treatments. Our findings indicated that HT significantly decreased CO_2 assimilation (P_n) by 45% and stomatal conductance (g_s) by 35.2% compared to the AT treatment. By contrast, the intercellular CO_2 concentration (Ci) showed higher levels (about 6%) in HT plants than in AT plants. Fruit quality analyses revealed that the fruit weight and equatorial diameter decreased by 39% and 13%, respectively, in the HT treatment compared to the AT treatment. By contrast, the firmness and total soluble solids (TSS) were higher in the HT treatment than in the AT treatment. Meanwhile, the titratable acidity showed no changes between temperature treatments. In our study, P_n reduction could be associated with stomatal and non-stomatal limitations under HT treatment. Although these findings improve our understanding of the impact of HTs on fruit growth and quality in *V. corymbosum*, further biochemical and molecular studies are need.

Keywords: CO_2 assimilation; leaf temperature; fruit weight; equatorial diameter; total soluble solids

1. Introduction

Climate change is expected to intensify extreme climate events such as infrequent and erratic precipitation, water deficit, and extreme temperatures [1,2]. In extreme temperatures, a high temperature (HT) negatively affects plant growth and crop yields [3–5]. Researchers have used polyethylene chambers to study the effects of HTs on crops, which is a reliable method of imposing heat treatments in field experiments [6–8]. It has been reported that HTs trigger morphological changes such as a reduced leaf area, shoot and root growth inhibition, leaf senescence, and sunburn on leaves and fruits [9–11]. At the physiological level, it has been reported that photosynthesis is the primary process affected by HT, altering Calvin–Benson cycle activity, increasing photorespiration, decreasing electron transport, photochemical reactions, and chlorophyll biosynthesis, and inducing the inactivation of Rubisco activase [12–14]. Otherwise, it has been reported that HTs lead to reactive oxygen species (ROS) overproduction, damaging cellular organelles and modifying structurally the thylakoid membranes in the chloroplasts at the cellular level [15,16]. However, the plant responses to HTs depends on its duration and intensity, the period of the day (day/night temperature) and the cultivar, phenological stage, and experimental conditions [15]. Currently, most studies have analyzed the responses to HTs in herbaceous annual species such as *Arabidopsis thaliana*, *Triticum aestivum*, *Oryza sativa*, *Solanum lycopersicum*, and *Solanum tuberosum* under controlled growth conditions, with significantly less information about woody species [6,9,17,18].

Highbush blueberry (*V. corymbosum* L.) is a shrub species whose fruits are characterized by their high levels of phenolic compounds and antioxidant capacity, with great human health benefits such as antidiabetic, antibacterial, anticarcinogenic, and cardiovascular protective properties [19–21]. Highbush blueberry production and fruit demand have increased worldwide during the last ten years [22,23]. Chile now cultivates 18,000 ha with this fruit crop, with Legacy (3217 ha), Duke (2524 ha), and Brigitta (2222 ha) being the main cultivars, established between El Maule and La Araucanía Region [24]. It has been reported that the optimal growth temperature ranges between 25 and 30 °C for *V. corymbosum* [25]. A previous study classified Bluecrop, Brigitta, Gulfcoast, and O'Neal as HT-tolerant cultivars; meanwhile, Duke and Ridge were classified as HT-sensitive cultivars, based on their morphological and physiological responses at 35/30 °C (day/night) during 90 days under greenhouse conditions [26]. On the other hand, Chen et al. [27] reported that Brigitta was the most sensitive cultivar, followed by Duke and Misty after 6 h at 35 °C under controlled conditions. In their study, Estrada et al. [28] found differential physiological and morphological responses among the Bluegold, Elliott, and Liberty cultivars when plants were exposed to temperatures 10 °C above ambient temperature. It has been reported that the fruits of *V. corymbosum* cv. Elliott can be very sensitive to HT, provoking softening, shriveling, and necrosis [29]. Thus, the responses of *V. corymbosum* to HTs might be dependent on multiple factors, including the cultivar, temperature level applied, treatment duration, phenological stage, and other experimental conditions. Although previous studies have reported the responses of some *V. corymbosum* cultivars to HTs under controlled conditions, less is documented under field conditions and no prior studies address the effects of HTs on cv. Legacy and fruit-related parameters. Therefore, this study aimed to evaluate the responses of *V. corymbosum* cv. Legacy regarding its physiological performance and fruit quality parameters to diurnal HTs under field conditions.

2. Results

2.1. Environmental Conditions during the Experiment

Our results showed that the diurnal mean temperature (between 09:00 to 18:00 h) was 21.9 ± 0.91 °C for the ambient temperature (AT) treatment and 26.9 ± 0.91 °C for the high-temperature (HT) treatment, resulting in a significant difference of 5.03 ± 0.12 °C throughout the experiment (Figure 1A). During nighttime, non-significant changes were detected in the temperature between the AT and HT treatments. The maximum temperature registered was 37.7 °C for AT and 42.3 °C for the HT treatment on 25 December. Concerning relative humidity (HR), the diurnal mean was 55.5% for AT and 41.2% for HT (Figure 1B). The minimum RH registered was 22.5% for AT and 27.1% for HT on 25 December. During the nighttime, both treatments showed high RH values (above 95%), with no significant differences between treatments.

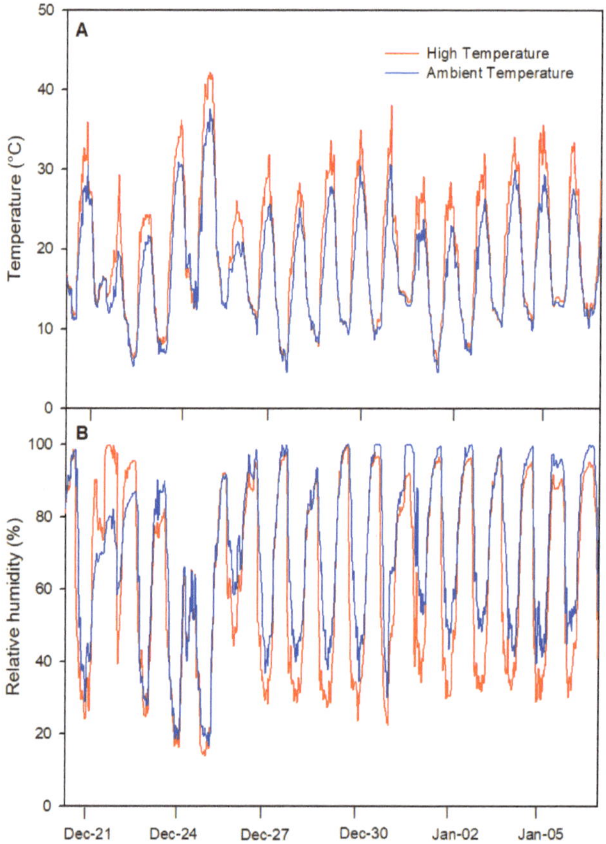

Figure 1. (**A**) Daily temperature (T°) and (**B**) relative humidity (RH) during the experiment: (i) ambient temperature (control) and (ii) high temperature (5 °C ± 1 °C above the ambient temperature).

2.2. Soil Water Content, Plant Water Status, and Leaf Temperature

Our results revealed no significant changes in the soil water content (SWC) in *V. corymbosum* plants between treatments during the experiment, showing a mean of 32.5% (Figure 2). The stem water potential (Ψ_w), which was measured only at the end of the experiment, reached −0.32 ± 0.03 MPa for the AT treatment and −0.33 ± 0.03 MPa for the HT treatment. Regarding leaf temperature, the HT treatment led to an increase in the leaf temperature (by about 5 °C) in *V. corymbosum* plants compared to the AT treatment throughout the experiment, which was expected considering the difference in temperature between the AT and HT treatments (Figure 3).

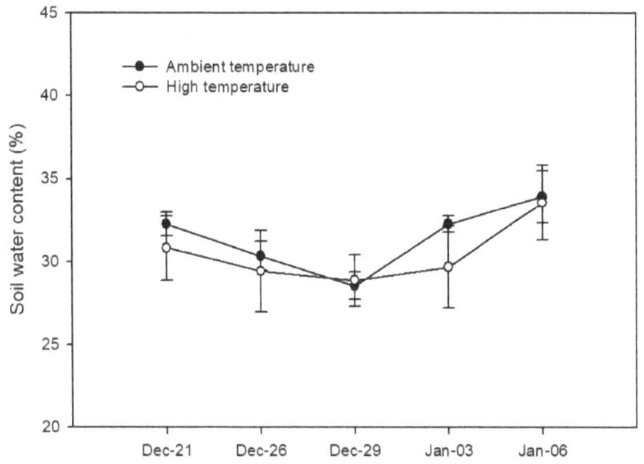

Figure 2. Soil water content (SWC, %) in *V. corymbosum* cv. Legacy plants subjected to two temperature treatments: (i) ambient temperature (control) and (ii) high temperature (5 °C ± 1 °C above the ambient temperature). The SWC was measured in the morning between 08:00 and 10:00 h. The value represents the means ± SE (n = 8).

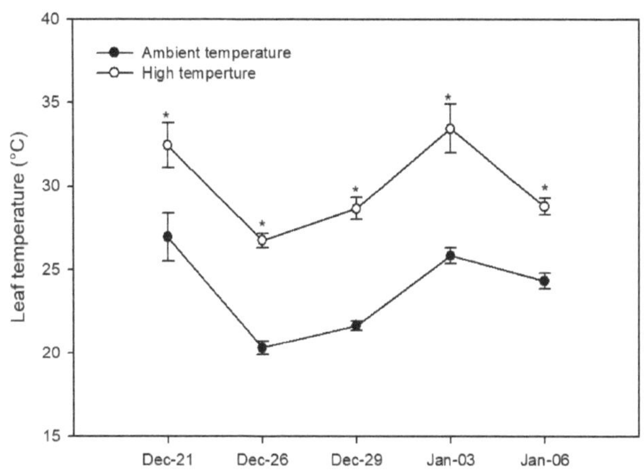

Figure 3. Leaf temperature in *V. corymbosum* cv. Legacy plants subjected to two temperature treatments: (i) ambient temperature (control) and (ii) high temperature (5 °C ± 1 °C above the ambient temperature). Leaf temperature was measured in the morning between 08:00 and 10:00 h. Asterisks indicate significant differences according to Student's t-test ($p \leq 0.05$). The value represents the means ± SE (n = 8).

2.3. *Gas-Exchange in V. corymbosum Plants*

In our study, the HT treatment decreased CO_2 assimilation (P_n) by 45%, stomatal conductance (g_s) by 35.2%, and transpiration (E) by 42% compared to the AT treatment in *V. corymbosum* plants (Figure 4). By contrast, we observed that the intercellular CO_2 concentration (Ci) showed higher levels (about 6%) in *V. corymbosum* plants under the HT treatment than when under the AT treatment (Figure 4).

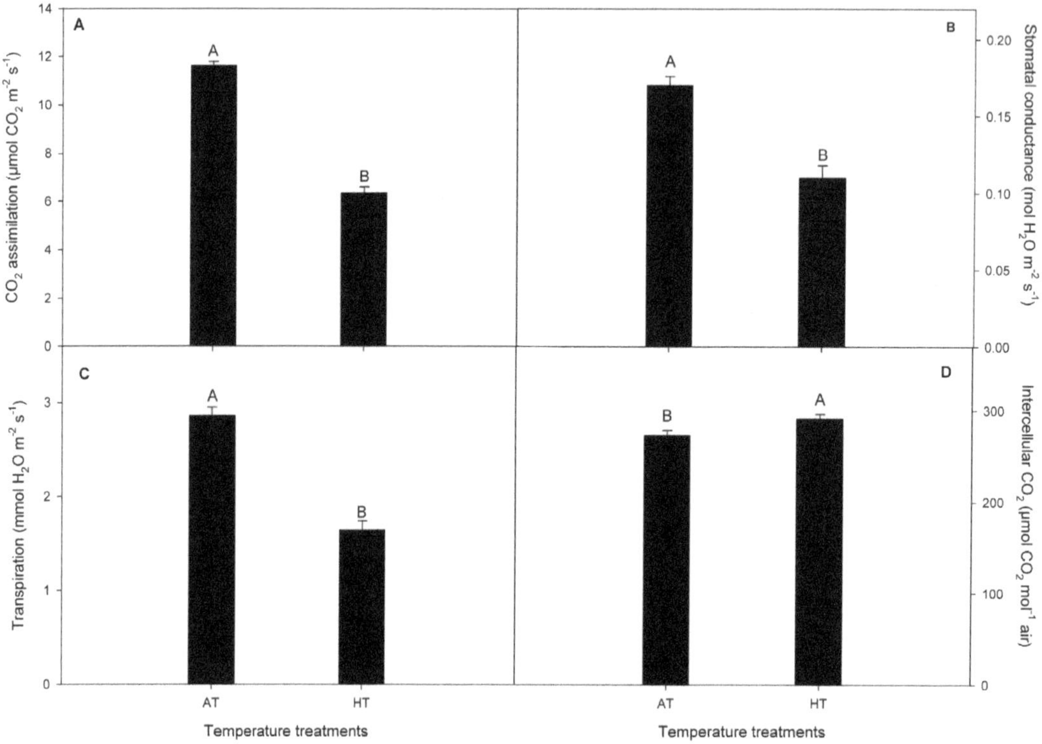

Figure 4. (**A**) CO_2 assimilation (P_n), (**B**) stomatal conductance (g_s), (**C**) transpiration (E), and (**D**) intercellular CO_2 concentration (Ci) in *V. corymbosum* cv. Legacy plants subjected to two temperature treatments: (i) ambient temperature (control) and (ii) high temperature (5 °C ± 1 °C above of ambient temperature). Different uppercase letters indicate significant differences between temperature treatments according to Student's *t*-test ($p \leq 0.05$). The value represents the mean ± SE (n = 8).

2.4. Fruit Quality

Our results revealed a higher fresh weight (39%) in the fruits of *V. corymbosum* plants grown under the AT treatment than those grown under the HT treatment (Figure 5). Similarly, the equatorial diameter was increased by 13% in the fruits of plants exposed to the AT treatment (Figure 5). By contrast, the fruits of plants subjected to the HT treatment had a higher firmness and total soluble solids compared to plants grown under the AT treatment. On the other hand, the titratable acidity of fruits did not vary between the temperature treatments (Figure 5).

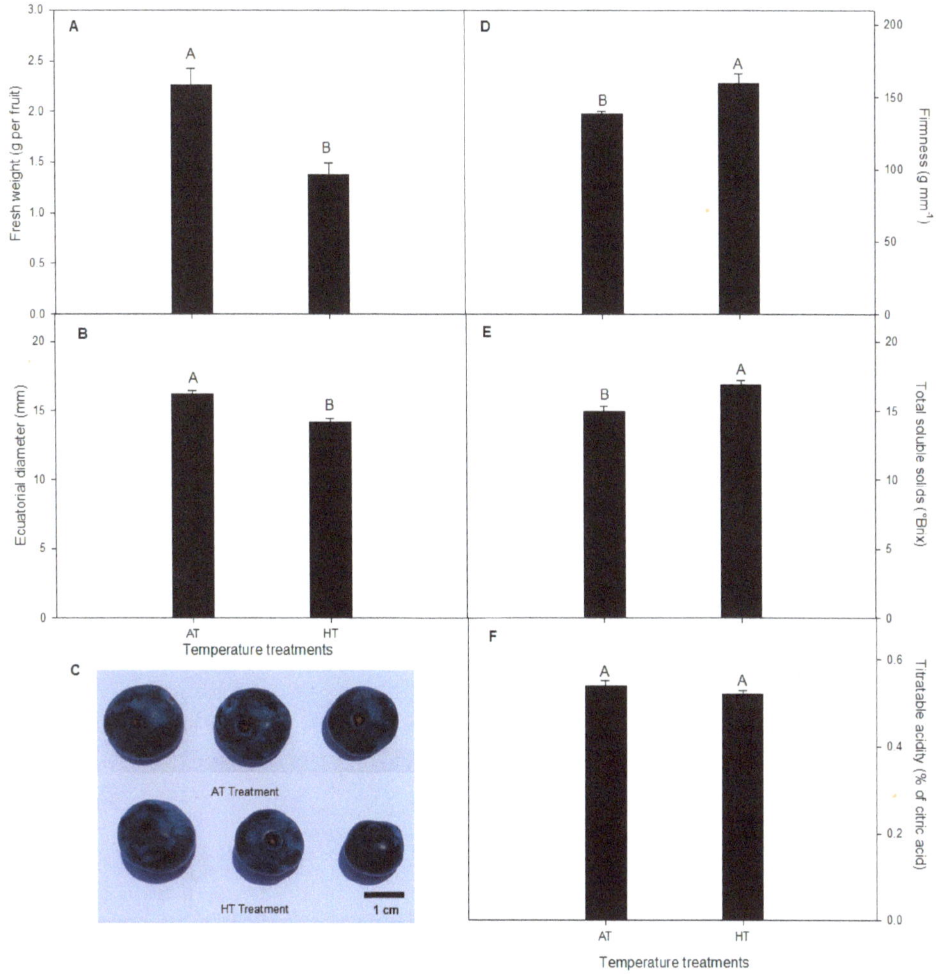

Figure 5. Fruit quality parameters: (**A**) fresh weight, (**B**) equatorial diameter, (**C**) fruits, (**D**) firmness, (**E**) total soluble solids, and (**F**) titratable acidity in *V. corymbosum* cv. Legacy plants subjected to two temperature treatments: (i) ambient temperature (control) and (ii) high temperature (5 °C ± 1 °C above of ambient temperature). Different uppercase letters indicate significant differences between temperature treatments according to Student's *t*-test ($p \leq 0.05$). The value represents the mean ± SE ($n = 50$).

2.5. *Multivariate Analysis Based on Physiological Responses and Quality Parameters Measured in V. corymbosum under Different Treatments*

To visualize and understand the relationships between the physiological responses and quality parameters evaluated as a function of the different treatments applied, we carried out a principal component analysis (PCA) (Figure 6). The biplot shows individual replicates in each treatment combined with the measured variables. This analysis revealed a separation between the AT and HT treatments. The sum of PC1 and PC2 explained around 82.3% of the observed variability, which was responsible for separating the HT and AT treatments into distinct groups. The HT treatment positively influenced the quality parameters of firmness and TSS, but had a negative relationship with parameters such as P_n, g_s, E, and the fruit fresh weight. In general, it can be seen that HTs induce a reduction in the physiological responses analyzed, which consequently leads to a reduction in the biometric parameters of the fruit, as shown in Figures 4 and 5. In addition, HTs are directly related to an increase in firmness and TSS, which can be explained by the lower fruit fresh weight found in this treatment, which implies a higher concentration of solutes in these fruits. Firmness, Ci and g_s were the three characteristics that contributed most to PC1 and PC2. Titratable acidity was the characteristic that least influenced the separation of PC1 and PC2.

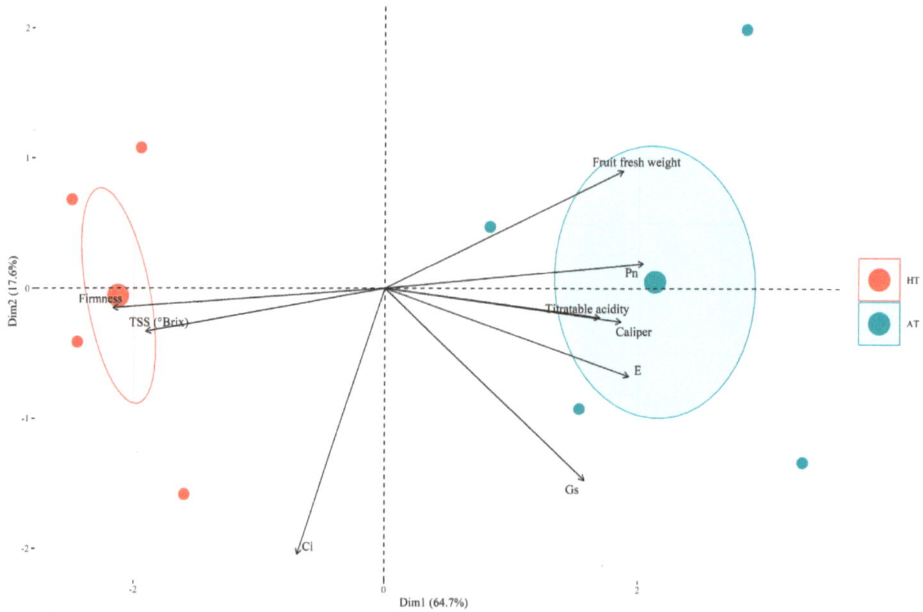

Figure 6. Principal component analysis (PCA) biplot of data derived from physiological responses and quality parameters measured in *V. corymbosum* fruit under different treatments: ambient temperature (AT) and high temperature (HT). All variables that are grouped are positively correlated with each other. The greater the distance between the variable and the origin, the better represented that variable is in relation to dimension 1 (first principal component) and dimension 2 (second principal component). Negatively correlated variables are displayed on opposite sides of the biplot origin. Abbreviations: CO_2 assimilation (P_n), stomatal conductance (g_s), transpiration (E), intercellular CO_2 concentration (Ci), and total soluble solids (TSS).

3. Discussion

3.1. Physiological Responses of V. corymbosum cv. Legacy Plants

High temperatures (HTs) induce several morphological and physiological changes in plants, negatively affecting crop yields [3,5]. However, less is known about the effects of HTs in *V. corymbosum* under field conditions. In this study, we evaluated the physiological and fruit quality responses of *V. corymbosum* cv. Legacy to HTs in a field experiment, using a polyethylene chamber approach to increase air temperature according to Kim et al. [7]. In our study, the HT treatment increased the diurnal temperature by 5.03 ± 0.12 °C above ambient temperature (AT) throughout the experiment (Figure 1A). Meanwhile, the soil water content (SWC) and stem water potential (Ψ_w) were unaffected by the HT treatment in *V. corymbosum* cv. Legacy plants (Figure 2), discarding additional stressing factors in our experiment such as irrigation differences. By contrast, we found that the HT treatment increased the leaf temperature in *V. corymbosum* cv. Legacy plants, rising by 5 °C above the AT treatment (Figure 3). Concerning leaf gas exchange, *V. corymbosum* cv. Legacy plants experienced decreased P_n (45%), g_s (35.2%), and E (42%) under the HT treatment compared to the AT treatment (Figure 4). Our results agree with those of Hao et al. [26], who reported a significant decrease in P_n and g_s (by about 32%) in *V. corymbosum* cv. Duke subjected to HTs (35 °C), meaning that it was classified as a HT-sensitive cultivar. Meanwhile, they showed that HT-tolerant cultivars (Bluecrop, Brigitta, O'Neal, and Gulfcoast) did not exhibit significant changes in P_n at high temperatures. The authors showed that HT-tolerant cultivars exhibit higher transpiration rates compared to sensitive ones, dissipating heat by leaf transpiration. They suggested that the HT tolerance might be explained by changes in stomatal traits such as the stomatal distribution, density, aperture size, and shape, allowing higher leaf transpiration to dissipate heat. However, in our study, we observed a decrease in P_n and E, suggesting that Legacy might be classified as a sensitive cultivar. It has been widely reported that photosynthesis may be reduced by stomatal and non-stomatal limitations [30,31]. Chen et al. [32] reported that HTs negatively affect Calvin–Benson cycle enzymes such as ribulose bisphosphate regeneration (RuBP), rubisco activase (RCA), and ribulose-1,5-bisphosphate carboxylase/oxygenase (RuBisCo). Thus, in our study, the reduction in photosynthesis might be due to a negative effect on Calvin/Benson cycle enzymes. As we mentioned before, P_n and g_s were reduced in plants under the HT treatment. However, we also observed that *V. corymbosum* cv. Legacy plants had significantly higher Ci levels (about 6%) under the HT treatment compared to the AT treatment. Chen et al. [32] found a significant reduction in F_v/F_m in four *V. corymbosum* cultivars exposed to 40 °C, indicating photoinhibitory damage under HTs, which could be associated with the reduction in P_n in *V. corymbosum* cv. Legacy plants in our study. On the other hand, Hao et al. [26] reported that HT-sensitive *V. corymbosum* cvs. Duke and Blue Ridge exhibited swollen chloroplasts with significant damage, disordered grana lamella, and stromal lamella under HT treatment. In fact, Chen et al. [27] indicated that the ROS and lipid peroxidation levels increased in *V. corymbosum* plants exposed to HTs, suggesting significant damage to cellular membranes and/or thylakoid membranes. Therefore, the reduction in P_n levels could be associated with stomatal and non-stomatal limitations in *V. corymbosum* cv. Legacy plants exposed to HTs in our study.

3.2. *Fruit Quality Changes in V. corymbosum cv. Legacy Plants*

Our results revealed that the HT treatment significantly decreased the fresh weight by 39% and the equatorial diameter of fruits by 13% in *V. corymbosum* cv. Legacy compared to the AT treatment, suggesting that fruit growth was inhibited during the HT treatment. This growth inhibition could be explained by the lower P_n in *V. corymbosum* cv. Legacy plants under HTs in our study (Figures 4 and 5). By contrast, the firmness and total soluble solids were increased in *V. corymbosum* cv. Legacy fruits under the HT treatment compared with the AT treatment (Figure 5). Our findings agree with Bryla et al. [33], who reported that the fruit firmness and total soluble solids increase due to a reduction in fruit volume. Likewise, Faghih et al. [34] reported that smaller fruits are firmer than larger fruits due to their higher cell density. Therefore, the higher firmness and total soluble solids of fruits of plants exposed to the HT treatment could be explained by the lower fruit size (lower fruit weight and equatorial diameter) in our study.

3.3. *Principal Component Analysis (PCA)*

The PCA shows a negative relationship between HTs and physiological responses, which was to be expected (Figure 6). It is known that at high temperatures plants tend to partially close their stomata to avoid excessive water loss through transpiration (E). In this context, the availability of CO_2 can become limiting for photosynthesis, resulting in a reduction in carbohydrate production by the plant. This relationship would explain the positive correlation visible in the PCA between P_n, g_s, E, and fruit weight. The stem water potential was not altered under HTs when compared to the AT treatment at the end of the experiment. This highlights the success of this plant strategy to avoid excessive water loss under HTs. On the other hand, this generates a stomatal limitation that directly influences the final fruit production. Excessive heat can negatively affect the functioning of enzymes and degrade photosynthetic pigments and other components of the photosynthetic machinery, resulting in collateral damage to cell membranes and the inhibition of metabolic processes. Consequently, this affects the photosynthetic efficiency and reduces the biomass production of the plant. This would also explain the correlation between the physiological responses and production parameters such as the fruit fresh weight and equatorial diameter observed in the PCA.

4. Materials and Methods

4.1. Plant Material and Description of the Study Site

The field experiment was conducted at the Experimental Station of the Universidad Católica de Temuco (39°30′08″ S; 72° 47′59″ W), located in Lautaro, La Araucanía Region, Chile, during the 2022/2023 season. The plant material corresponds to three-year-old *V. corymbosum* cv. Legacy plants, which were transplanted in plastic pots containing 30 L of soil during the 2019 season. The soil was classified as the Temuco series (Andisol, Typic Hapludands) [35]. The soil texture analysis showed a silt loam surface (19.6% sand, 42.8% silt, 37.6% clay). The soil nutrient analysis showed an organic matter content of 17.21%, a pH of 5.75, P of 13 mg kg^{-1}, K of 216 mg kg^{-1}, Ca of 7.3 cmol+ kg^{-1}, Mg of 1.47 cmol+ kg^{-1}, and Na of 0.07 cmol+ kg^{-1}. Agronomic management, such as irrigation, pest control, pruning, and fertilization, were performed following commercial recommendations. Weeds were manually controlled.

4.2. Treatments and Experimental Conditions

Plants were subjected to two treatments for 20 days from fruit load set (17 December) to fruit harvest (7 January): (i) ambient temperature (AT; control) and (ii) high temperature (HT; 5 °C ± 1 °C above ambient temperature). A chamber was built of wood (1.5 × 1.5 × 1.5 m) and covered with transparent polyethylene (100 µm thick) to increase the temperature in the HT treatment (Figure 7), as suggested by Ávila-Valdés et al. [6] and Kim et al. [7]. The transmittance of the polyethylene was about 95%, as measured by a Polypen (PSI, Brno, Czech Republic). The chamber was equipped with a thermostatic electric heater and controlled by a temperature regulator (Cavadevices, Buenos Aires, Argentina) [6]. To avoid extreme temperatures inside the chamber, the top was removed daily. The environmental conditions (inside and outside of chamber) were continually monitored, with the air temperature and relative humidity recorded using data loggers (Elitech Technology, Inc., San Jose, CA, USA).

Figure 7. High-temperature chamber covered with transparent polyethylene (100 µm thick).

4.3. Soil Water Content and Plant Water Status

The soil water content (SWC) was determined twice a week throughout the whole experiment using a portable time-domain reflectometer (TDR) soil moisture meter (TDR-300, Spectrum Technologies Inc., Plainfield, IL, USA). Otherwise, the plant water status was determined by measuring the stem water potential (Ψ_w). For this, leaves were covered with aluminum foil in a plastic bag for 60 min before measurement [36]. The Ψ_w was determined at fruit harvest using a Scholander chamber Model 1000 (PMS, Instruments Co., Corvallis, OR, USA) between 08:00 and 10:00 h.

4.4. Leaf Temperature

During the experiment, the leaf temperature was monitored twice a week using a hand-held infrared thermometer Fluke 62 Max (Fluke Corporation, Everett, WA, USA), according to Barai et al. [37]. The measurements were performed on three attached fully expanded leaves per each plant between 08:00 and 10:00 h.

4.5. Gas Exchange Measurement

Gas exchange was determined at the fruit harvest time using a portable infrared gas analyzer (IRGA) (Li-6400; LI-COR, Inc., Lincoln, NE, USA) following the protocol of Reyes-Díaz et al. [38]. The analyzed parameters were CO_2 assimilation (P_n), stomatal conductance (g_s), transpiration (E), and intercellular CO_2 concentration (Ci). The CO_2 reference concentration was 400 µmol mol^{-1}, with a flow rate of 300 mL min^{-1} and 60% relative humidity inside the leaf chamber, and the temperature was maintained at 20 ± 2 °C. The measurement was performed in vivo on attached fully expanded leaves during the light period between 08:00 and 10:00 h. Five measurements per plant were performed.

4.6. Fruit Quality

When ripe (100% blue), fruits were harvested early in the morning (between 08:00 to 10:00 h), placed in 500 g clamshell containers, immediately stored in a portable refrigerator (4 °C), and transferred to the laboratory to determine the fruit quality parameters within 24 h after harvest. Fifty fruits from each plant were used to determine quality parameters such as the fresh weight (FW), equatorial diameter (ED), firmness, total soluble solids (TSS), and titratable acidity (TI). The FW was determined using a precision balance (Model BA2204B, Biobase Meihua Trading, Jinan, China). The ED and firmness were determined using a digital caliper (Mitutoyo Corp., Kawasaki, Japan) and a texture meter (FirmPro, Happyvolt, Santiago, Chile), respectively, as described by Retamal-Salgado et al. [39]. The firmness was expressed as the force in grams (g) necessary to deform the fruit in 1 mm (g mm^{-1}). The total soluble solids (TSS) were determined in the fruit juice using a thermo-compensated digital refractometer (ATAGO, Mod. PAL-BX I ACID F5, Saitama, Japan) and expressed as °Brix. The titratable acidity (TA) was determined by the volumetric titration method with sodium hydroxide (0.1 N), using an automatic titrator HI-84532 (HANNA Instruments, Woonsocket, RI, USA), and expressed as the percentage (%) of citric acid, according to Mazzoni et al. [40]. For TSS and TA, 4 samples per treatment were used, each sample consisting of a juice obtained by macerating 10 fruits.

4.7. Experimental Design and Statistical Analysis

The experiment was performed using a completely randomized design with four replicates for each treatment. Kolmogorov–Smirnov and Levene tests were used to verify the normality of data and the variance homogeneity. A t-Student test was used to compare treatments (ambient temperature and high temperature). The statistical analyses were performed using Sigma Stat v.2.0 (SPSS, Chicago, IL, USA). The dataset was subjected to principal component analysis (PCA), preserving as much statistical information as possible. The analysis was carried out using R software version R 4.3.1 (R Core Team, Statistical computing, Vienna, Austria, 2023).

5. Conclusions

The negative influence of stomatal limitations on the development of *V. corymbosum* under HTs is clear, as evidenced by the physiological responses found in this study. Stomatal and non-stomatal limitations related to damage due to excessive heat certainly also have a major impact on the growth and development of fruit under HTs, but in order to really measure their contribution, more biochemical and metabolic studies are needed on *V. corymbosum* under HTs.

Author Contributions: J.G.-V., E.J.-F. and K.Á. designed and coordinated the experiment; H.A.G. and J.G.-V. performed gas-exchange analyses; K.Á. performed field determinations; J.G.-V., K.Á. and A.R.-F. carried out the fruit quality and statistical analyses; P.F.-G. and A.N.-N. performed the PCA analysis; J.G.-V. formulated the draft of manuscript. J.G.-V., K.Á., H.A.G., L.A.B., G.C., C.R., E.J.-F., A.N.-N., P.F.-G. and M.M.R.-D. revised and improved the current version of the manuscript. All authors have read and agreed to the published version of the manuscript.

Funding: This research was funded by ANID/FONDECYT 1211856, ANID/FONDECYT 11220732, ANID/FONDAP/15130015 and ANID/FONDAP/1523A0001, ANID/Anillo ATE230007 projects of the National Agency for Research and Development (ANID, ex CONICYT), and Internal project from UC Temuco (2023PF-06-JG).

Data Availability Statement: All data supporting the findings of this study are available within the paper.

Conflicts of Interest: The authors declare no conflicts of interest.

References

1. IPCC. *Climate Change 2014. Fifth Assessment Synthesis Report (Longer Report) of Intergovernmental Panel on Climate Change*; Cambridge University Press: Cambridge, UK; New York, NY, USA, 2014.
2. IPCC. *Climate Change 2021: The Physical Science Basis. Contribution of Working Group I to the Sixth Assessment Report of the Intergovernmental Panel on Climate Change*; Masson-Delmotte, V., Zhai, P., Pirani, A., Connors, S.L., Péan, C., Berger, S., Caud, N., Chen, Y., Goldfarb, L., Gomis, M.I., et al., Eds.; Cambridge University Press: Cambridge, UK; New York, NY, USA, 2021.
3. Siebert, S.; Ewert, F.; Rezaei, E.; Kage, H.; Grab, R. Impact of heat stress on crop yield-on the importance of considering canopy temperature. *Environ. Res. Lett.* **2014**, *9*, 044012. [CrossRef]
4. Wang, L.; Ma, K.B.; Lu, Z.G.; Ren, S.X.; Jiang, H.R.; Cui, J.W.; Chen, G.; Teng, N.J.; Lam, H.M.; Jin, B. Differential physiological, transcriptomic and metabolomic responses of Arabidopsis leaves under prolonged warming and heat shock. *BMC Plant Biol.* **2020**, *20*, 86. [CrossRef] [PubMed]
5. Sato, H.; Mizoi, J.; Shinozaki, K.; Yamaguchi-Shinozaki, K. Complex plant responses to drought and heat stress under climate change. *Plant J.* **2024**, *177*, 1873–1892. [CrossRef] [PubMed]
6. Ávila-Valdés, A.; Quinet, M.; Lutts, S.; Martínez, J.P.; Lizana, C. Tuber yield and quality responses of potato to moderate temperature increase during Tuber bulking under two water availability scenarios. *Field Crops Res.* **2020**, *251*, 107786. [CrossRef]
7. Kim, J.; Slafer, G.; Savin, R. Are Are portable polyethylene tents reliable for imposing heat treatments in field-grown wheat? *Field Crops Res.* **2021**, *271*, 108206. [CrossRef]
8. Kim, J.; Savin, R.; Slafer, G. Quantifying pre- and post-anthesis heat waves on grain number and grain weight of contrasting wheat cultivars. *Field Crops Res.* **2024**, *307*, 109264. [CrossRef]
9. Rivero, R.; Mestre, T.; Mittler, R.; Rubio, F.; Garcia-Sánchez, F.; Martínez, V. The combined effect of salinity and heat reveales a specific physiological, biochemical and molecular response in tomato plants. *Plant Cell Environ.* **2014**, *38*, 1037–1258.
10. Fahad, S.; Bajwa, A.A.; Nazir, U.; Anjum, S.A.; Farooq, A.; Zohaib, A.; Sadia, S.; Nasim, W.; Adkins, S.; Saud, S.; et al. Crop Production under Drought and Heat Stress: Plant Responses and Management Options. *Front. Plant Sci.* **2017**, *8*, 1147. [CrossRef]
11. Djalovic, I.; Kundu, S.; Bahuguna, R.; Pareek, A.; Raza, A.; Singla-Pareek, S.; Prasad, P.; Varshney, R. Maize and heat stress: Physiological, genetic, and molecular insights. *Plant Genome* **2024**, *17*, e20378. [CrossRef]
12. Cai, Y.; Tarin, M.; Fan, L.; Xie, D.; Rong, J.; He, T.; Chen, L.; Zheng, Y. Responses of photosynthesis, chloroplast ultrastructure, and antioxidant system of Morinda officinalis how. to exogenous 2, 4-epibrassinolide treatments under high temperature stress. *Appl. Ecol. Environ. Res.* **2020**, *18*, 3981–4004. [CrossRef]
13. Lal, M.K.; Tiwari, R.K.; Gahlaut, V. Physiological and molecular insights on wheat responses to heat stress. *Plant Cell Rep.* **2022**, *41*, 501–518. [CrossRef] [PubMed]
14. Zahra, N.; Hafeez, M.; Ghaffar, A.; Kausar, A.; Zeidi, M.; Siddique, K.; Farooq, M. Plant photosynthesis under heat stress: Effects and management. *Environ. Exp. Bot.* **2023**, *206*, 105178. [CrossRef]
15. Pospíšil, P. Production of Reactive Oxygen Species by Photosystem II as a Response to Light and Temperature Stress. *Front. Plant Sci.* **2016**, *7*, 1950. [CrossRef] [PubMed]

16. Wang, Q.-L.; Chen, J.-H.; He, N.-Y.; Guo, F.-Q. Metabolic reprogramming in chloroplasts under heat stress in plants. *Int. J. Mol. Sci.* **2018**, *19*, 849. [CrossRef] [PubMed]
17. Khan, A.; Min, L.; Ma, Y.; Zeeshan, M.; Jin, S.; Zhang, X. High-temperature stress in crops: Male sterility, yield loss and potential remedy approaches. *Plant Biotechnol. J.* **2023**, *21*, 680–697. [CrossRef] [PubMed]
18. Li, Y.; Jiang, F.; Niu, L.; Wang, G.; Yin, J.; Song, X.; Ottosen, C.; Rosenqvist, E.; Mittler, R.; Wu, Z.; et al. Synergistic regulation at physiological, transcriptional and metabolic levels in tomato plants subjected to a combination of salt and heat stress. *Plant J.* **2024**, *177*, 1656–1675. [CrossRef]
19. Ribera, A.E.; Reyes-Diaz, M.; Alberdi, M.; Zuñiga, G.E.; Mora, M.L. Antioxidant compounds in skin and pulp of fruits change among genotypes and maturity stages in highbush blueberry (*Vaccinium corymbosum* L.) grown in southern Chile. *J. Soil Sci. Plant Nutr.* **2010**, *10*, 509–536. [CrossRef]
20. Onuh, J.O.; Dawkins, N.L.; Aluko, R.E. Cardiovascular disease protective properties of blueberry polyphenols (*Vaccinium corymbosum*): A concise review. *Food Prod. Process Nutr.* **2023**, *5*, 27. [CrossRef]
21. Shi, J.; Xiao, Y.; Jia, C.; Zhang, H.; Gan, Z.; Li, X.; Yang, M.; Yin, Y.; Zhang, G.; Hao, J.; et al. Physiological and biochemical changes during fruit maturation and ripening in highbush blueberry (*Vaccinium corymbosum* L.). *Food Chem.* **2023**, *410*, 135299.
22. FAS-USDA. Blueberries around the Globe—Past, Present, and Future. 2021. Available online: https://www.fas.usda.gov/sites/default/files/2021-10/GlobalBlueberriesFinal_1.pdf (accessed on 8 April 2024).
23. Lagos, L.O.; Souto, C.; Lillo-Saavedra, M. Daily crop evapotranspiration and diurnal dynamics of the surface energy balance of a drip-irrigated blueberry (*Vaccinium corymbosum*) orchard. *Irrig. Sci.* **2024**, *42*, 1–13. [CrossRef]
24. INE. *Censo Agropecuario y Forestal 2021*; Instituto Nacional de Estadística (INE): Santiago, Chile, 2021. Available online: https://www.ine.cl/estadisticas/economia/agricultura-agroindustria-y-pesca/censos-agropecuarios (accessed on 5 March 2024).
25. Hancock, J. Highbush blueberry breeding. *Latvian J. Agron.* **2009**, *12*, 35–38.
26. Hao, L.; Guo, L.; Li, R.; Cheng, Y.; Huang, L.; Zhou, H.; Xu, M.; Li, F.; Zhang, X.; Zheng, Y. Responses of phosotynthesis to high temperature stress associated with changes in leaf structure and biochemistry of blueberry (*Vaccinium corymbosum* L.). *Sci. Hortic.* **2019**, *246*, 251–264. [CrossRef]
27. Chen, W.; Cen, W.; Chen, L.; Di, L.; Li, Y.; Guo, W. Differential sensitivity of four highbush blueberry (*Vaccinium corymbosum*) cultivars to heat stress. *Pak. J. Boyt.* **2012**, *44*, 853–860.
28. Estrada, F.; Escobar, A.; Romero-Bravo, S.; González-Talice, J.; Poblete-Echeverría, C.; Caligari, P.; Lobos, G.A. Fluorescence phenotyping in blueberrry breeding for genotype selection under drought conditions, with or with heat stress. *Sci. Hortic.* **2015**, *181*, 147–161. [CrossRef]
29. Yang, F.-H.; Bryla, D.; Strik, B. Critical Temperatures and Heating Times for Fruit Damage in Northern Highbush Blueberry. *HortScience* **2019**, *54*, 2231–2239. [CrossRef]
30. Flexas, J.; Medrano, H. Drought-inhibition of Photosynthesis in C3 Plants: Stomatal and Non-stomatal Limitations Revisited. *Ann. Bot.* **2002**, *89*, 183–189. [CrossRef]
31. Zhou, J.; Jiang, X.; Agathokleous, E.; Lu, X.; Yang, Z.; Li, R. High temperature inhibits photosynthesis of chrysanthemum (*Chrysanthemum morifolium* Ramat.) seedlings more than relative humidity. *Front. Plant Sci.* **2023**, *14*, 1272013. [CrossRef]
32. Chen, J.H.; Tang, M.; Jin, X.-Q.; Li, H.; Chen, L.-S.; Wang, Q.-L.; Sun, A.-Z.; Yi, Y.; Guo, F.-Q. Regulation of Calvin–Benson cycle enzymes under high temperature stress. *aBIOTECH* **2022**, *3*, 65–77. [CrossRef]
33. Bryla, D.; Yorgey, B.; Shireman, A. Irrigation management effects on yield and fruit quality of highbush blueberry. *Acta Hortic.* **2009**, *810*, 649–656. [CrossRef]
34. Faghih, S.; Zamani, Z.; Fatahi, R.; Omidi, M. Influence of kaolin application on most important fruit and leaf characteristics of two apple cultivars under sustained deficit irrigation. *Biol. Res.* **2021**, *54*, 1. [CrossRef]
35. CIREN. Estudio Agrológico. Descripciones de suelos, materiales y símbolos IX Región. Centro de Información de Recursos Naturales. *Publicación* **2022**, *122*, 343.
36. Begg, J.E.; Turner, N.C. Water potential gradients in field tobacco. *Plant Physiol.* **1970**, *46*, 343–346. [CrossRef] [PubMed]
37. Barai, K.; Calderwood, L.; Wallhead, M.; Vanhanen, H.; Hall, B.; Drummond, F.; Zhang, Y.-J. High Variation in Yield among Wild Blueberry Genotypes: Can Yield Be Predicted by Leaf and Stem Functional Traits? *Agronomy* **2022**, *12*, 617. [CrossRef]
38. Reyes-Díaz, M.; Meriño-Gergichevich, C.; Alarcón, E.; Alberdi, M.; Horst, W.J. Calcium sulfate ameliorates the effect of aluminum toxicity differentially in genotypes of highbush blueberry (*Vaccinium corymbosum* L.). *J. Soil Sci. Plant Nutr.* **2011**, *11*, 59–78. [CrossRef]
39. Retamal-Salgado, J.; Batías, R.; Wilckens, R.; Paulino, L. Influence of microclimatic conditions under high tunnels on the physiological and productive responses in blueberry 'O'Neal'. *Chil. J. Agric. Res.* **2015**, *75*, 291–297. [CrossRef]
40. Mazzoni, L.; Balducci, F.; Di Vittori, L.; Scalzo, J.; Capocasa, F.; Zhong, C.; Forbes-Hernández, T.; Giampieri, F.; Battino, M.; Mezzetti, B. Yield and nutritional quality of highbush blueberry genotypes trialled in a Mediterranean hot summer climate. *J. Sci. Food Agric.* **2020**, *100*, 3675–3686. [CrossRef]

plants

MDPI

Article

Overexpression of *AcWRKY31* Increases Sensitivity to Salt and Drought and Improves Tolerance to Mealybugs in Pineapple

Myat Hnin Wai [1,2,3,4,†], Tiantian Luo [1,2,3,†], S. V. G. N. Priyadarshani [5], Qiao Zhou [1,2,3], Mohammad Aqa Mohammadi [1,2,3], Han Cheng [1,2,3], Mohammad Aslam [1,2,3], Chang Liu [1,2,3], Gaifeng Chai [1,2,3], Dongping Huang [1,2,3], Yanhui Liu [1,2,3], Hanyang Cai [1,2,3], Xiaomei Wang [1,2,3,6], Yuan Qin [1,2,3,*] and Lulu Wang [1,2,3,*]

[1] College of Agriculture, Fujian Provincial Key Laboratory of Haixia Applied Plant Systems Biology, Pingtan Science and Technology Research Institute, Fujian Agriculture and Forestry University, Fuzhou 350002, China; myathninwai.edu@gmail.com (M.H.W.); luotiantian1022@163.com (T.L.); zhouqiao0606@163.com (Q.Z.); mohammadaqam85@gmail.com (M.A.M.); c.hanngu922@gmail.com (H.C.); aslampmb1@gmail.com (M.A.); changliu08@163.com (C.L.); 15117155242@163.com (G.C.); z35068120000419@163.com (D.H.); yanhuiliu520@gmail.com (Y.L.); caihanyang123@163.com (H.C.); wangxiaomei159@163.com (X.W.)

[2] College of Life Science, Fujian Provincial Key Laboratory of Haixia Applied Plant Systems Biology, Pingtan Science and Technology Research Institute, Fujian Agriculture and Forestry University, Fuzhou 350002, China

[3] College of Horticulture, Fujian Provincial Key Laboratory of Haixia Applied Plant Systems Biology, Pingtan Science and Technology Research Institute, Fujian Agriculture and Forestry University, Fuzhou 350002, China

[4] Department of Botany, Mandalay University of Distance Education, Ministry of Education, Mandalay 05024, Myanmar

[5] Department of Applied Sciences, Faculty of Humanities and Sciences, Sri Lanka Institute of Information Technology, New Kandy Road, Malabe 10115, Sri Lanka; nirosha.p@sliit.lk

[6] Horticulture Research Institute, Guangxi Academy of Agricultural Sciences, Nanning Investigation Station of South Subtropical Fruit Trees, Ministry of Agriculture, Nanning 530007, China

* Correspondence: yuanqin@fafu.edu.cn (Y.Q.); luluwanghn@163.com (L.W.)

† These authors contributed equally to this work.

Citation: Wai, M.H.; Luo, T.; Priyadarshani, S.V.G.N.; Zhou, Q.; Mohammadi, M.A.; Cheng, H.; Aslam, M.; Liu, C.; Chai, G.; Huang, D.; et al. Overexpression of *AcWRKY31* Increases Sensitivity to Salt and Drought and Improves Tolerance to Mealybugs in Pineapple. *Plants* **2024**, *13*, 1850. https://doi.org/10.3390/plants13131850

Academic Editors: Violetta Katarzyna Macioszek, Iwona Ciereszko and Andrzej K. Kononowicz

Received: 29 May 2024
Revised: 29 June 2024
Accepted: 30 June 2024
Published: 5 July 2024

Abstract: Pineapple is a globally significant tropical fruit, but its cultivation faces numerous challenges due to abiotic and biotic stresses, affecting its quality and quantity. WRKY transcription factors are known regulators of stress responses, however, their specific functions in pineapple are not fully understood. This study investigates the role of *AcWRKY31* by overexpressing it in pineapple and *Arabidopsis*. Transgenic pineapple lines were obtained using *Agrobacterium*-mediated transformation methods and abiotic and biotic stress treatments. Transgenic *AcWRKY31-OE* pineapple plants showed an increased sensitivity to salt and drought stress and an increased resistance to biotic stress from pineapple mealybugs compared to that of WT plants. Similar experiments in *AcWRKY31-OE*, *AtWRKY53-OE*, and the *Arabidopsis* Atwrky53 mutant were performed and consistently confirmed these findings. A comparative transcriptomic analysis revealed 5357 upregulated genes in *AcWRKY31-OE* pineapple, with 30 genes related to disease and pathogen response. Notably, 18 of these genes contained a W-box sequence in their promoter region. A KEGG analysis of RNA-Seq data showed that upregulated DEG genes are mostly involved in translation, protein kinases, peptidases and inhibitors, membrane trafficking, folding, sorting, and degradation, while the downregulated genes are involved in metabolism, protein families, signaling, and cellular processes. RT-qPCR assays of selected genes confirmed the transcriptomic results. In summary, the *AcWRKY31* gene is promising for the improvement of stress responses in pineapple, and it could be a valuable tool for plant breeders to develop stress-tolerant crops in the future.

Keywords: pineapple; WRKY transcription factor; *AcWRKY31*; biotic stress; abiotic stress

1. Introduction

Plant breeders are currently facing a huge challenge developing new, high-yielding, and stress-tolerant crop varieties that can cope with environmental cues. Plants encounter various stresses throughout their life cycle, such as salinity, drought, temperature fluctuations, and pathogen infections [1]. To adapt to challenging environments, plants attempt to regulate their physiological and biochemical mechanisms in a timely manner [2,3]. At the same time, plant breeders are working on the development of stress-tolerant varieties to meet global demand.

The detrimental effects of abiotic stress on plants primarily arise from unfavorable alterations in the natural environment due to changes in temperature, salinity, etc. These changes promote corresponding physiological and biochemical responses in plants, ultimately causing significant harm to their inherent traits and phenotypic characteristics, leading to crop losses [4]. High levels of plant osmotic pressure and high levels of salinity can impair plant growth and development [5]. Drought is another significant factor that exerts detrimental effects on plant processes [6]. Additionally, drought stress has been associated with increased plant root biomass [7]. When the water content decreases in plant environments, the water potential in leaves also declines, which impairs photosynthesis and metabolism, and, in severe cases, leads to plant mortality [8]. Biotic stress, primarily caused by harmful pathogens, can penetrate plant cuticles, invade stomata, or exploit natural wounds to disrupt normal organ functions, leading to reduced photosynthetic rates, water imbalances, and nutrient transport obstacles [9,10].

Ananas comosus (L.) Merr. (pineapple) is an economically important crop that is distinctly exposed to disease conditions caused by bacteria, fungi, viruses, and nematodes, as well as abiotic stresses, leading to yield reductions [11]. *Dysmicoccus brevipes* (pineapple mealybug) is one of the most serious pests, and it weakens plants by sap-sucking and acts as a vector for the pineapple mealybug wilt-associated virus (PMWaV) [12]. Another destructive pathogen, *Sclerotinia sclerotiorum*, infects over 400 plant species, impacting crop yield and quality [13]. These pathogens pose a significant threat to various plants, including dicotyledons, such as sunflower, soybean, and rape, as well as monocotyledons, like onion and tulip [14]. Consequently, advanced biotechnologies play an essential role in the improvement of plant resistance to environmental stresses, identifying stress-resistant genes and facilitating the development of new plant varieties.

Among the transcription factor (TF) families involved in stress regulation, WRKY TFs are known for their diverse regulatory mechanisms. Typical WRKY proteins efficiently bind to W-box elements to modulate downstream gene transcription and can form protein complexes by interacting with other activated elements, enhancing their transcriptional binding ability. Structurally, WRKY TFs consist of two main domains: the DNA-binding N-terminal domain and the C-terminal zinc finger structure [15]. The WRKY DNA binding domain exhibits variations within the family according to the basic features of conservative heptapeptides, such as WRKYGMK, WRKYGKK, WRKYGQK, WSKYGQK, WKRYGQK, WVKYGQK, and WKKYGQK [16]. Zinc finger structures are primarily composed of C_2H_2 and C_2HC motifs and are crucial constituents of WRKY TFs. WRKY TFs are classified into subfamilies I, II, and III by the number of their domains and their zinc finger-like structure [17]. The data from an evolutionary analysis showed that subfamily II can divide into sub-subfamilies, like IIa, IIb, IIc, IId, and IIe [18]. Furthermore, some WRKY families also contain the proline enrichment domain, glutamate enrichment domain, and a leucine structure [19].

The role of WRKY TFs in regulating responses to abiotic stress has been studied in various species. The *IbWRKY47* gene is upregulated in salt stress and confers salinity resistance to sweet potatoes [20]. *MiR156/SPL* induces the expression of the salt-tolerant gene *MdWRKY100*, improving the salt tolerance of *Begonia* [21]. *SbWRKY50* participates in sweet sorghum plant responses to salt stress by controlling its ion balance according to the binding of the upstream promoters of *AtSOS1* and *AtHKT1* [22,23]. In chrysanthemum, *CmWRKY17* is negatively regulated by salt stress [24]. Transcripts, proteins, metabolite levels, hormones, ROS, small RNAs, epigenetic modifications, post-translational modifications, and environmental cues are the main factors regulating the seed germination of *Arabidopsis* and rice [25]. *GhWRKY68* regulates the ABA-mediated pathway and reduce resistance to drought and salt in cotton [26]. The overexpression of the *BdWRKY36* transgenic line can reduce ROS accumulation by the activation of *NtNCED1, NtDREB3,* and *NtLEA5* in the ABA biosynthesis pathway and lead to significant resistance to drought stress in tobacco [27]. In pepper, *CaWRKY6* activates *CaWRKY40* and makes it more resistant to heat and humidity stress [28]. *GmWRKY27* inhibits the *GmNAC29* promoter through an independent inhibitory effect and inhibits the expression of *GmNAC29* with *GmMYB174* to improve plant resistance to salt and drought stress [29]. In pineapple, 54 WRKY genes have been investigated in this family [30]. However, there has been less research regarding functional analyses of the pineapple WRKY family's genes. Huang *et al.* reported that the ectopic overexpression of *AcWRKY31* negatively affects tolerance to drought and salt stresses in rice [31]. The first-ever study conducted with transgenic pineapple showed the ability of overexpressed *AcWRKY28* in pineapple to confer salt stress tolerance [32]. Through the regulation of bromelain and oxidative stress, the effect of the DA-6 and COS PGRs can create a tolerance to drought in pineapple [33].

Ecological stresses have a significant impact on the fruit development, quality, and yield of pineapple plants [34]. Conventional pineapple breeding for new variety development is time-consuming for commercial production, and self-incompatibility further complicates this process. Molecular breeding and biotechnological tools make breeding programs easier and more efficient. Transgenic *AcWRKY31-OE* pineapple and *Arabidopsis* plants were successfully developed using the *Agrobacterium*-mediated transformation method. The results of this study showed that the overexpression of *AcWRKY31* increased sensitivity to drought and salt tolerance in both pineapple and *Arabidopsis* and increased tolerance to biotic stress caused by pineapple mealybugs in pineapple. In *Arabidopsis,* the overexpression of *AcWRKY31* increased resistance to *Sclerotinia sclerotiorum* infection. The findings of our study on stress-responsive genes provide insight into molecular breeding for developing improved pineapple varieties to boost the economic growth of producers.

2. Results

2.1. The Expression Profiles and Subcellular Localization of AcWRKY31

By analyzing the gene expression patterns, we can preliminarily identify the developmental processes involved and provide a theoretical foundation for subsequent phenotypic and functional analyses. To examine the expression patterns of *AcWRKY31* in different pineapple tissues, we downloaded and analyzed the transcriptome data from various tissues, including root, leaf, flower, calyx, petal, stamen, pistil, ovule, and fruit tissues [35,36]. The expression level of *AcWRKY31* was shown to be the highest in the pistil, followed by the ovule and root (Figure 1A). In addition, an RT-qPCR analysis confirmed the accuracy of the expression level of *AcWRKY31* in different pineapple tissues and revealed high expression levels in the pistil and ovule (Figure 1B). Based on the tobacco transient transformation, we found that *AcWRKY31* is localized in the nucleus (Figure 1C). These expression patterns suggest that *AcWRKY31* serves an important function in pineapple pistil and ovule development as well as in root development for water absorption.

2.2. Phylogenetic Analysis and Sequence Alignment of AcWRKY31

To characterize the pineapple gene *AcWRKY31,* the CDS sequence and genome sequence were downloaded, and the gene structure is depicted using GSDS 2.0 (Figure 2A). A phylogenetic tree of the pineapple *AcWRKY31* and its homologous genes in *Arabidopsis* and rice was constructed. The phylogenetic tree indicated that *AcWRKY31* shares a high degree of similarity with *AtWRKY53* in *Arabidopsis* and *OsWRKY113* in rice (Figure 2B). In addition, the alignment of multiple protein sequences showed that *AcWRKY31* belongs to group III of the pineapple WRKY TF family, which contains a conserved WRKYGQK domain and C_2HC-type zinc finger motif (C-X7-C-X23-H-X-C) (Figure 2C).

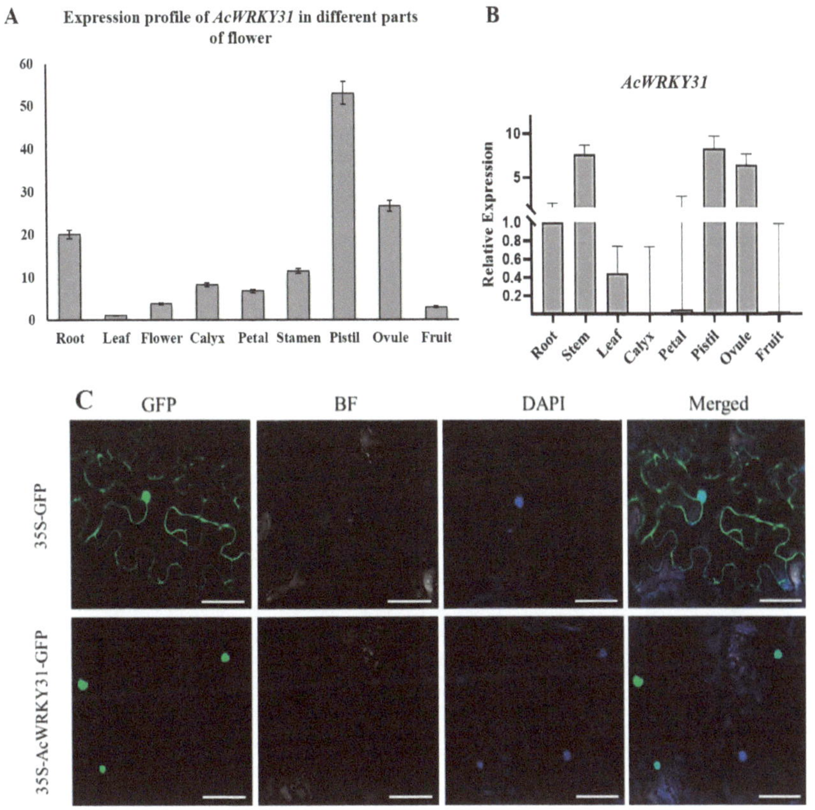

Figure 1. Expression levels of *AcWRKY31* in different parts of pineapple tissues; (**A**) Expression profiles of RNA-sequencing; (**B**) Relative expression levels *AcWRKY31* from RT-qPCR analysis. The error bars indicate ±SD (n = 3) (**C**) Subcellular localization of *AcWRKY31* that located in nucleus (Bar = 50 μm; Notes: blue and green color dots indicate the nucleus of cells under microscope).

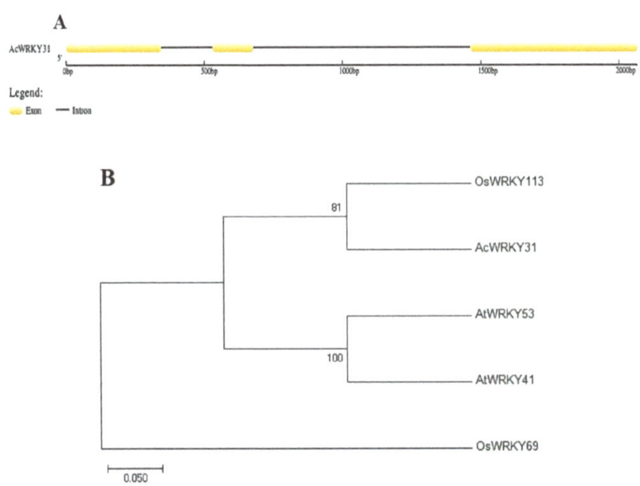

Figure 2. *Cont.*

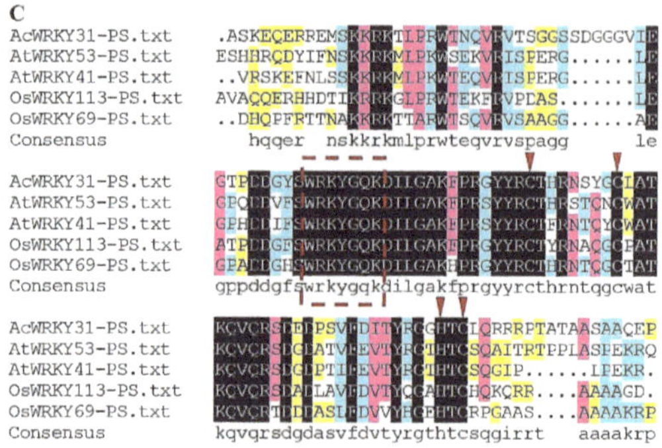

Figure 2. Bioinformatics analysis of *AcWRKY31*. (**A**) Gene structure of *AcWRKY31*. (**B**) Phylogenetic tree of *AcWRKY31* and its homologous genes in *Arabidopsis* and rice. The unrooted phylogenetic tree was generated using MEGA 7.0 software with neighbor-joining procedure and the following parameters: Poisson model, pairwise gap deletion, and 1000 bootstraps. Bar = 0.05 indicates the distance scale. (**C**) Multiple sequence alignment of *AcWRKY31* and its homologous protein sequences (Note: the red dotted line refers to the WRKYGQK domain, and red triangles refer to the C_2HC-type zinc finger motif).

2.3. Transformation and Identification of Overexpressed AcWRKY31 in Pineapple

To investigate the regulations and functions of *AcWRKY31*, a genetic transformation was performed using an *Agrobacterium*-mediated transformation system as described in Priyadarshani et al. with a few modifications [37]. Transgenic callus and plants were selected rigorously (Figure 3A–G) in the selection medium. The screened plants were further confirmed through PCR testing and sequencing. A total of 12 independent positive *AcWRKY31*-overexpressed pineapple plants were verified (Figure 3H). The WT plants and 12 *AcWRKY31-OE* transgenic pineapple plants were used to extract the total RNA, and RT-qPCR testing was performed to confirm the expression levels of *AcWRKY31*. The result showed that the expression level of *AcWRKY31* was increased more than twice in all of the overexpressed transgenic plants compared to the wild-type plants (Figure 3I).

Figure 3. Transformation and identification process of overexpression of *AcWRKY31* in pineapple; (**A**) *Agrobacterium* transformation in pineapple callus; (**B**) Infected callus in 1st selection medium;

(**C**) Plant regeneration from resistant callus in normal medium; (**D**) Plant regenerated callus in 2nd selection medium; (**E**) Resistant pineapple plantlets in normal medium; (**F**) Plantlets that died and changed color in 3rd selection medium; (**G**) Completed and resistant pineapple plantlets after undergoing selection three times; (**H**) Confirmation of positive transgenic pineapple by PCR method; (**I**) The expression level of *AcWRKY31-OE* lines and WT pineapple by RT-qPCR analysis. The error bars indicate ±SD (n = 3), and the asterisks indicate the significant differences based on Student's t-test (** $p < 0.01$).

2.4. Overexpression of AcWRKY31 Reduces Tolerance to Salt and Drought Stresses in Pineapple

To investigate the function of *AcWRKY31* in pineapple, *AcWRKY31-OE* transgenic plants were subjected to salt and drought stress. After three weeks of treatment with 400 mM NaCl and 400 mM mannitol solutions, the leaves of the *AcWRKY31-OE* pineapple plants were wilted, dried, and drooped, while the leaves of the wild-type (WT) pineapple plants remained upright (Figure 4A,B). After undergoing one month of recovery in a greenhouse, the *AcWRKY31-OE* transgenic pineapple plants were completely dried and noticeably lighter compared to the WT plants. According to the results of the RT-qPCR analysis, enzyme- and salt-stress-related marker genes, such as *AcPOD1*, *AcABI5*, *AcABA1*, and *AcPR1* genes, in the *AcWRKY31-OE* plants showed reduced relative expression levels compared to those of the WT plants (Figure 4C). Similarly, the relative expression levels of hormone- and drought-stress-related marker genes (*AcCAT1*, *AcPR1*, *AcLOX4*, *AcABA1*, *AcRD22*, and *AcDREB2A*) in the *AcWRKY31-OE* plants were lower than then those to the WT pineapple (Figure 4D). These findings indicate the overexpression of *AcWRKY31* enhances the sensitivity to salt and drought stress in pineapple.

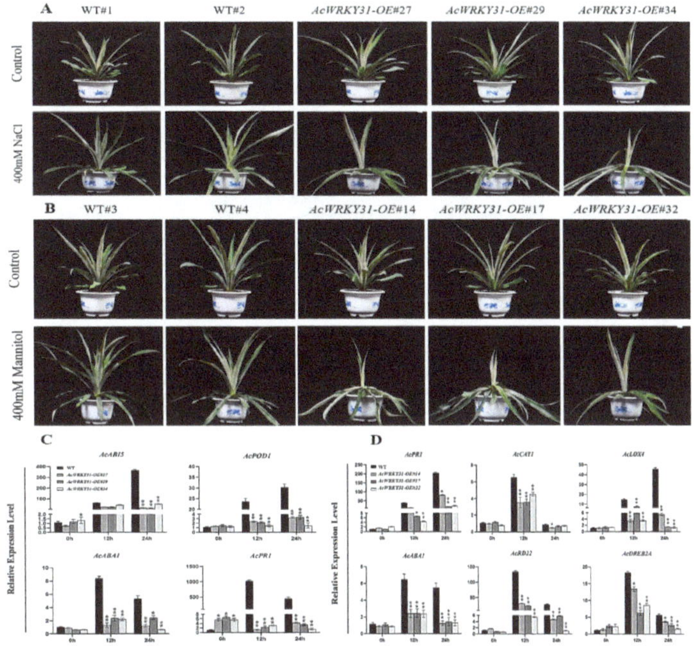

Figure 4. Abiotic stress treatments on transgenic pineapple; (**A**) Comparative phenotypic characters of WT and transgenic pineapple after 3 weeks of salt treatment; (**B**) Comparative phenotypic characters of WT and transgenic pineapple after 3 weeks of mannitol treatment; (**C**,**D**) The expression level of salt- and drought-stress-related genes by RT-qPCR analysis in *AcWRKY31-OE* lines and WT pineapple under abiotic treatments. The error bars indicate ±SD (n = 3), and the asterisks indicate significant differences based on Student's t-test (** $p < 0.01$, * $p < 0.05$).

2.5. Overexpression of AcWRKY31 Increases the Resistance to Pineapple Mealybug

To investigate the function of *AcWRKY31* that response to biotic stress, four-month-old *AcWRKY31-OE* and WT pineapple plants were inoculated with *Dysmicoccus brevipes*. There were no significant differences in *AcWRKY31-OE* and WT plants in the first two months after inoculation. However, after 3 months of inoculation, we observed a significantly higher density of colonies on the crown meristem and dorsal surfaces of the WT leaves compared to that of those the *AcWRKY31-OE* pineapple plants. In addition, the size of the colonies was significantly larger in the WT plants (Figure 5A). In the WT plants, there were approximately 200 colonies on the dorsal surface and less than 100 colonies on the ventral surface of the pineapple leaves, and the smallest colony sizes were dotted while the maximum colony diameter was 0.8 cm. In the transgenic pineapple leaves, the number of mealybug colonies did not exceed 60 on the dorsal surface and was less than 15 on the ventral surfaces. Most of the colonies were tiny, and the largest colonies were only 0.2 cm in diameter (Figure 5B,C). Interestingly, we found that the dorsal surface was more susceptible to mealybug infection in both the WT and transgenic plants compared to the ventral leaf surface. To understand the relationship between the higher infection rate on the dorsal side compared to the ventral side, free-hand transverse sections were cut, and the cellular arrangement was observed under a microscope. We found a soft epidermal cell layer and thick layers of water storage cells on the dorsal side of both the WT and transgenic leaves. On the ventral side of the leaves, rigid epidermal cells, chlorenchyma cells, fibers, and vascular bundles were observed (Figure 5D). According to the cell arrangements in both the WT and transgenic leaves, mealybugs easily penetrate the cell layers and infect the dorsal surfaces rather than the ventral surfaces. Based on the phenotypic characteristics and statistical analysis data of the colonies on the leaf surfaces, we can conclude that the overexpression of *AcWRKY31* enhanced resistance to *Dysmicoccus brevipes* in the pineapple.

Figure 5. *Cont.*

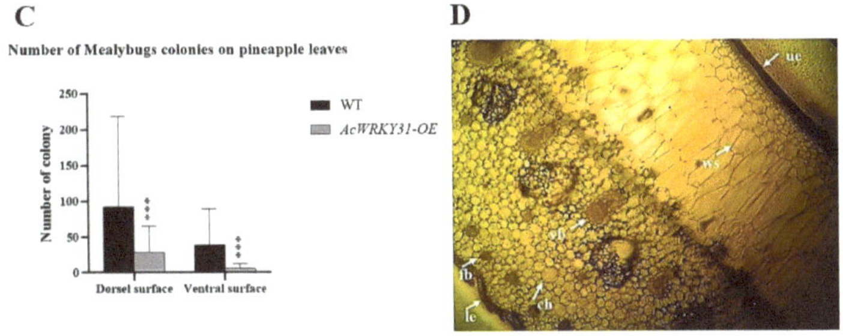

Figure 5. Inoculation of *Dysmicoccus brevipes* (pineapple mealybugs) on pineapple leaves; (**A**) Comparison of infected leaves between WT and *AcWRKY31-OE* pineapple plants after 3 months of inoculation, bars = 1 cm; (**B**) Comparison of infected mealybugs and number of colonies on dorsiventral surfaces in both WT and transgenic leaves (Bar= 1 cm) (**C**) Statistical results of mealybugs' colonies on dorsiventral surfaces in both WT and transgenic lines (the error bars indicate ±SD (n = 3)) (**D**) Free-hand section of pineapple leaves to understand the cell arrangement on both surfaces (ue: upper epidermal layer, le: lower epidermal layer, ch: chlorenchyma cell, ws: water storage cell, vb: vascular bundle, and fb: fiber; Note: cell arrangements are the same in both WT and transgenic lines). The asterisks indicate significant differences (*** $p < 0.001$).

2.6. RNA-Seq Analysis of AcWRKY31-OE Pineapple

To study the functions and mechanisms of *AcWRKY31* regulation in pineapple, a transcriptomic analysis was conducted by using the total RNA from the WT and *AcWRKY31-OE* transgenic pineapple. In this analysis, we identified 5357 upregulated genes and 662 downregulated genes from the transgenic pineapple plants compared to the WT plants. The putative functions of these differentially expressed genes (DEGs) are provided in Supplementary Tables S2.1 and S2.2. The Gene Ontology (GO) enrichment analysis of DEGs showed that 5357 upregulated genes were significantly enriched in the RNA metabolic process, tRNA aminoacylation, the establishment of localization in cell, intracellular transport, protein folding, RNA splicing via transesterification reaction, cellular response to stimulus, transcription according to RNA polymerase II, ncRNA processing, etc. (Figure 6A). The Kyoto Encyclopedia of Genes and Genomes (KEGG) pathway analysis indicated that the upregulated DEGs were mostly involved in the process of translation, protein kinases, peptidases and inhibitors, membrane trafficking, folding, sorting and degradation, endocytosis and translation factors, etc. (Figure 6B). Among the upregulated genes, a total of 30 genes were identified relating to diseases and pathogen response, and further analysis showed that 18 of them contained the conserved WRKY binding domain (W-box TGAC) in the promoter region. These 18 genes typically belong to the CC-NBS-LRR and TIR-NBS-LRR classes of the disease resistance protein family (Supplementary Table S2.3). The GO analysis of 662 downregulated genes revealed their involvement in 19 GO terms, primarily concentrated in transport, the establishment of localization, alpha-amino acid metabolic process, transmembrane transport and the organic substance catabolic process, the carbohydrate catabolic process, etc. (Figure 6C). In addition, the KEGG analysis pointed out that these downregulated genes were activated in seven pathways, especially in mitochondrial biogenesis, metabolism, glycosyltransferases, oxidative phosphorylation, protein families, signaling and cellular processes, energy metabolism, and transporters (Figure 6D). Furthermore, three differentially expressed genes (*Aco010322*: MYB-like transcription factor family protein, *Aco014286*: Disease resistance protein CC-NBS-LRR class family, and *Aco024514*: Tetratricopeptide repeat TPR-like superfamily protein; three that are downregulated (*Aco001317*: Mitochondrial substrate carrier family protein *Aco002224*: Beta-1,4-N-acetylglucosaminyltransferase family protein, and *Aco010847*: Glutaredoxin family protein from downregulated were selected and performed RT-qPCR analysis to confirm the accuracy of the RNA-sequencing data. The results showed that the expression profiles for these six genes are consistent and aligned with the RNA-seq data (Supplementary Figure S3).

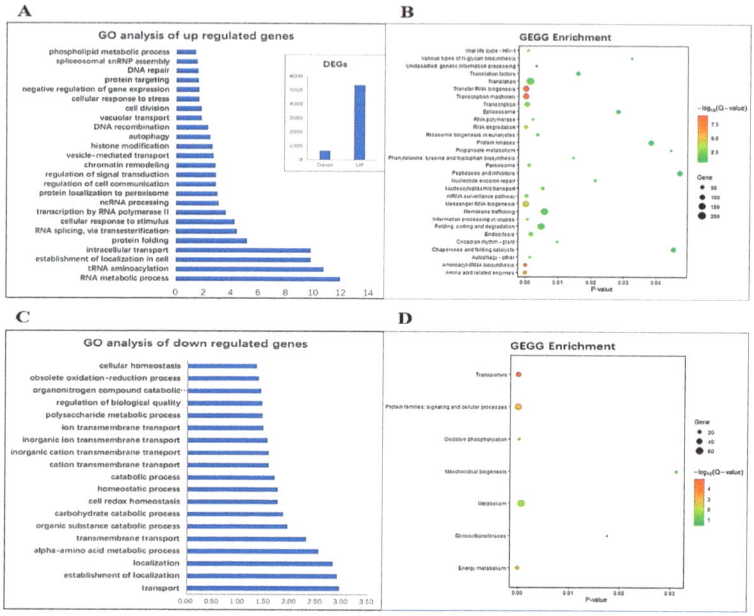

Figure 6. Transcriptomic analysis of WT and *AcWRKY31-OE* transgenic pineapple plants. (**A**) GO analysis of the upregulated genes; (**B**) KEGG pathway analysis for upregulated genes; (**C**) GO analysis of the downregulated genes; (**D**) KEGG pathway analysis for downregulated genes.

2.7. *AcWRKY31-OE Transgenic Arabidopsis Reduces Tolerance to Salt and Drought Stresses*

To confirm the heterologous functions of *AcWRKY31-OE* in *Arabidopsis*, we transformed the overexpressed vector *35S:AcWRKY31-GFP* gene and its homologous gene *35S:AtWRKY53-GFP* into *Arabidopsis* by the floral dip method. The overexpressed *AcWRKY31* and its homologous gene (*AtWRKY53*) in *Arabidopsis* seeds underwent salt and drought treatments. Under the treatment of 100 and 150 mM NaCl, both the WT and transgenic *Arabidopsis* seeds could germinate completely; however, the fresh weight and root length of *AcWRK31-OE Arabidopsis* were significantly reduced compared to the WT plants (Figure 7A). Likewise, a notably adverse impact was observed for the results of the *AcWRK31-OE Arabidopsis* plants subjected to 200 mM and 250 mM mannitol treatments (Figure 7B). Furthermore, the same findings were observed in *AtWRKY53-OE Arabidopsis* under the stress of the salt and drought treatments (Supplementary Figures S4 and S5). Sun and Yu showed that the *Atwrky53* mutant negatively regulates drought tolerance by mediating stomata movement [38]. According to the phenotypic characters of this experiment, *AcWRKY31* plays a negative role in the regulation of drought and salt stress response in *Arabidopsis*. To verify the further molecular mechanism of *AcWRKY31-OE Arabidopsis* in salt and drought stress, an RT-qPCR analysis was performed after the treatments. The expression levels of enzyme-related and stress-related genes, such as *AtPOD1*, *AtLOX4*, *AtABI5*, and *AtPR1*, were significantly decreased in all of the transgenic lines after the NaCl and mannitol treatments (Figure 7C,D). The results showed that the overexpression of *AcWRKY31* enhanced the sensitivity to salt and drought stress in *Arabidopsis*. Based on the RT-qPCR assay of ABA stress treatment on WT pineapple, we can speculate that *AcWRKY31* may be involved in the ABA-mediated stress response process. To prove this speculation, ABA stress treatments were performed in *AcWRKY31-OE*, *AtWRKY53-OE*, and in *Atwrky53 Arabidopsis* avn showed consistent findings that the ABA was induced in both overexpression lines (Supplementary Figure S6).

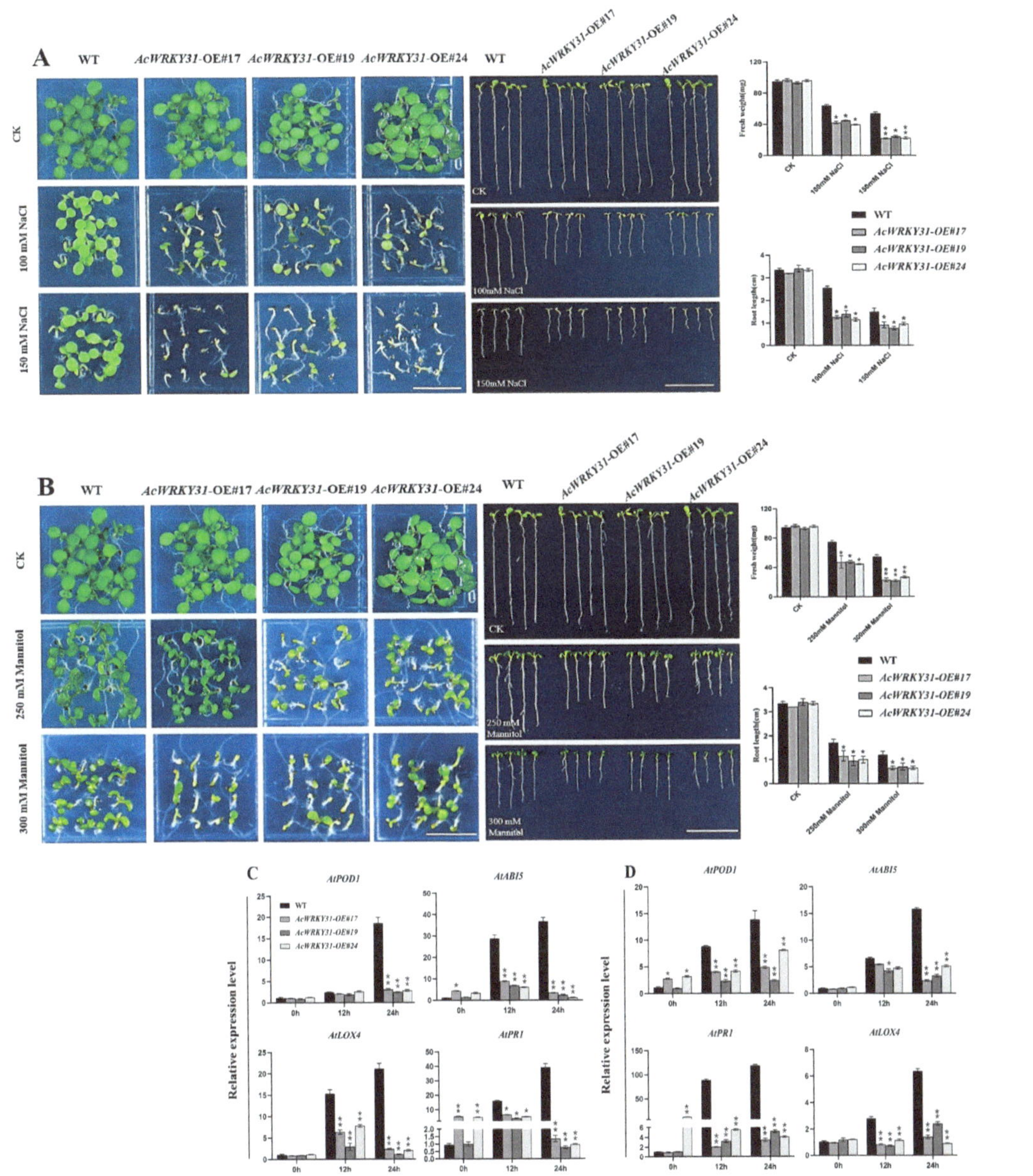

Figure 7. Salt and drought treatments in *AcWRKY31-OE Arabidopsis*; (**A**) Observation of germination rate, fresh weight, and root length of *AcWRKY31-OE* and WT *Arabidopsis* under NaCl treatment (bar = 1 cm); (**B**) Observation of germination rate, fresh weight, and root length of *AcWRKY31-OE* and WT *Arabidopsis* under mannitol treatment (bar = 1 cm); (**C**,**D**) The expression of abiotic-stress-related genes in the WT and *AcWRKY31* transgenic *Arabidopsis* plants in response to salt and drought stresses. Statistical results of fresh weight and root length after 7 days; the error bars indicate ±SD (n = 3), and the asterisks indicate the significant differences based on Student's t-test (** $p < 0.01$, * $p < 0.05$).

2.8. *AcWRKY31-OE Transgenic Arabidopsis Increases Resistance to the Sclerotinia sclerotiorum Pathogen*

To investigate the function of *AcWRKY31* in response to pathogenic infection, leaves of *AcWRKY31-OE* transgenic and WT *Arabidopsis* were inoculated with *Sclerotinia sclerotiorum*. After 24 h of infection, DAB staining was performed to observe the infected area, and the results revealed that the infected area of the WT leaves was significantly larger than that of three *AcWRKY31-OE* lines (Figure 8A,B). Disease- and hormone-related marker genes, such as *AtICS1*, *AtPDF1.2*, *AtPR1*, *AtERF1*, *AtLOX4*, and *AtABI5*, were selected for an RT-qPCR assay after infection [39–41]. The RT-qPCR results showed the expression levels of these selected marker genes were significantly increased, indicating that the overexpression of *AcWRKY31* enhanced the plants' sensitivity to pathogenic disease (Figure 8C). In addition, we inoculated the same pathogen on *Atwrky53*, *Atwrky53-com*, and *Atwrky53pAcWRKY31* leaves. The affected areas of the *Atwrky53pAcWRKY31* leaves were smaller than those of the *Atwrky53* and *Atwrky53-com*, and the expression levels of marker genes were also increased in the *Atwrky53pAcWRKY31* leaves (Figure 9A–C). Therefore, these results provided proof that the functions of *AcWRKY31* and its homologous gene *AtWRKY53* can increase the resistance to biotic stress in both pineapple and *Arabidopsis*.

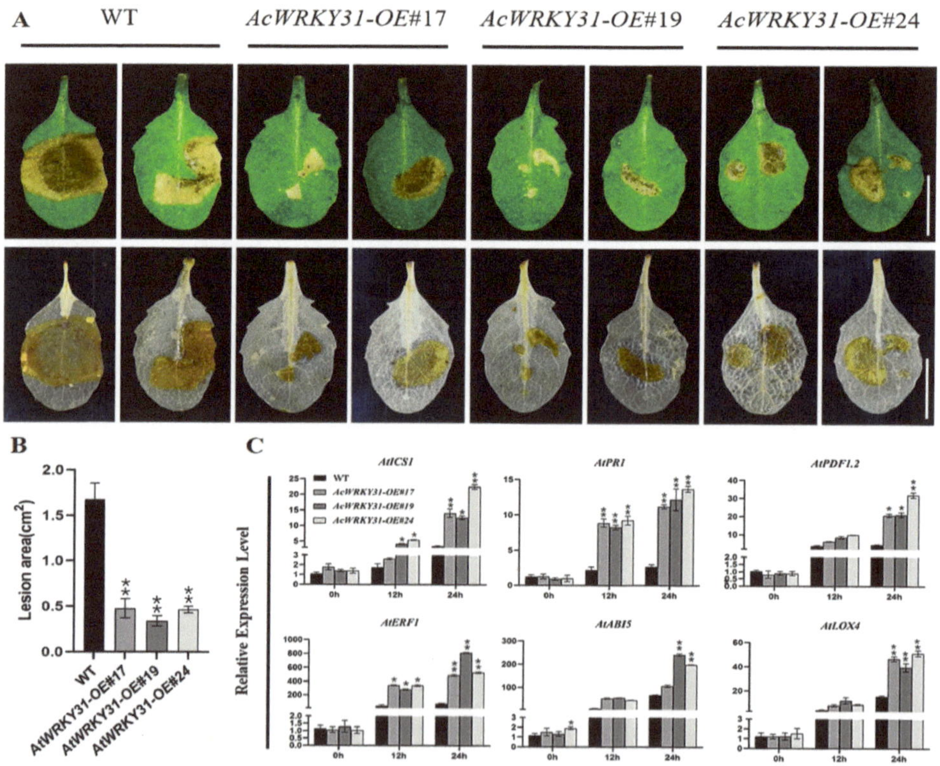

Figure 8. Inoculation of pathogen treatment on *AcWRKY31-OE Arabidopsis*. (**A**) The infection of *Sclerotinia sclerotiorum* on WT and *AcWRKY31-OE Arabidopsis* lines for 24 h and relative plaque area after DAB staining (bar = 1 cm); (**B**) The statistical data of lesion area; (**C**) The expression level of disease-related genes by RT-qPCR in *AcWRKY31-OE* and WT plants. The error bars indicate ±SD (n = 3), and the asterisk indicates significant differences based on Student's t-test (** $p < 0.01$, * $p < 0.05$).

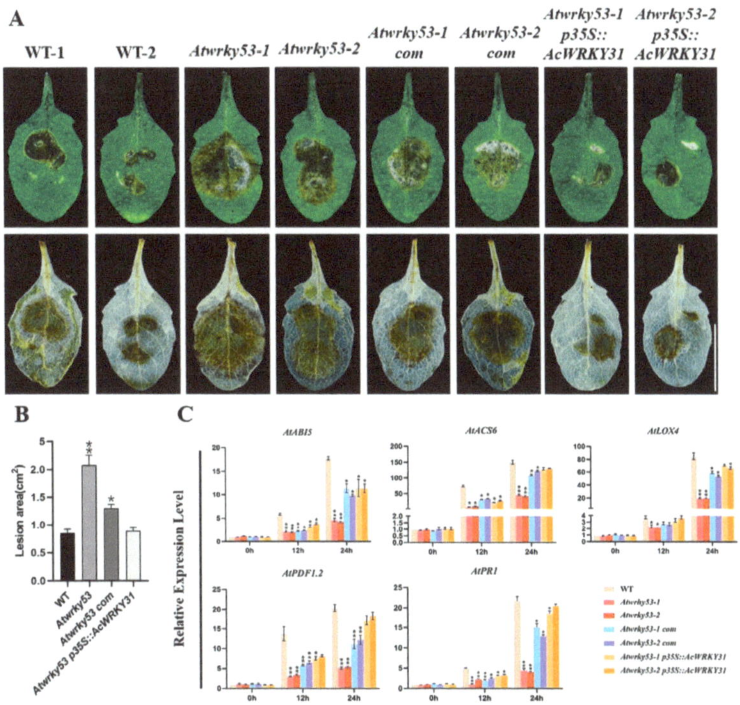

Figure 9. Inoculation of pathogen treatment on *Atwrky53 Arabidopsis*. (**A**) The infection of *Sclerotinia sclerotiorum* on WT, *Atwrky53* mutant, *Atwrky53* complementary lines, and *Atwrky53p35S:AcWRKY31 Arabidopsis* lines for 24 h and relative plaque area after DAB staining (bar = 1 cm); (**B**) The statistical data of lesion area; (**C**) The expression levels of disease-related genes by RT-qPCR in transgenic lines and WT plants. The error bars indicate ±SD (n = 3), and the asterisks indicate significant differences based on Student's t-test (** $p < 0.01$, * $p < 0.05$).

3. Discussion

Environmental factors such as drought, salt, pathogens, and pests have a significant impact on crop yield and quality. Pineapple, an important economic fruit crop, is susceptible to various ecological stresses during its lifespan. WRKY transcription factors (TFs) are involved in biotic and abiotic stress in different crops, such as pepper [42], cotton [43], chrysanthemum [24], and rice [44]. In pineapple, 54 WRKY genes have been identified and reported by the authors of [30,34]. However, only limited research has been conducted to elucidate the functions of WRKY genes in pineapple [31,32]. Therefore, the current study was focused on cloning and transforming the *AcWRKY31* gene to investigate its functions in pineapple.

WRKY transcription factors (TFs) can be divided into three groups (I, II, and III) based on their motifs and domains. Members of group III contain a conserved WRKYGQK domain and a C_2HC zinc finger structure (C-X7-C-X23-H-X-C). Previous studies have shown that group III WRKY genes in Chinese rose [45], *Sorghum* [46], barley [47], maize [48], rice [49], *Arabidopsis* [50], and pineapple [30] share these same domains and motifs. In our study, *AcWRKY31* exhibited high expression profiles and relative expression levels in the pistils and ovules (Figure 1A,B). The phylogenetic analysis and the alignment of multiple protein sequences revealed that the *AcWRKY31* gene and its homologous genes in rice and *Arabidopsis* belong to the third subfamily from the WRKY transcription factor family, containing a conserved WRKYGQK domain and a C_2HC zinc finger structure (C-X7-C-X23-H-X-C). Additionally, *AcWRKY31* is exclusively localized in the nucleus and demonstrates transcriptional activation activity (Figure 2). Similarly, our expression profile findings of *AcWRKY31* were consistent with the transcriptomic profile and displayed the same gene structure and functions.

In recent years, abiotic stresses, particularly salt and drought, have severely affected crops [51,52]. Many researchers have investigated and reported on the role of WRKY genes in different crops. For instance, *GhWRKY68* reduces resistance to salt and drought by affecting the germination rate, survival rate, and stomatal closure of transgenic tobacco. [26]. *AtWRKY53* is involved in drought stress and leaf senescence in *Arabidopsis* by mediating stomatal movement and ABA induction [38,42]. In rice, *OsWRKY46*, *OsWRKY64*, and *OsWRKY113* are upregulated in BR IRGA 409 and involve iron toxicity [53]. Our current study demonstrated that *AcWRKY31-OE* transgenic pineapples exhibited adverse phenotypic characteristics compared to wild-type pineapples during drought and salt stress. The expression levels of enzyme- and stress-related marker genes, like *AcPOD1*, *AcABI5*, *AcABA1*, and *AcPR1* for salt stress and *AcCAT1*, *AcPR1*, *AcLOX4*, *AcABA1*, *AcRD22*, and *AcDREB2A* for drought stress, were reduced in all transgenic pineapple lines compared to WT plants (Figure 4). In addition, the heterologous overexpression of *AcWRKY31* in *Arabidopsis* also showed an increased sensitivity to salt and drought compared to the WT plants. Therefore, our findings indicate that the overexpression of *AcWRKY31* enhances the sensitivity to abiotic stresses, especially salt and drought, in both pineapple and *Arabidopsis*.

Plants must adapt to drastic environmental changes throughout their lifespan by altering their biochemical and physiological pathways. Among those challenges, attacks from pathogens, insects, pests, and weeds can cause significant damage to plants and reduce crop yields. Many researchers have reported the effects of WRKY TFs on different biotic stresses, for example, the overexpression of *AtWRKY28* conferred resistance to *Sclerotinia* and *Botrytis cinerea* in *Arabidopsis* [54], the overexpression of *AtWRKY75* promoted resistance to *Sclerotinia* and *Pseudomonas syringae* [55], and *AtWRKY70 Arabidopsis* increased the resistance to SA-mediated powdery mildew but rendered plants sensitive to JA-mediated black spot [56]. The *AtWRKY22* genes promote susceptibility to aphids and modulate salicylic acid and jasmonic acid signaling [57]. In pineapple, Zhou et al. stated that *AcWRKY28* mediated the activation of *AcCPK* genes and conferred drought and salt stress tolerance in transgenic pineapple and resistance to *Sclerotinia* in transgenic *Arabidopsis* [32]. In rice, *OsWRKY6* positively regulated resistance to *Xanthomonas oryzae* [58]. The overexpression of *OsWRKY31* and *OsWRKY13* enhanced the resistance to rice blast in rice [59,60]. The pineapple mealybug (*Dysmicoccus brevipes*) is one of the most common insects that affect plant quality and yield by sap-sucking and acts as a vector for the pineapple mealybug wilt-associated virus (PMWaV) [12]. Our result showed that the overexpression of *AcWRKY31* significantly enhances the resistance to mealybugs and *Sclerotinia sclerotiorum* infection. The RT-qPCR analysis of disease-related and hormone-related marker genes (*AtICS1*, *AtPDF1.2*, *AtPR1*, *AtERF1*, *AtLOX4*, and *AtABI5)* also indicated that *AcWRKY31* improved pineapple tolerance to biotic stress.

The WRKY family is also considered to play a crucial role in hormone signaling pathways, including ABA, BRs, ETH, JA, and SA, and can participate in plant defenses [61,62]. The ABA signaling pathway is mainly involved in the regulation of the abiotic stress response in plants [63]. The nuclear protein *GmWRKY12* was responsive to drought, salt, ABA, and salicylic acid (SA) stress [64]. The overexpression of *GhWRKY1* in *Arabidopsis* constitutively activated ABA biosynthesis genes, signaling genes, responsive genes, and drought-related maker genes and led to an enhanced tolerance to drought [65]. Under an exogenous hormone treatment, the expression level of *AcWRKY31* was reduced at 4 h, 8 h, and 12 h, then it instantly increased at 24 h and 48 h under an ABA treatment (Supplementary Figure S1). In addition, the phenotype of *AcWRKY31-OE Arabidopsis* lines in the ABA treatment showed a significantly reduced resistance to ABA than WT plants. During the salt and drought stress experiments, it was found that the expression levels of marker genes in ABA-related signaling pathways were significantly increased. This expression change may be due to the involvement of *AcWRKY31* in the ABA signaling pathway, which coordinates metabolic and physiological processes within plants and leads to changes in gene expression to adapt to alternative environments, but further research is needed. Therefore, we speculate that *AcWRKY31* may regulate plant stress response through ABA-mediated signaling pathways.

Altogether, based on transcriptome and experimental results, we conclude that *AcWRKY31* plays distinct roles in pineapple under both biotic and abiotic stress. Its reduction in plant response to drought and salt stress may be mediated by the ABA signaling pathway, while its increase in plant resistance to biological stress may be achieved by regulating the expression of disease-resistance-related genes (Figure 10).

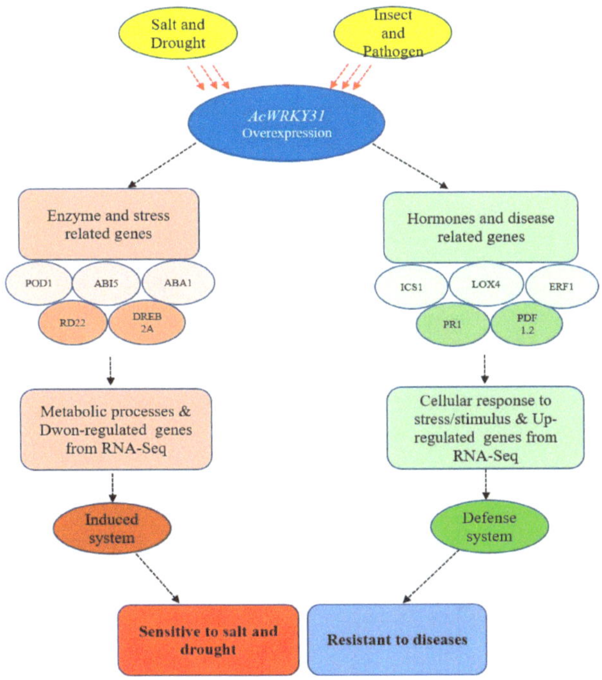

Figure 10. A role model of *AcWRKY31-OE* under biotic and abiotic stresses. *AcWRKY31* gene regulates enzyme- and stress-related genes and then negatively responds to salt and drought stresses while it positively responds to disease by regulating hormone- and disease-related genes.

4. Materials and Methods

4.1. Plant Materials and Pineapple Callus Induction

The pineapple (Tainong 11) plants and *Arabidopsis* (Columbia-0) seeds used in this experiment were obtained from Qin Lab, Center of Genomics Biotechnology, Fujian Agriculture and Forestry University, Fujian, China, where the experiments took place. For pineapple callus induction, we used the pineapple micro-propagation methods of Priyadarshani [66] with some modifications, especially for sterilization processes. Briefly, the explants such as leaf bases and crown meristems were soaked in disinfectant solution overnight to reduce contaminations and then washed with running tap water. In addition, the explants were surface sterilized with 75% ethanol for 1 min and then sterilized with 3–5% sodium hypochlorite solution for 8–10 min. Furthermore, the explants were cut into sizes of 0.5–1.5 cm and then sterilized with 0.2% of $HgCl_2$ solution for 10 min. The explants were washed with sterilized double-distilled water 3–5 times after each sterilization. After sterilizations, the samples were air-dried and cultured in full MS solid medium (with pH of −5.8) containing 30% sucrose, 4 mg/L of BAP, and 0.2 mg/L of NAA for callus induction. The callus was subcultured in full MS solid medium containing 1 mg/L of BAP and 0.2 mg/L of NAA. All cultures were kept in a tissue culture room under the following controlled conditions: 8 h/16 h dark and light cycle, light intensity of 3000 lux, and temperature of ±26 °C. All the sterilization and cultural processes were carried out in an aseptic condition.

4.2. Vector Construction, Gene Transformation of *AcWRKY31*, and Transgenic Plant Selection in Pineapple

The peptide sequences of *AcWRKY31* (*Aco000358.1*) and its homologous genes from rice were downloaded from the *Phytozome13* database "https://phytozome-next.jgi.doe.gov/ (accessed on 20 September 2021)" and those of *Arabidopsis* were downloaded from TAIR "https://www.arabidopsis.org/ (accessed on 20 September 2021)". The CDS sequence (containing 1098 bps) and genome sequence (containing 2068 bps) of *AcWRKY31* were downloaded from the *Phytozome13* database, and the gene structure was displayed by GSDS 2.0. The protein sequences of *AcWRKY31* and its two homologous genes from rice and *Arabidopsis* were used, and the alignment of multiple sequences was carried out using DNAMAN V6.0 software. The specific primers were designed using SnapGene^{TM}1.1.3 software. The total RNA from young pineapple leaves was extracted and cloned to the cDNA. The cDNA and gene-specific primers of *AcWRKY31* were used for PCR amplification to obtain the target DNA. After purification of the PCR products, the target DNA was cloned into the pENTER/D-TOPO vector and then recombined into the final pGWB502 vector using LR Clonase II enzyme mix (Thermo Fisher Scientific, Waltham, MA USA). The plasmids from the final vectors were extracted and transformed into *Agrobacterium* (GV3101) strains. The *Agrobacterium* with targeted genes was transformed into pineapple callus. In the transformation process, positive pineapple callus selection and plant screening were carried out as explained by Priyadarshani [37,67] with some modifications, especially in regards to selection time. Time for selection and antibiotic concentration were optimized according to the gene. Three repeated selections were carried out alternatively in the selection and normal medium. Following selection, the resilient plantlets were excised and cultivated on an MS solid medium supplemented with 0.2 mg/L NAA to stimulate root growth. Putative transgenic plants underwent DNA extraction for subsequent PCR amplification. DNA samples yielding positive PCR outcomes were forwarded to Bosun Biotechnology (Shanghai, China) Co., LTD for sequencing.

4.3. Analysis of Salt and Drought Stress in *AcWRKY31*-Overexpressed Pineapple

To analyze the response of transgenic and wild-type (WT) pineapple plants to salt and drought stress, plantlets at comparable growth stages were selected for experimentation. WT and transgenic pineapple plantlets were transplanted into pots with composted soil and placed in a greenhouse at 30 °C ± 2 °C for four months until they grew into adult pineapple plants. After four months, six *AcWRKY31-OE* transgenic plants and WT pineapple plants were treated with 100 mL of 400 mM NaCl solution for salt stress and 400 mM mannitol solution for drought stress. The phenotypic characteristics were noted and recorded daily. In addition, leaf samples of treated WT and *AcWRKY31-OE* transgenic pineapple were collected at 0 h, 12 h, and 24 h, respectively. After irrigating for 3 weeks, the phenotypic characters of treated WT and transgenic pineapple plants were compared. For the molecular assays, the total RNA was isolated from these treated samples and reverse-transcribed into cDNA for further experiments. Selected marker genes were analyzed using RT-qPCR with specific primers (Supplementary Table S1).

4.4. Analysis of Biotic Stress in *AcWRKY31-OE* Transgenic Pineapple

To investigate the role of *AcWRKY31* in biotic stress response, pineapple mealybugs (*Dysmicoccus brevipes*) from a previously infected pineapple plant in the greenhouse were used to inoculate the wild-type (WT) and transgenic pineapple plants. A known number of mealybugs were inoculated onto the leaves of four-month-old, healthy wild-type (WT) and transgenic pineapple plants. Plants were maintained in the greenhouse at 30 °C ± 2 °C and were observed weekly. After four months, disease symptoms, plant growth, and insect colonization were visually inspected and recorded. The statistical data of mealybugs colonies were counted and calculated using IBM SPSS Statistics 24.0.0 software.

4.5. Genetic Transformation of AcWRKY31 and AtWRKY53 in Arabidopsis

To confirm and validate the functions of *AcWRKY31*, we transformed *AcWRKY31*, its homologous gene *AtWRKY53*, and knockoff mutant *Atwrky53* into *Arabidopsis* using the floral dip method [68]. Inflorescences were immersed in *Agrobacterium* solution containing the targeted plasmids: p35S:*AcWRKY31* and p35S: *AtWRKY53* [68,69]. Subsequently, the infected plants were allowed to grow until the seeds were ready for harvesting. The seeds from the T_0 generation were germinated in soil and selected by spraying them with Basta (1:1000 dilution of water) to obtain transgenic plants. DNA was extracted from selected putative plants by using the CTAB method [70] to identify the positive plants for T_1 generation through PCR. The second-generation seeds were obtained by repeated screening and identification, and they were used for subsequent studies.

4.6. Analysis of Abiotic Treatments in Transgenic Arabidopsis

The previous study reported that the expression of WRKY TFs in pineapple can be affected by temperature stresses (4 °C and 45 °C), drought, and salt stresses [31]. To understand the regulation of heterologous overexpression of *AcWRKY31* under abiotic stress, the wild-type and T_2 generation of transgenic *Arabidopsis* seeds were treated with 100 mM–150 mM of NaCl for salt stress, 250 mM–300 mM of mannitol for drought stress, and 0.25–0.75 μM of ABA for ABA stress. After 7 days, the germination rates, fresh weight, and root length were investigated. To analyze the expression level of *AcWRKY31*, RT-qPCR analysis was performed for salt and drought stress with stress-related marker genes. All the experiments were conducted with three replicates.

4.7. Analysis of Sclerotinia sclerotiorum Inoculation on Transgenic Arabidopsis

To examine the functions of *AcWRKY31* in response to pathogens, WT and three independent lines of *AcWRKY31-OE*, *AtWRKY53-OE*, *Atwrky53*, *Atwrky53-com*, and *Atwrky53-AcWRKY31* T_2-generation *Arabidopsis* seeds were used. The preserved fungal strain of *Sclerotinia sclerotiorum* was sub-cultured on potato dextrose agar medium for the production of new hyphae. Briefly, five seeds from each line were planted in plastic pots with composted soil for germination. Leaves from 3-week-old transgenic and WT *Arabidopsis* plants were inoculated with the same quantity of *Sclerotinia sclerotiorum.* The infected area was analyzed after 24 h with DAB staining to analyze the accumulation of reactive oxygen species. For DAB staining, leaves were immersed in 50 mL of DAB aqueous solution (0.05 g DAB; ddH_2O 50 mL) overnight in darkness. Dyed leaves were de-colorized using 75% ethanol in a water bath for 15 min. The lysis areas were recorded by taking photographs and measured using ImageJ 1.51j/Java 1.8.0-112(64-bit) software [71]. The data were analyzed using the Student's *t*-test and GraphPad Prism 8.3.0 (538). All the experiments were repeated three times.

4.8. Analysis of Transcriptomic Data

Total RNA was extracted from three biological replicates of wild-type (WT) and *AcWRKY31*-overexpressing (*AcWRKY31-OE*) pineapple leaves using the Trizol method (Invitrogen, Carlsbad, CA, USA) following the manufacturer's instructions. Raw reads and adapter sequences were processed to remove ambiguities and adapters using TRIMMOMATIC v 0.3 software [72]. The RNA libraries were sequenced by the Novogene company on a HiSeq2500 sequencing instrument using 150 bp paired-end protocols. The processed reads were aligned to the reference genome using Tophat v2.1.1 software with default parameters. Then, the transcripts were quantified and assembled according to Cufflinks, and the differently expressed genes (DEGs) were identified. Differential expression was determined based on a log2 fold change $-1\leq$ and $\geq+1$ and a value of FDR ≤ 0.05) [73]. Kyoto Encyclopedia of Genomes and Genes (KEGG) and Gene Ontology (GO) analyses of DEGs were carried out using TBtools v1.09 software [74] and R package UpSet v1.0.0. The promoter region analysis was performed through the online PlantCARE database (http://bioinformatics.psb.ugent.be/webtools/plantcare/html/) (accessed on 13 March 2024). RT-qPCR analysis of six selected genes from up- and downregulated genes was performed to actuate the function of *AcWRKY31* with the RNA-Seq data. The raw data sequence in Fastq format was deposited in the China National Center for Bioinformation Database (https://bigd.big.ac.cn; BioProject ID: PRJCA019756) (accessed on 13 September 2023).

4.9. *Quantitative Real-Time PCR Analysis*

The cDNA was synthesized from extracted total RNA according to the instructions of AMV reverse-transcriptase kit (Takara, Shiga, Japan) [75]. RT-qPCR was carried out using SYBR Premix Ex Taq II (Takara, Japan) system and Bio-Rad Real-Time PCR system. The reaction was conducted in a total volume of 20 µL with 10 µL of enzyme mix, 7.5 µL of nuclease-free water, 1.5 µL of cDNA, and 0.5 µL of each specific forward and reverse primer. The RT-qPCR was performed by using given conditions (95 °C—30 s, 95 °C—5 s, 60 °C—34 s, and 95 °C—15 s) with 40 cycles [76]. The experiment was performed in triplicate. According to the comparative threshold period ($2^{-\Delta\Delta Ct}$) protocol, the level of relative expression in stress-related genes was analyzed and confirmed [77].

5. Conclusions

The outcomes of our current research highlight the notable expression patterns of *AcWRKY31*, particularly in the pistil and ovule of pineapple. Notably, our investigations reveal that the upregulation of *AcWRKY31* in both pineapple and *Arabidopsis* plants is correlated with a reduced tolerance to salt and drought stress, alongside a heightened resistance to pests and pathogens. These findings underscore the intricate response mechanisms of pineapple to diverse stressors. Consequently, our study yields significant insights crucial for the development of advanced biotechnological interventions and breeding approaches aimed at enhancing the resilience and productivity of pineapple varieties.

Supplementary Materials: The following supporting information can be downloaded at: https://www.mdpi.com/article/10.3390/plants13131850/s1. Figure S1: Relative expression levels of *AcWRKY31* in different hormone treatments (A) Abscisic Acid (ABA) (B) Ethylene (ETH) (C) Jasmonate Acid (JA) and (D) Salicylic Acid (SA). Figure S2: Analysis of transcriptional activation activity of *AcWRKY31* (A) The growth of *AcWRKY31*-BD/AD, negative and positive control yeast on SD-Leu-Trp. (B) The growth of *AcWRKY31*-BD/AD and negative-positive co-transformed yeast on SD-Leu-Trp-Ade-His+X-α-gal medium. Figure S3: Relative expression levels of Up Regulated Genes and Down Regulated Genes from RNA sequence analysis of WT and transgenic pineapples lines, the error bars indicate ±SD (n = 3). Figure S4: Salt treatment on *AcWRKY53-OE Arabidopsis* (A) Observation of germination rate and root length, bar = 1 cm (B) Statistical results of fresh weight and root length of *AcWRKY53-OE* and WT *Arabidopsis* after 7 days of NaCl treatment, the error bars indicate ±SD (n = 3) the asterisks indicate the different significances based on Student's t-test (** $p < 0.01$, * $p < 0.05$). Figure S5: Drought treatment on *AcWRKY53-OE Arabidopsis* (A) Observation of germination rate and root length, bar = 1 cm (B) Statistical results of fresh weight and root length of *AcWRKY53-OE* and WT *Arabidopsis* after 7 days of Mannitol treatment, the error bars indicate ±SD (n = 3) the asterisks indicate the different significances based on Student's t-test (** $p < 0.01$, * $p < 0.05$). Figure S6: ABA treatment on *Arabidopsis* (A) Observation of germination rate and root length of *AcWRKY53-OE* and WT *Arabidopsis* (bar = 1 cm) (B) Observation of germination rate and root length of *AtWRKY53* mutant, complementary and WT *Arabidopsis* (bar = 1 cm); Statistical results of fresh weight and root length after 7 days (Notes: the error bars indicate ±SD (n = 3) the asterisks indicate the different significances based on Student's t-test (** $p < 0.01$, * $p < 0.05$)). Table S1: Primer sequences. Table S2: Up and down- regulated Differential Expressed Genes.

Author Contributions: Conceptualization, M.H.W. and T.L.; methodology, M.H.W., Y.Q. and L.W.; software, H.C. (Han Cheng), Y.L. and X.W.; validation, Q.Z. and M.A.M.; formal analysis, C.L., G.C. and D.H.; resources, Y.Q.; data curation, M.H.W. and T.L.; writing—original draft preparation, M.H.W.; writing—review and editing, S.V.G.N.P., H.C. (Hanyang Cai), M.A., Y.Q. and L.W.; supervision, Y.Q. and L.W.; project administration, Y.Q. and L.W.; funding acquisition, Y.Q. and L.W. All authors have read and agreed to the published version of the manuscript.

Funding: This work was supported by the Natural Science Foundation of Guangxi (2022GXNSFBA035523) and the China Postdoctoral Science Foundation (2022MD713731) through a grant to LW, "the Science and Technology Major Project of Guangxi" (Gui Ke AA22068096), "Project of Guangxi featured fruit innovation team on pineapple breeding and cultivation post under the national modern agricultural industry technology system" (nycytxgxcxtd-17-05), the National Natural Science Foundation of China (32170352 and 31970333), the Excellent Youth Foundation of Fujian Agriculture and Forestry University through a grant to H.C. (xjq202108), and the Project for Scientific and Technological Innovation of Fujian Agriculture and Forestry University (CXZX2020050A).

Data Availability Statement: All analytical data of this study are described in the article and its additional files.

Acknowledgments: A special thanks to the editor and reviewers for their suggestions and helpful comments on our original manuscript. Special thanks to foundations for their financial supports on this research.

Conflicts of Interest: The authors declare no conflicts of interest.

Abbreviations

WT	Wild-type
TFs	Transcription factors
OE	Overexpression
DEGs	Differentially expressed genes
PCR	Polymerase chain reaction
RT-qPCR	Quantitative real-time polymerase chain reaction
PGRs	Plant growth regulators
DA-6	Diethyl aminoethyl hexanoate
COS	Chitosan oligosaccharide
ABA	Abscisic acid
BRs	Brassinosteroids
ETH	Ethylene
JA	Jasmonate acid
SA	Salicylic acid
MS	Murishige and Skoog medium
BAP	6-Benzylaminopurine
NAA	Naphthalene acetic acid
GFP	Green fluorescent protein
DNA	Deoxyribonucleic acid
cDNA	Complementary DNA
NaCl	Sodium chloride
ROS	Relative oxygen species
Cotton:	
GhHBA	High level of beta-amylase activity
Tobacco:	
NtNCED1	9-Cis-Epoxycarotenoid Dioxygenase 1
NtDREB3	Dehydration Response Element-Binding protein 3
NtLEA5	Late Embryogenesis Abundant-like 5
Pineapple:	
AcPOD	Peroxidase
AcABI5	ABA Insensitive 5
AcABA1	ABA Deficient 1

AcPR1	Pathogenesis-Related Gene 1
AcCAT1	Catalase 1
AcLOX4	Lipoxygenase 4
AcRD22	Responsive to Desiccation 22
AcDREB2A	Dehydration Response Element-Binding protein 2
AcCPK	Calcium-dependent Protein Kinase
Arabidopsis:	
AtSOS1	Salt Overly Sensitive 1
AtHKT1	High-Affinity K+ Transporter 1
AtICS1	Isochorismate Synthase 1 (SA related)
AtPDF1.2	Plant Defensin 1.2
AtPR1	Pathogenesis-Related Gene 1
AtLOX4	Lipoxygenase 4 (JA related)
AtERF1	Ethylene Response Factor 1
AtABI5	ABA Insensitive 5

References

1. Sinha, A.K.; Jaggi, M.; Raghuram, B.; Tuteja, N. Mitogen-activated protein kinase signaling in plants under abiotic stress. *Plant Signal. Behav.* **2011**, *6*, 196–203. [CrossRef]
2. Rodziewicz, P.; Swarcewicz, B.; Chmielewska, K.; Wojakowska, A.; Stobiecki, M. Influence of abiotic stresses on plant proteome and metabolome changes. *Acta Physiol. Plant.* **2013**, *36*, 1–19. [CrossRef]
3. Gong, Z.; Xiong, L.; Shi, H.; Yang, S.; Herrera-Estrella, L.R.; Xu, G.; Chao, D.Y.; Li, J.; Wang, P.Y.; Qin, F.; et al. Plant abiotic stress response and nutrient use efficiency. *Sci. China Life Sci.* **2020**, *63*, 635–674. [CrossRef] [PubMed]
4. Boisson-Dernier, A.; Roy, S.; Kritsas, K.; Grobei, M.A.; Jaciubek, M.; Schroeder, J.I.; Grossniklaus, U. Disruption of the pollen-expressed FERONIA homologs ANXUR1 and ANXUR2 triggers pollen tube discharge. *Development* **2009**, *136*, 3279–3288. [CrossRef] [PubMed]
5. Arif, Y.; Singh, P.; Siddiqui, H.; Bajguz, A.; Hayat, S. Salinity induced physiological and biochemical changes in plants: An omic approach towards salt stress tolerance. *Plant Physiol. Biochem.* **2020**, *156*, 64–77. [CrossRef]
6. Shao, H.B.; Chu, L.Y.; Shao, M.A.; Jaleel, C.A.; Mi, H.M. Higher plant antioxidants and redox signaling under environmental stresses. *Comptes Rendus Biol.* **2008**, *331*, 433–441. [CrossRef] [PubMed]
7. Gu, R.; Song, X.F.; Liu, X.F.; Yan, L.Y.; Zhou, Z.Y.; Zhang, X.L. Genome-wide analysis of CsWOX transcription factor gene family in cucumber (*Cucumis sativus* L.). *Sci. Rep.* **2020**, *10*, 6216. [CrossRef]
8. Evans, M.; Kermicle, J.L. Interaction between maternal effect and zygotic effect mutations during maize seed development. *Genetics* **2001**, *159*, 303–315. [CrossRef] [PubMed]
9. Ranjan, A.; Westrick, N.M.; Jain, S.; Piotrowski, J.S.; Ranjan, M.; Kessens, R.; Stiegman, L.; Grau, C.R.; Conley, S.P.; Smith, D.L.; et al. Resistance against Sclerotinia sclerotiorum in soybeans involves reprogramming of the phenylpropanoid pathway and up-regulation of antifungal activity targeting ergosterol biosynthesis. *Plant Biotechnol. J.* **2019**, *17*, 1567–1581. [CrossRef]
10. Caceres, M.; Hidalgo, W.; Stashenko, E.; Torres, R.; Ortiz, C. Essential Oils of Aromatic Plants with Antibacterial, Anti-Biofilm and Anti-Quorum Sensing Activities against Pathogenic Bacteria. *Antibiotics* **2020**, *9*, 147. [CrossRef]
11. Santa-Cecília, L.V.C.; Prado, E.; Souza, B. Probing behavior of Dysmicoccus brevipes mealybug in pineapple plants1. *Pesqui. Agropecu. Trop.* **2016**, *46*, 458–463. [CrossRef]
12. Rohrbach, K.G.; Johnson, M.W. Pests, diseases and weeds. In *The Pineapple: Botany, Production and Uses*; CABI Publishing: Wallingford, UK, 2003; pp. 203–251.
13. Bolton, M.D.; Thomma, B.P.; Nelson, B.D. *Sclerotinia sclerotiorum* (Lib.) de Bary: Biology and molecular traits of a cosmopolitan pathogen. *Mol. Plant Pathol.* **2006**, *7*, 1–16. [CrossRef]
14. Zhou, F.; Zhang, X.L.; Li, J.L.; Zhu, F.X. Dimethachlon Resistance in *Sclerotinia sclerotiorum* in China. *Plant Dis.* **2014**, *98*, 1221–1226. [CrossRef] [PubMed]
15. Phukan, U.J.; Jeena, G.S.; Shukla, R.K. WRKY Transcription Factors: Molecular Regulation and Stress Responses in Plants. *Front. Plant Sci.* **2016**, *7*, 760. [CrossRef]
16. Xu, Y.H.; Sun, P.W.; Tang, X.L.; Gao, Z.H.; Zhang, Z.; Wei, J.H. Genome-wide analysis of WRKY transcription factors in *Aquilaria sinensis* (Lour.) Gilg. *Sci. Rep.* **2020**, *10*, 3018. [CrossRef]
17. Xie, Z.; Zhang, Z.L.; Zou, X.; Huang, J.; Ruas, P.; Thompson, D.; Shen, Q.J. Annotations and functional analyses of the rice WRKY gene superfamily reveal positive and negative regulators of abscisic acid signaling in aleurone cells. *Plant Physiol.* **2005**, *137*, 176–189. [CrossRef]
18. Rushton, P.J.; Bokowiec, M.T.; Han, S.; Zhang, H.; Brannock, J.F.; Chen, X.; Laudeman, T.W.; Timko, M.P. Tobacco transcription factors: Novel insights into transcriptional regulation in the Solanaceae. *Plant Physiol.* **2008**, *147*, 280–295. [CrossRef]
19. Chen, L.; Song, Y.; Li, S.; Zhang, L.; Zou, C.; Yu, D. The role of WRKY transcription factors in plant abiotic stresses. *Biochim. Biophys. Acta* **2012**, *1819*, 120–128. [CrossRef] [PubMed]

20. Chang, X.; Yang, Z.; Zhang, X.; Zhang, F.; Huang, X.; Han, X. Transcriptome-wide identification of WRKY transcription factors and their expression profiles under different stress in *Cynanchum thesioides*. *PeerJ* **2022**, *10*, e14436. [CrossRef]
21. Ma, Y.; Xue, H.; Zhang, F.; Jiang, Q.; Yang, S.; Yue, P.; Wang, F.; Zhang, Y.; Li, L.; He, P.; et al. The miR156/SPL module regulates apple salt stress tolerance by activating MdWRKY100 expression. *Plant Biotechnol. J.* **2021**, *19*, 311–323. [CrossRef]
22. Chen, C.; Shang, X.; Sun, M.; Tang, S.; Khan, A.; Zhang, D.; Yan, H.; Jiang, Y.; Yu, F.; Wu, Y.; et al. Comparative Transcriptome Analysis of Two Sweet Sorghum Genotypes with Different Salt Tolerance Abilities to Reveal the Mechanism of Salt Tolerance. *Int. J. Mol. Sci.* **2022**, *23*, 2272. [CrossRef] [PubMed]
23. Song, Y.; Li, J.; Sui, Y.; Han, G.; Zhang, Y.; Guo, S.; Sui, N. The sweet sorghum SbWRKY50 is negatively involved in salt response by regulating ion homeostasis. *Plant Mol. Biol.* **2020**, *102*, 603–614. [CrossRef] [PubMed]
24. Li, P.; Song, A.; Gao, C.; Wang, L.; Wang, Y.; Sun, J.; Jiang, J.; Chen, F.; Chen, S. Chrysanthemum WRKY gene CmWRKY17 negatively regulates salt stress tolerance in transgenic chrysanthemum and Arabidopsis plants. *Plant Cell Rep.* **2015**, *34*, 1365–1378. [CrossRef] [PubMed]
25. Zhao, J.; He, Y.; Zhang, H.; Wang, Z. Advances in the molecular regulation of seed germination in plants. *Seed Biol.* **2024**, *3*, e006. [CrossRef]
26. Jia, H.; Wang, C.; Wang, F.; Liu, S.; Li, G.; Guo, X. GhWRKY68 reduces resistance to salt and drought in transgenic *Nicotiana benthamiana*. *PLoS ONE* **2015**, *10*, e0120646, Erratum in *PLoS ONE* **2019**, *14*, e0213540. [CrossRef] [PubMed]
27. Sun, J.; Hu, W.; Zhou, R.; Wang, L.; Wang, X.; Wang, Q.; Feng, Z.; Li, Y.; Qiu, D.; He, G.; et al. The *Brachypodium distachyon* BdWRKY36 gene confers tolerance to drought stress in transgenic tobacco plants. *Plant Cell Rep.* **2015**, *34*, 23–35. [CrossRef]
28. Cai, H.; Yang, S.; Yan, Y.; Xiao, Z.; Cheng, J.; Wu, J.; Qiu, A.; Lai, Y.; Mou, S.; Guan, D.; et al. CaWRKY6 transcriptionally activates CaWRKY40, regulates *Ralstonia solanacearum* resistance, and confers high-temperature and high-humidity tolerance in pepper. *J. Exp. Bot.* **2015**, *66*, 3163–3174. [CrossRef]
29. Wang, F.; Chen, H.W.; Li, Q.T.; Wei, W.; Li, W.; Zhang, W.K.; Ma, B.; Bi, Y.D.; Lai, Y.C.; Liu, X.L.; et al. GmWRKY27 interacts with GmMYB174 to reduce the expression of GmNAC29 for stress tolerance in soybean plants. *Plant J.* **2015**, *83*, 224–236. [CrossRef] [PubMed]
30. Xie, T.; Chen, C.; Li, C.; Liu, J.; Liu, C.; He, Y. Genome-wide investigation of WRKY gene family in pineapple: Evolution and expression profiles during development and stress. *BMC Genom.* **2018**, *19*, 490. [CrossRef]
31. Huang, Y.; Chen, F.; Chai, M.; Xi, X.; Zhu, W.; Qi, J.; Liu, K.; Ma, S.; Su, H.; Tian, Y.; et al. Ectopic Overexpression of Pineapple Transcription Factor AcWRKY31 Reduces Drought and Salt Tolerance in Rice and Arabidopsis. *Int. J. Mol. Sci.* **2022**, *23*, 6269. [CrossRef]
32. Zhou, Q.; Priyadarshani, S.V.G.N.; Qin, R.; Cheng, H.; Luo, T.; Wai, M.H.; Mohammadi, M.A.; Liu, Y.; Liu, C.; Cai, H.; et al. AcWRKY28-mediated activation of AcCPK genes confers salt tolerance in pineapple (*Ananas comosus*). *Hortic. Plant J.* **2024**, *10*, 398–412. [CrossRef]
33. Huang, X.; Rao, G.; Peng, X.; Xue, Y.; Hu, H.; Feng, N.; Zheng, D. Effect of plant growth regulators DA-6 and COS on drought tolerance of pineapple through bromelain and oxidative stress. *BMC Plant Biol.* **2023**, *23*, 180. [CrossRef] [PubMed]
34. Ming, R.; Wai, C.M.; Guyot, R. Pineapple Genome: A Reference for Monocots and CAM Photosynthesis. *Trends Genet.* **2016**, *32*, 690–696. [CrossRef]
35. Gao, Y.; Yao, Y.; Chen, X.; Wu, J.; Wu, Q.; Liu, S.; Guo, A.; Zhang, X. Metabolomic and transcriptomic analyses reveal the mechanism of sweet-acidic taste formation during pineapple fruit development. *Front. Plant Sci.* **2022**, *13*, 971506. [CrossRef] [PubMed]
36. Xie, T.; Zhang, J.; Luan, A.; Zhang, W.; Wu, J.; Cai, Z.; He, Y. Comparative transcriptome analysis of a fan-shaped inflorescence in pineapple using RNA-seq. *Genomics* **2021**, *113*, 3653–3665. [CrossRef]
37. Priyadarshani, S.; Cai, H.; Zhou, Q.; Liu, Y.; Cheng, Y.; Xiong, J.; Patson, D.L.; Cao, S.; Zhao, H.; Qin, Y. An Efficient Agrobacterium Mediated Transformation of Pineapple with GFP-Tagged Protein Allows Easy, Non-Destructive Screening of Transgenic Pineapple Plants. *Biomolecules* **2019**, *9*, 617. [CrossRef] [PubMed]
38. Sun, Y.; Yu, D. Activated expression of AtWRKY53 negatively regulates drought tolerance by mediating stomatal movement. *Plant Cell Rep.* **2015**, *34*, 1295–1306. [CrossRef]
39. Cai, J.; Liu, T.; Li, Y.; Ow, D.W. A C-terminal fragment of Arabidopsis OXIDATIVE STRESS 2 can play a positive role in salt tolerance. *Biochem. Biophys. Res. Commun.* **2021**, *556*, 23–30. [CrossRef]
40. Xiang, S.; Wu, S.; Zhang, H.; Mou, M.; Chen, Y.; Li, D.; Wang, H.; Chen, L.; Yu, D. The PIFs Redundantly Control Plant Defense Response against Botrytis cinerea in Arabidopsis. *Plants* **2020**, *9*, 1246. [CrossRef]
41. Grebner, W.; Stingl, N.E.; Oenel, A.; Mueller, M.J.; Berger, S. Lipoxygenase6-dependent oxylipin synthesis in roots is required for abiotic and biotic stress resistance of Arabidopsis. *Plant Physiol.* **2013**, *161*, 2159–2170. [CrossRef]
42. Zentgraf, U.; Laun, T.; Miao, Y. The complex regulation of WRKY53 during leaf senescence of Arabidopsis thaliana. *Eur. J. Cell Biol.* **2010**, *89*, 133–137. [CrossRef] [PubMed]
43. Ren, X.; Chen, Z.; Liu, Y.; Zhang, H.; Zhang, M.; Liu, Q.; Hong, X.; Zhu, J.K.; Gong, Z. ABO3, a WRKY transcription factor, mediates plant responses to abscisic acid and drought tolerance in Arabidopsis. *Plant J.* **2010**, *63*, 417–429. [CrossRef] [PubMed]
44. Jaglo-Ottosen, K.R.; Gilmour, S.J.; Zarka, D.G.; Schabenberger, O.; Thomashow, M.F. Arabidopsis CBF1 overexpression induces COR genes and enhances freezing tolerance. *Science* **1998**, *280*, 104–106. [CrossRef] [PubMed]

45. Yan, X.; Zhao, J.; Huang, W.; Liu, C.; Hao, X.; Gao, C.; Deng, M.; Wen, J. Genome-Wide Identification of WRKY Transcription Factor Family in Chinese Rose and Response to Drought, Heat, and Salt Stress. *Genes* **2024**, *15*, 800. [CrossRef]
46. Baillo, E.H.; Hanif, M.S.; Guo, Y.; Zhang, Z.; Xu, P.; Algam, S.A. Genome-wide Identification of WRKY transcription factor family members in sorghum (*Sorghum bicolor* (L.) Moench). *PLoS ONE* **2020**, *15*, e0236651. [CrossRef] [PubMed]
47. Zheng, J.; Zhang, Z.; Tong, T.; Fang, Y.; Zhang, X.; Niu, C.; Li, J.; Wu, Y.; Xue, D.; Zhang, X. Genome-Wide Identification of WRKY Gene Family and Expression Analysis under Abiotic Stress in Barley. *Agronomy* **2021**, *11*, 521. [CrossRef]
48. Wei, K.F.; Chen, J.; Chen, Y.F.; Wu, L.J.; Xie, D.X. Molecular phylogenetic and expression analysis of the complete WRKY transcription factor family in maize. *DNA Res.* **2012**, *19*, 153–164. [CrossRef]
49. Jeyasri, R.; Muthuramalingam, P.; Satish, L.; Adarshan, S.; Lakshmi, M.A.; Pandian, S.K.; Chen, J.-T.; Ahmar, S.; Wang, X.; Mora-Poblete, F.; et al. The Role of OsWRKY Genes in Rice When Faced with Single and Multiple Abiotic Stresses. *Agronomy* **2021**, *11*, 1301. [CrossRef]
50. Song, Y.; Gao, J. Genome-wide analysis of WRKY gene family in Arabidopsis lyrata and comparison with *Arabidopsis thaliana* and *Populus trichocarpa*. *Chin. Sci. Bull.* **2014**, *59*, 754–765. [CrossRef]
51. Fei, X.; Hou, L.; Shi, J.; Yang, T.; Liu, Y.; Wei, A. Patterns of Drought Response of 38 WRKY Transcription Factors of *Zanthoxylum bungeanum* Maxim. *Int. J. Mol. Sci.* **2018**, *20*, 68. [CrossRef]
52. Cheng, Y.; Wang, Y.; Sun, J.; Liao, Z.; Ye, K.; Hu, B.; Dong, C.; Li, Z.; Deng, F.; Wang, L.; et al. Unveiling the genomic blueprint of salt stress: Insights from *Ipomoea pes-caprae* L. *Seed Biol.* **2023**, *2*, 21. [CrossRef]
53. Viana, V.E.; Marini, N.; Finatto, T.; Ezquer, I.; Busanello, C.; Dos Santos, R.S.; Pegoraro, C.; Colombo, L.; Costa de Oliveira, A. Iron excess in rice: From phenotypic changes to functional genomics of WRKY transcription factors. *Genet. Mol. Res.* **2017**, *16*, gmr16039694. [CrossRef] [PubMed]
54. Windram, O.; Madhou, P.; McHattie, S.; Hill, C.; Hickman, R.; Cooke, E.; Jenkins, D.J.; Penfold, C.A.; Baxter, L.; Breeze, E.; et al. Arabidopsis defense against *Botrytis cinerea*: Chronology and regulation deciphered by high-resolution temporal transcriptomic analysis. *Plant Cell* **2012**, *24*, 3530–3557. [CrossRef] [PubMed]
55. Encinas-Villarejo, S.; Maldonado, A.M.; Amil-Ruiz, F.; de los Santos, B.; Romero, F.; Pliego-Alfaro, F.; Munoz-Blanco, J.; Caballero, J.L. Evidence for a positive regulatory role of strawberry (*Fragaria* × *ananassa*) Fa WRKY1 and Arabidopsis At WRKY75 proteins in resistance. *J. Exp. Bot.* **2009**, *60*, 3043–3065. [CrossRef] [PubMed]
56. Li, J.; Brader, G.; Kariola, T.; Palva, E.T. WRKY70 modulates the selection of signaling pathways in plant defense. *Plant J.* **2006**, *46*, 477–491. [CrossRef] [PubMed]
57. Kloth, K.J.; Wiegers, G.L.; Busscher-Lange, J.; van Haarst, J.C.; Kruijer, W.; Bouwmeester, H.J.; Dicke, M.; Jongsma, M.A. AtWRKY22 promotes susceptibility to aphids and modulates salicylic acid and jasmonic acid signaling. *J. Exp. Bot.* **2016**, *67*, 3383–3396. [CrossRef]
58. Hwang, S.H.; Yie, S.W.; Hwang, D.J. Heterologous expression of OsWRKY6 gene in *Arabidopsis* activates the expression of defense-related genes and enhances resistance to pathogens. *Plant Sci.* **2011**, *181*, 316–323. [CrossRef] [PubMed]
59. Zhang, J.; Peng, Y.; Guo, Z. Constitutive expression of pathogen-inducible OsWRKY31 enhances disease resistance and affects root growth and auxin response in transgenic rice plants. *Cell Res.* **2008**, *18*, 508–521. [CrossRef] [PubMed]
60. Qiu, D.; Xiao, J.; Xie, W.; Liu, H.; Li, X.; Xiong, L.; Wang, S. Rice Gene Network Inferred from Expression Profiling of Plants Overexpressing OsWRKY13, a Positive Regulator of Disease Resistance. *Mol. Plant* **2008**, *1*, 538–551. [CrossRef]
61. Dietz, K.J.; Vogel, M.O.; Viehhauser, A. AP2/EREBP transcription factors are part of gene regulatory networks and integrate metabolic, hormonal, and environmental signals in stress acclimation and retrograde signaling. *Protoplasma* **2010**, *245*, 3–14. [CrossRef]
62. Liu, Z.Q.; Yan, L.; Wu, Z.; Mei, C.; Lu, K.; Yu, Y.T.; Liang, S.; Zhang, X.F.; Wang, X.F.; Zhang, D.P. Cooperation of three WRKY-domain transcription factors WRKY18, WRKY40, and WRKY60 in repressing two ABA-responsive genes ABI4 and ABI5 in Arabidopsis. *J. Exp. Bot.* **2012**, *63*, 6371–6392. [CrossRef]
63. Khoso, M.A.; Hussain, A.; Ritonga, F.N.; Ali, Q.; Channa, M.M.; Alshegaihi, R.M.; Meng, Q.; Ali, M.; Zaman, W.; Brohi, R.D.; et al. WRKY transcription factors (TFs): Molecular switches to regulate drought, temperature, and salinity stresses in plants. *Front. Plant Sci.* **2022**, *13*, 1039329. [CrossRef]
64. Shi, W.Y.; Du, Y.T.; Ma, J.; Min, D.H.; Jin, L.G.; Chen, J.; Chen, M.; Zhou, Y.B.; Ma, Y.Z.; Xu, Z.S.; et al. The WRKY Transcription Factor GmWRKY12 Confers Drought and Salt Tolerance in Soybean. *Int. J. Mol. Sci.* **2018**, *19*, 4087. [CrossRef] [PubMed]
65. Hu, Q.; Ao, C.; Wang, X.; Wu, Y.; Du, X. GhWRKY1-like, a WRKY transcription factor, mediates drought tolerance in *Arabidopsis* via modulating ABA biosynthesis. *BMC Plant Biol.* **2021**, *21*, 458. [CrossRef]
66. Priyadarshani, S.V.G.N.; Hu, B.; Li, W.; Ali, H.; Jia, H.; Zhao, L.; Ojolo, S.P.; Azam, S.M.; Xiong, J.; Yan, M.; et al. Simple protoplast isolation system for gene expression and protein interaction studies in pineapple (*Ananas comosus* L.). *Plant Methods* **2018**, *14*, 95. [CrossRef]
67. He, Y.; Luan, A.; Wu, J.; Zhang, W.; Lin, W. Overcoming key technical challenges in the genetic transformation of pineapple. *Trop. Plants* **2023**, *2*, 6. [CrossRef]
68. Zhang, X.; Henriques, R.; Lin, S.S.; Niu, Q.W.; Chua, N.H. Agrobacterium-mediated transformation of *Arabidopsis thaliana* using the floral dip method. *Nat. Protoc.* **2006**, *1*, 641–646. [CrossRef] [PubMed]
69. Clough, S.J.; Bent, A.F. Floral dip a simplified method for Agrobacterium-mediated transformation of *Arabidopsis thaliana*. *Plant J.* **1998**, *16*, 735–743. [CrossRef] [PubMed]

70. Aydemir, B.C.; Ozmen, C.Y.; Kibar, U.; Mutaf, F.; Buyuk, P.B.; Bakir, M.; Ergul, A. Salt stress induces endoplasmic reticulum stress-responsive genes in a grapevine rootstock. *PLoS ONE* **2020**, *15*, e0236424. [CrossRef]
71. Bowler, C.; Benvenuto, G.; Laflamme, P.; Molino, D.; Probst, A.V.; Tariq, M.; Paszkowski, J. Chromatin techniques for plant cells. *Plant J.* **2004**, *39*, 776–789. [CrossRef]
72. Bolger, A.M.; Lohse, M.; Usadel, B. Trimmomatic: A flexible trimmer for Illumina sequence data. *Bioinformatics* **2014**, *30*, 2114–2120. [CrossRef] [PubMed]
73. Trapnell, C.; Roberts, A.; Goff, L.; Pertea, G.; Kim, D.; Kelley, D.R.; Pimentel, H.; Salzberg, S.L.; Rinn, J.L.; Pachter, L. Differential gene and transcript expression analysis of RNA-seq experiments with TopHat and Cufflinks. *Nat. Protoc.* **2012**, *7*, 562–578. [CrossRef] [PubMed]
74. Chen, C.; Chen, H.; Zhang, Y.; Thomas, H.R.; Frank, M.H.; He, Y.; Xia, R. TBtools: An Integrative Toolkit Developed for Interactive Analyses of Big Biological Data. *Mol. Plant* **2020**, *13*, 1194–1202. [CrossRef] [PubMed]
75. Roth, R.; Madhani, H.D.; Garcia, J.F. Total RNA Isolation and Quantification of Specific RNAs in Fission Yeast. *Methods Mol. Biol.* **2018**, *1721*, 63–72. [CrossRef] [PubMed]
76. Singh, M.; Goel, S.; Meeley, R.B.; Dantec, C.; Parrinello, H.; Michaud, C.; Leblanc, O.; Grimanelli, D. Production of Viable Gametes without Meiosis in Maize Deficient for an ARGONAUTE Protein. *Plant Cell* **2011**, *23*, 443–458. [CrossRef]
77. Sang, J.; Han, X.; Liu, M.; Qiao, G.; Jiang, J.; Zhuo, R. Selection and validation of reference genes for real-time quantitative PCR in hyperaccumulating ecotype of Sedum alfredo under different heavy metals stresses. *PLoS ONE* **2013**, *8*, e82927. [CrossRef]

plants

Article

Melatonin-Induced Chromium Tolerance Requires Hydrogen Sulfide Signaling in Maize

Xiaoxiao Yang [1,2], Qifeng Shi [3], Xinru Wang [3], Tao Zhang [1], Ke Feng [3], Guo Wang [3], Juan Zhao [3,*], Xiangyang Yuan [3,*] and Jianhong Ren [1,*]

[1] College of Life Sciences, Shanxi Agricultural University, Jinzhong 030800, China; nwafu_yxx@163.com (X.Y.); sxau_zt@163.com (T.Z.)
[2] State Key Laboratory of Soil Erosion and Dryland Farming on the Loess Plateau, College of Life Sciences, Northwest A&F University, Xianyang 712100, China
[3] College of Agriculture, Shanxi Agricultural University, Jinzhong 030800, China; sxau_sqf@126.com (Q.S.); sxau_wxr@126.com (X.W.); sxau_fk@126.com (K.F.); sxau_wg@126.com (G.W.)
* Correspondence: sxau_zj@126.com (J.Z.); yuanxiangyang200@sxau.edu.cn (X.Y.); renjh@sxau.edu.cn (J.R.)

Abstract: Both melatonin and hydrogen sulfide (H_2S) mitigate chromium (Cr) toxicity in plants, but the specific interaction between melatonin and H_2S in Cr detoxification remains unclear. In this study, the interaction between melatonin and H_2S in Cr detoxification was elucidated by measuring cell wall polysaccharide metabolism and antioxidant enzyme activity in maize. The findings revealed that exposure to Cr stress (100 μM $K_2Cr_2O_7$) resulted in the upregulation of *L-/D*-cysteine desulfhydrase (*LCD/DCD*) gene expression, leading to a 77.8% and 27.3% increase in endogenous H_2S levels in maize leaves and roots, respectively. Similarly, the endogenous melatonin system is activated in response to Cr stress. We found that melatonin had a significant impact on the relative expression of *LCD/DCD*, leading to a 103.3% and 116.7% increase in endogenous H_2S levels in maize leaves and roots, respectively. In contrast, NaHS had minimal effects on the relative mRNA expression of serotonin-Nacetyltransferase (*SNAT*) and endogenous melatonin levels. The production of H_2S induced by melatonin is accompanied by an increase in Cr tolerance, as evidenced by elevated gene expression, elevated cell wall polysaccharide content, increased pectin methylesterase activity, and improved antioxidant enzyme activity. The scavenging of H_2S decreases the melatonin-induced Cr tolerance, while the inhibitor of melatonin synthesis, p-chlorophenylalanine (*p*-CPA), has minimal impact on H_2S-induced Cr tolerance. In conclusion, our findings suggest that H_2S serves as a downstream signaling molecule involved in melatonin-induced Cr tolerance in maize.

Keywords: antioxidant; cell wall polysaccharide; crosstalk; H_2S; melatonin

Citation: Yang, X.; Shi, Q.; Wang, X.; Zhang, T.; Feng, K.; Wang, G.; Zhao, J.; Yuan, X.; Ren, J. Melatonin-Induced Chromium Tolerance Requires Hydrogen Sulfide Signaling in Maize. *Plants* **2024**, *13*, 1763. https://doi.org/10.3390/plants13131763

Academic Editors: Violetta Katarzyna Macioszek, Iwona Ciereszko and Andrzej K. Kononowicz

Received: 26 May 2024
Revised: 22 June 2024
Accepted: 25 June 2024
Published: 26 June 2024

1. Introduction

Chromium (Cr) is a prevalent heavy metal pollutant in agricultural areas, predominantly derived from industrial activities such as Cr ore smelting, leather tanning, electroplating, metal processing, and the production of paints and pigments. These activities contribute to the release of Cr-laden wastewater and exhaust gases, leading to contamination of soil and water sources [1,2]. This contamination poses a threat to food production safety, as it can enter the human body through the food chain, subsequently impacting human health [3]. The toxicity of Cr is dependent on its oxidation state, with trivalent Cr (Cr (III)) and hexavalent Cr (Cr (VI)) being the predominant forms found in soil. Cr (VI) is known to be more toxic than Cr (III) [4]. In China, a considerable portion of cultivable land, amounting to 1.26%, is presently at a notable risk of Cr pollution [5]. The severity of this contamination has led to the abandonment of 0.13% of arable land [6]. Therefore, an effective strategy to alleviate Cr toxicity in crop plants is essential.

The plant cell wall serves as the primary barrier preventing heavy metal ions from entering the cell, with cell wall polysaccharides being key components involved in the sequestration of heavy metals [7]. Cell wall polysaccharides, such as cellulose, hemicellulose, and pectin, possess functional groups such as phosphoric acid, aldehyde, and carboxyl groups that facilitate the binding of heavy metal ions [8,9]. These groups interact with metal cations to immobilize them within the cell wall, thereby limiting the penetration of metal ions into the cytoplasm. In addition, it was observed that the presence of reactive oxygen species (ROS) in plants was heightened in response to heavy metal stress, resulting in significant oxidative stress and subsequent cell damage. The research indicated that plants possess an antioxidant defense mechanism capable of regulating ROS levels [10,11]. This defense mechanism comprises the following two primary systems: enzymatic, which includes superoxide dismutase (SOD), catalase (CAT), and peroxidase (POD); and non-enzymatic, which encompasses ascorbic acid (ASA), glutathione (GSH), and flavonoids [12].

Melatonin was initially identified in the pineal gland of cattle in 1958, and subsequent research has revealed its widespread presence in both plants and animals [13]. In animals, melatonin serves various physiological and biochemical roles, such as regulating sleep patterns, antioxidative effects, involvement in stem cell differentiation, and anti-aging properties [14]. In plants, melatonin functions as a significant growth regulator and shares structural similarities with auxin, as an indole compound [15]. Melatonin plays a significant role in various physiological processes related to plant growth and development, encompassing plant morphogenesis, seed germination, and fruit ripening [16,17]. Additionally, melatonin is involved in responses to both biological and abiotic stresses, including salt stress, water stress, heavy metal toxicity, and pathogen stress [18–20]. Specifically, melatonin enhances plant tolerance to heavy metal stress through mechanisms such as regulation of the ASA–GSH cycle, augmentation of antioxidant enzyme activity, modulation of polyamine metabolism, and control of chelate synthesis [21,22]. Moreover, research has confirmed that melatonin plays a role in feedback mechanisms within plants, specifically in regulating various REDOX networks, such as ROS and active nitrogen (RNS), with a particular emphasis on the functions of nitric oxide (NO) and hydrogen peroxide (H_2O_2) [23].

In plants, H_2S plays a significant role in the growth, development, and response mechanisms of plants to both biological and abiotic stresses [24]. This is achieved through intricate interactions with plant hormones, ROS, and various signaling molecules [25,26]. Research has demonstrated that H_2S interacts with NO, H_2O_2, and mitogen-activated protein kinases (MAPKs) to increase plant resistance. Specifically, in the case of broad beans, H_2S is implicated in the closure of stomata under salt stress conditions, acting as a downstream signal of H_2O_2 [27]. Furthermore, H_2S serves as a downstream signaling molecule of both NO and H_2O_2 in facilitating white clover's tolerance to drought stress regulated by spermidine [28]. In cucumbers, the protein MAPK4 functions as a downstream element of H_2S, modulating the transcription of genes responsive to low temperatures and controlling stomatal movement during periods of low temperature stress [29]. These investigations indicate that H_2S engages in crosstalk with additional signaling molecules, exerting a significant regulatory influence in response to abiotic stress. Nevertheless, the potential involvement of H_2S in mediating melatonin-induced tolerance to Cr stress remains uncertain.

Maize (*Zea mays* L.), a significant staple and feed crop, exhibits diverse applications in the fields of nutrition, livestock sustenance, and various industrial sectors [30]. However, the crop's propensity to accumulate toxic metals poses a constraint on its overall productivity within the maize production domain [31]. H_2S has been demonstrated to play a role in melatonin-induced salt stress tolerance as a downstream signal [32], though its involvement in melatonin-induced Cr stress tolerance remains uncertain. Additionally, H_2S and melatonin share common physiological functions in the regulation of plant tolerance to environmental stress, particularly in their involvement in the plant ROS signaling pathway [12,33]. In this study, we hypothesized that H_2S may serve as a downstream signaling molecule involved in melatonin-induced Cr tolerance in maize. This study specifically examines the impact of H_2S and melatonin on the regulation of cell wall Cr-binding ability and mitigation of oxidative stress.

2. Results

2.1. Response of Endogenous H_2S and Melatonin to Cr Stress

Treatment with $K_2Cr_2O_7$ resulted in a significant upregulation of genes involved in the synthesis of H_2S (*LCD1, LCD2, DCD1,* and *DCD2*) and melatonin (*T5H, SNAT,* and *ASMT*), leading to an increase in endogenous levels of both H_2S and melatonin. Additionally, exogenous melatonin notably enhanced the transcription of genes related to L-/D-cysteine desulphurase (L-/D-CD) synthesis, and the endogenous content of H_2S (Figure 1). Conversely, NaHS did not have a significant impact on the expression of melatonin biosynthesis genes or the endogenous levels of melatonin (Figure 2).

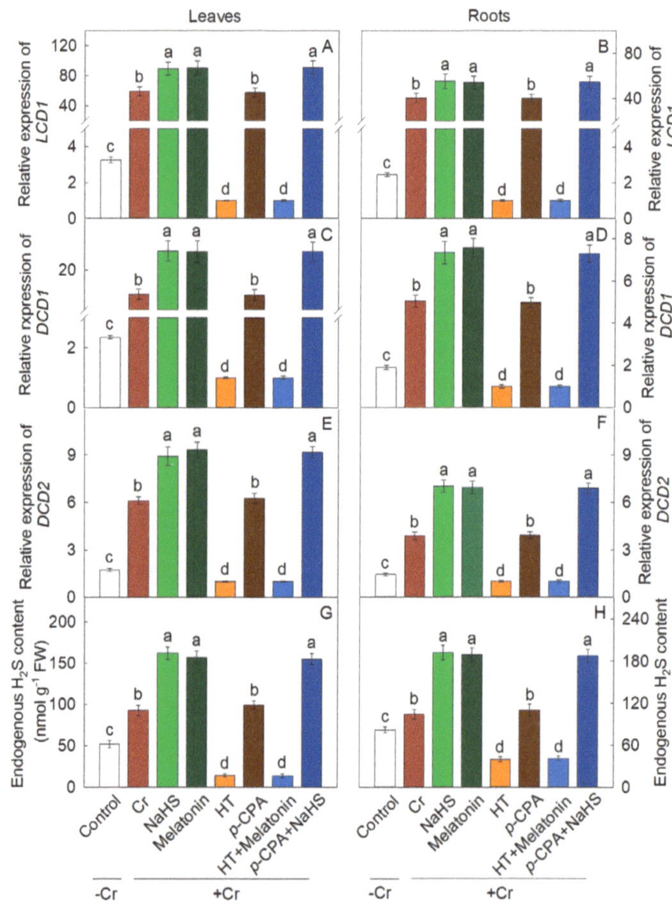

Figure 1. The effect of NaHS and melatonin on the expression of H_2S biosynthesis genes (**A–F**) and H_2S levels (**G,H**) in maize leaves and roots in response to Cr-induced stress. Different letters indicate statistically significant differences. Values are presented as mean ± SE (n = 3).

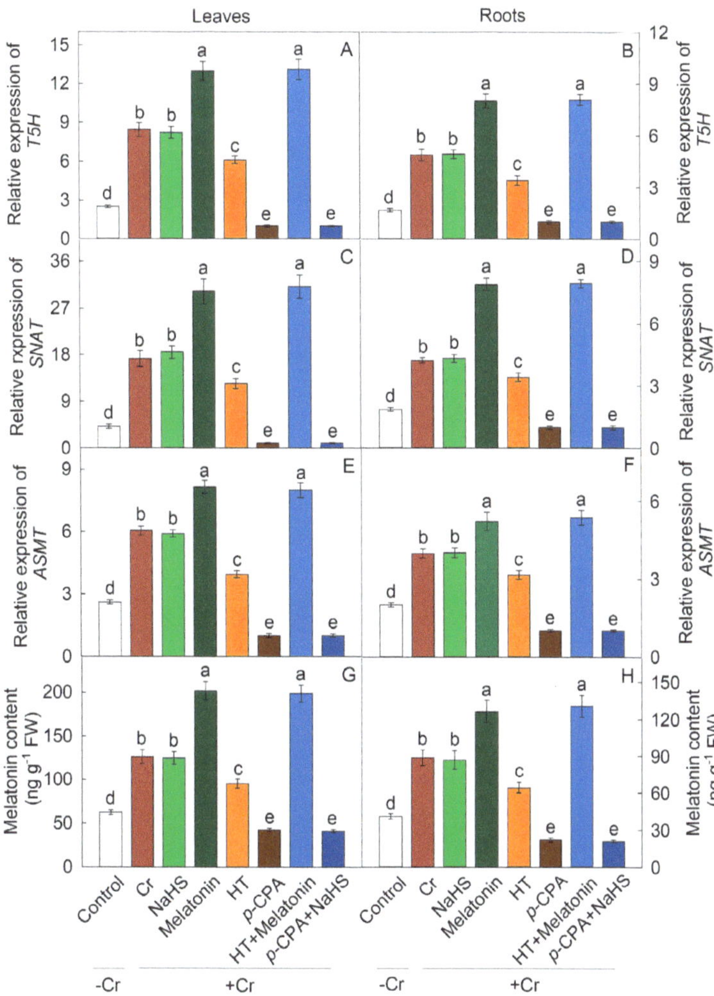

Figure 2. The effect of NaHS and melatonin on the expression of melatonin biosynthesis genes (**A–F**) and melatonin levels (**G,H**) in maize leaves and roots in response to Cr-induced stress. Different letters indicate statistically significant differences. Values are presented as mean ± SE (*n* = 3).

2.2. *Effects of Exogenous NaHS and Melatonin on Plant Growth*

The inhibitory effect of Cr stress on maize growth was significantly mitigated by the addition of exogenous NaHS and melatonin, resulting in a 88.3% and 90.3% increase in the shoot dry weight (SDW), respectively, compared to the Cr stress alone. Conversely, the application of hydroxylamine (H_2S synthesis inhibitor, HT) and *p*-chlorophenylalanine (melatonin synthesis inhibitor, *p*-CPA) resulted in a decrease in the SDW. The inhibitory effect of HT on SDW enhancement mediated by melatonin was noted, whereas the promotion of SDW by NaHS remained unaffected by *p*-CPA (Figure 3). Likewise, the alteration in root dry weight exhibited a consistent trend.

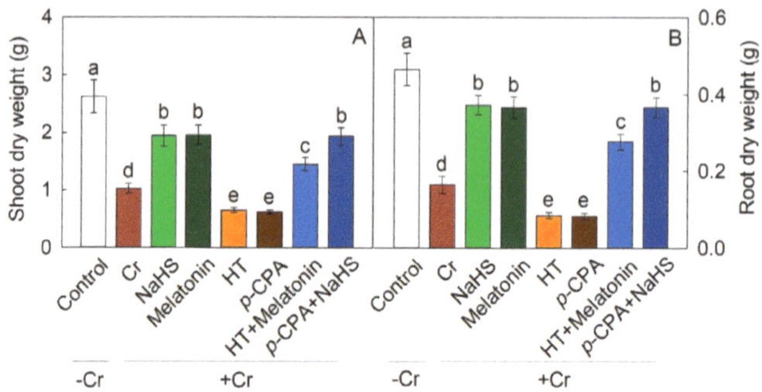

Figure 3. Effects of NaHS and melatonin on maize shoot dry weight (**A**) and root dry weight (**B**) under Cr stress. Different letters indicate statistically significant differences. Values are presented as mean ± SE ($n = 3$).

2.3. *Effects of Exogenous NaHS and Melatonin on Cr Accumulation*

The application of NaHS or melatonin resulted in a significant reduction in Cr accumulation in maize leaves and roots. Following 7 days of Cr stress, the Cr content in maize leaves and roots decreased with the addition of NaHS and melatonin, respectively. Conversely, the Cr content increased with HT and *p*-CPA treatment. HT was found to inhibit the decrease in Cr content caused by melatonin, whereas *p*-CPA did not affect the reduction in Cr content induced by NaHS (Figure 4).

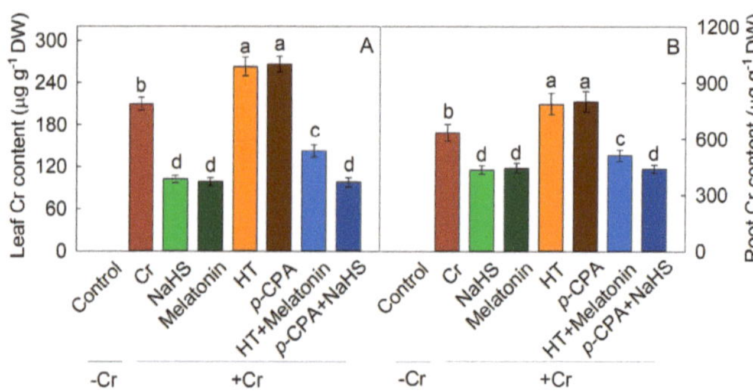

Figure 4. Effects of NaHS and melatonin on Cr accumulation in maize leaves (**A**) and roots (**B**). Different letters indicate statistically significant differences. Values are presented as mean ± SE ($n = 3$).

2.4. *Effects of Exogenous NaHS and Melatonin on Cell Wall Polysaccharides' Metabolism under Cr Stress*

The addition of NaHS or melatonin resulted in a significant elevation of pectin, hemicellulose 1 (HC1), and hemicellulose 2 (HC2) levels in both maize roots and leaves (Figure 5A–F). Furthermore, the expression of pectin and HC biosynthesis genes, such as *UGDH*, *GAE*, *GAUT*, *CSL*, and *XYL*, were upregulated by NaHS or melatonin (Figure 6). Additionally, it was observed that HT hindered the melatonin-induced elevation of pectin, HC1, and HC2 levels, while the increase induced by NaHS in these components was not impacted by *p*-CPA (Figure 5A–F). The application of Cr led to a rise in pectin methylase (PME) activity. The addition of NaHS and melatonin resulted in an increase in PME activity, whereas HT and *p*-CPA caused a decrease in PME activity (Figure 5G,H). Additionally, the expression level of the *PME* gene was significantly increased by melatonin or NaHS (Figure 6). HT hindered the increase in PME activity induced by melatonin, while the increase in PME activity caused by NaHS was unaffected by *p*-CPA.

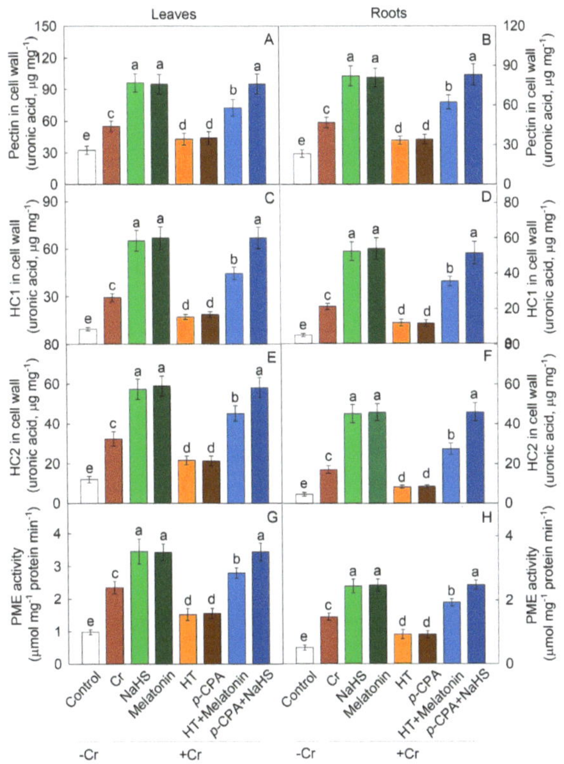

Figure 5. The Effects of NaHS and melatonin on the levels of pectin (**A**,**B**), hemicellulose 1 (HC1) (**C**,**D**), hemicellulose 2 (HC2) (**E**,**F**), and PME activity (**G**,**H**) in maize leaves and roots. Different letters indicate statistically significant differences. Values are presented as mean ± SE (*n* = 3).

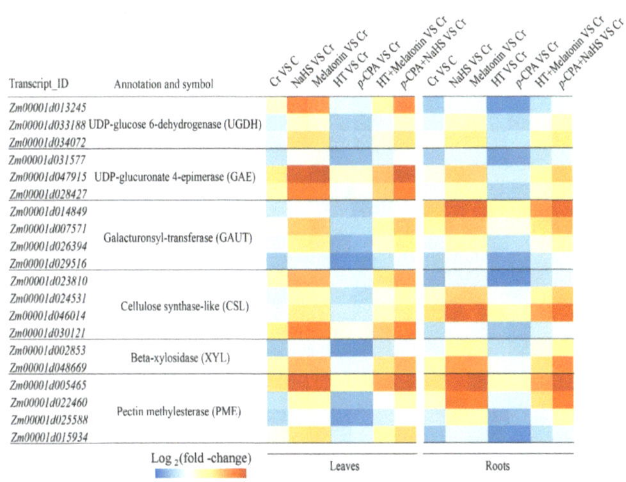

Figure 6. The effect of NaHS and melatonin on the expression of genes associated with cell wall polysaccharide metabolism. Scale bar denotes $\log_2$ (fold-change).

2.5. Effects of NaHS and Melatonin on Oxidative Damage

Cr stress resulted in elevated levels of $O_2^{\bullet -}$, H_2O_2, and malondialdehyde (MDA) in both maize leaves and roots. Conversely, treatment with exogenous NaHS or melatonin effectively mitigated the levels of $O_2^{\bullet -}$, H_2O_2, and MDA. In contrast, treatments with HT and *p*-CPA led to an increase in ROS content. HT was observed to hinder the reduction of $O_2^{\bullet -}$, H_2O_2, and MDA induced by melatonin, while the reduction induced by NaHS remained unaffected by *p*-CPA (Figure 7). Likewise, the transcriptional expression of genes involved in the generation of ROS exhibited a comparable trend (Figure 8). These findings suggest that both NaHS and melatonin have the ability to mitigate ROS levels in maize seedlings, with H_2S acting as a downstream signal of melatonin to alleviate oxidative damage caused by Cr stress.

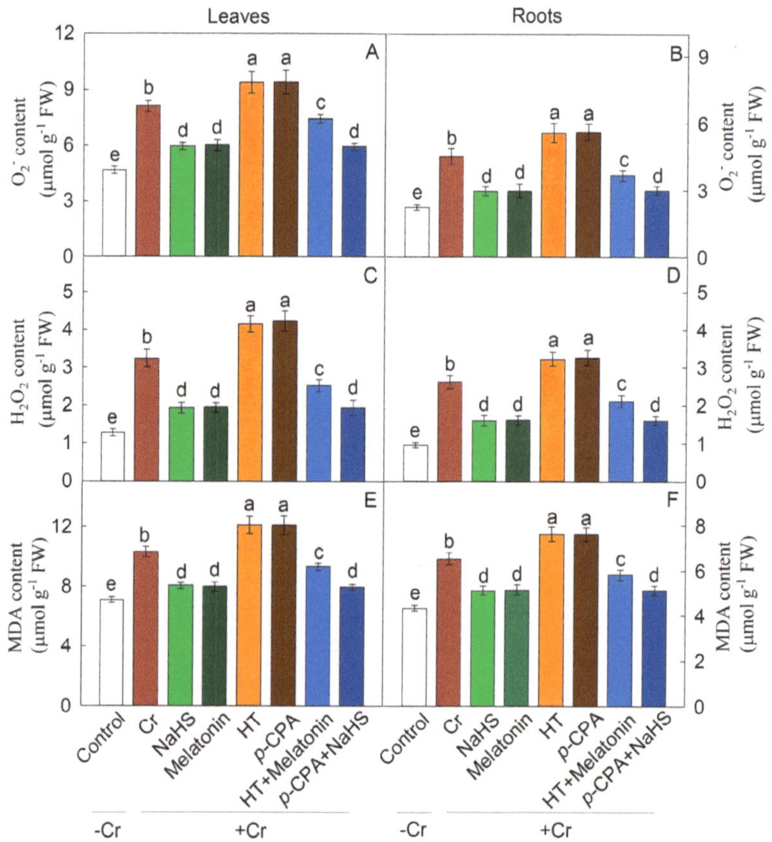

Figure 7. The effect of NaHS and melatonin on the levels of $O_2^{\bullet -}$ (**A**,**B**), H_2O_2 (**C**,**D**), and malondialdehyde (MDA) (**E**,**F**) in maize leaves and roots. Different letters indicate statistically significant differences. Values are presented as mean ± SE (n = 3).

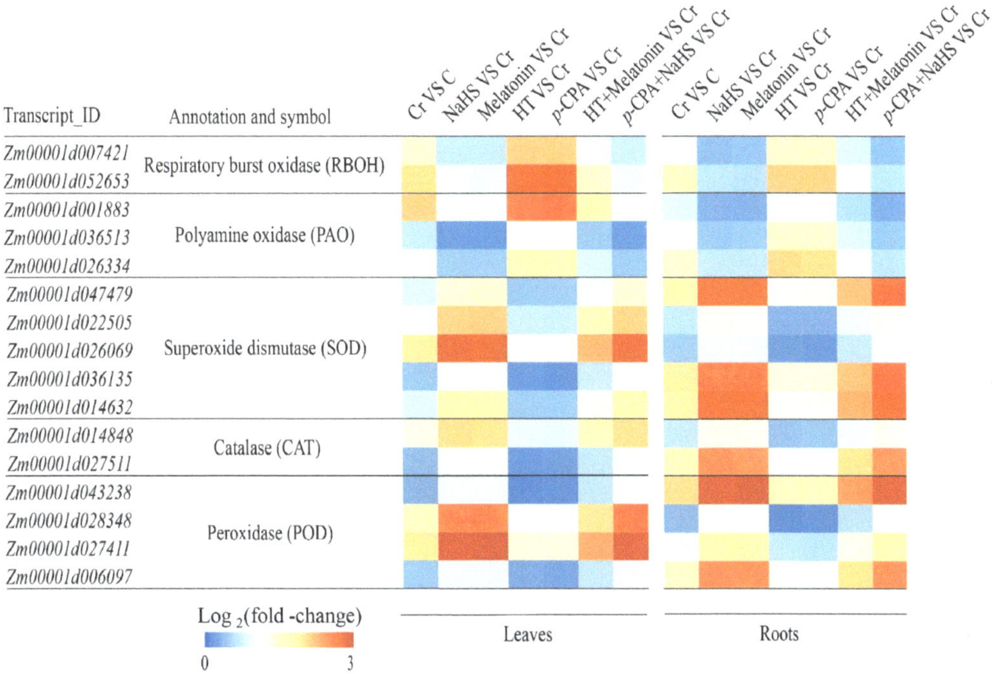

Figure 8. The effect of NaHS and melatonin on the expression of antioxidase synthesis genes. Scale bar denotes log_2 (fold-change).

2.6. *Effects of NaHS and Melatonin on the Antioxidant System*

Cr treatment resulted in a substantial increase in the expression of antioxidant enzymes' synthesis genes (SOD, POD, and CAT), as well as an enhancement in enzyme activity. The addition of exogenous NaHS or melatonin notably augmented the activity of these enzymes, whereas treatments with HT and *p*-CPA resulted in a decrease in antioxidant enzyme activity. HT was found to inhibit the enhancement of antioxidase activity induced by melatonin, whereas the increase in antioxidase activity induced by NaHS remained unaffected by *p*-CPA (Figure 9). Likewise, the transcriptional expression of genes responsible for the synthesis of antioxidant enzyme exhibited a comparable trend (Figure 8). The levels of ASA and GSH were notably elevated in response to Cr stress. When compared to exposure to $K_2Cr_2O_7$ alone, treatment with exogenous NaHS or melatonin effectively increases the levels of ASA and GSH. The presence of HT hindered the elevation of ASA and GSH levels caused by melatonin, whereas the increase in ASA and GSH levels induced by NaHS remained unaffected by *p*-CPA (Figures 10 and 11).

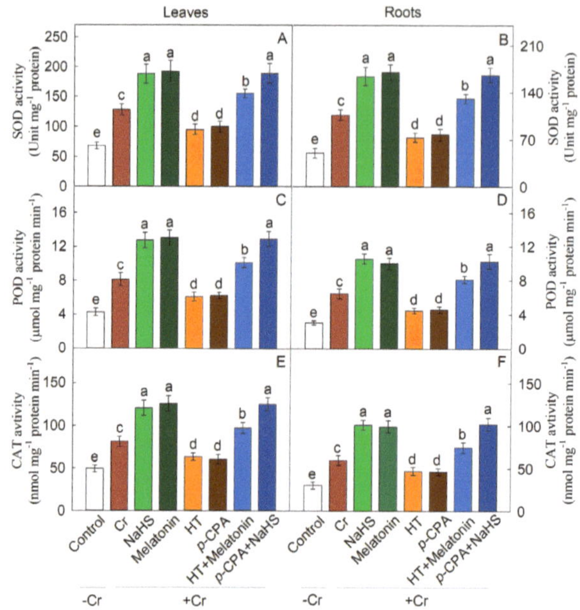

Figure 9. The effect of NaHS and melatonin on the enzymatic activities of antioxidants in maize leaves and roots. SOD: superoxide dismutase (**A**,**B**); POD: peroxidase (**C**,**D**); CAT: catalase (**E**,**F**). Different letters indicate statistically significant differences. Values are presented as mean ± SE (n = 3).

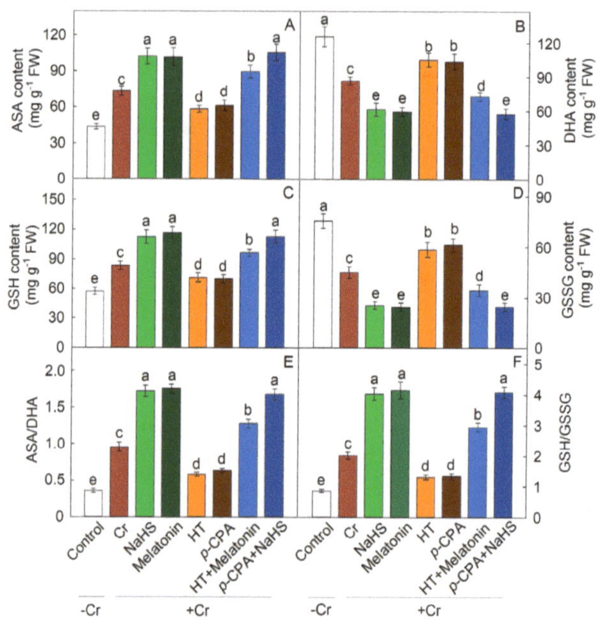

Figure 10. The effect of NaHS and melatonin on the levels of ascorbate (ASA) (**A**), dehydroascorbate (DHA) (**B**), glutathione (GSH) (**C**), oxidized glutathione (GSSG) (**D**), ASA/DHA ratio (**E**), and GSH/GSSG ratio (**F**) in maize leaves exposed to Cr stress. Different letters indicate statistically significant differences. Values are presented as mean ± SE (n = 3).

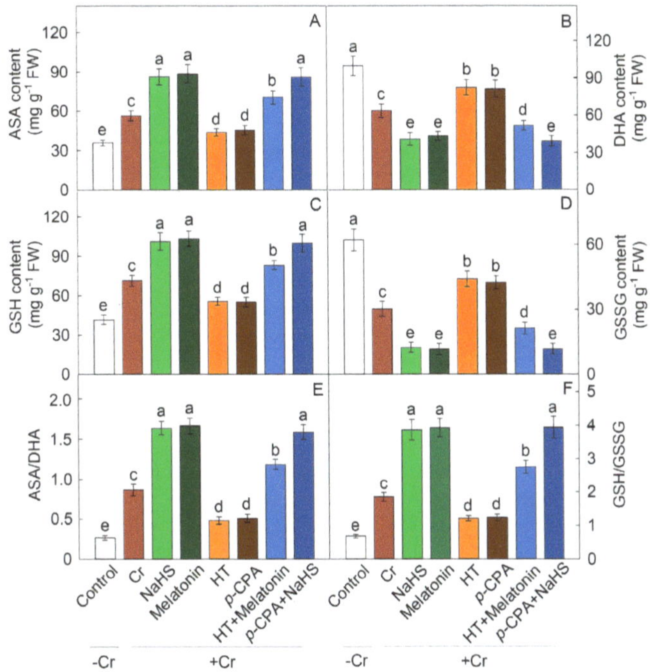

Figure 11. The effect of NaHS and melatonin on the levels of ascorbate (ASA) (**A**), dehydroascorbate (DHA) (**B**), glutathione (GSH) (**C**), oxidized glutathione (GSSG) (**D**), ASA/DHA ratio (**E**), and GSH/GSSG ratio (**F**) in maize roots exposed to Cr stress. Different letters indicate statistically significant differences. Values are presented as mean ± SE (n = 3).

3. Discussion

Previous research has indicated that H_2S [34,35] and melatonin [36,37] have the potential to improve stress tolerance in plants. This study demonstrates that exposure to Cr stress triggers the production of endogenous H_2S and melatonin. The growth of maize seedlings was notably impeded under Cr stress conditions, but supplementation with exogenous NaHS or melatonin alleviated this growth inhibition. The introduction of HT reversed the melatonin-induced tolerance to Cr stress, whereas *p*-CPA did not affect the H_2S-induced tolerance to Cr stress. Through the assessment of polysaccharide content in the cell wall, levels of ROS, and antioxidant enzyme activity, it is hypothesized that H_2S may serve as a downstream signal in mediating the tolerance of maize seedlings to Cr stress induced by melatonin.

The enzymes of LCD and DCD play crucial roles in H_2S synthesis in plants [38]. This study observed a significant increase in endogenous H_2S content and the expression levels of *LCD* and *DCD* in response to Cr stress. Additionally, the levels of melatonin and the expression of *SNAT* and *ASMT* were also found to increase significantly under Cr stress. Exogenous melatonin was found to markedly enhance the transcriptional activity of genes involved in the synthesis of L-/D-cysteine desulphurase (L-/D-CD) and to increase the levels of endogenous H_2S. Conversely, NaHS did not show a significant impact on the transcriptional regulation of genes associated with melatonin synthesis or the endogenous levels of melatonin. These findings suggest that H_2S may function as a downstream signaling molecule in mediating the tolerance of maize seedlings to Cr stress induced by melatonin.

The primary function of the root cell wall is to sequester heavy metals, thereby preventing their intrusion into the cytoplasm [39]. Among the components of the cell wall, HCs and pectin are recognized as the key elements responsible for binding heavy metals [8,40]. Higher levels of pectin and/or HC content led to increased accumulation of toxic metals in the cell walls of rice and rapeseed [9,41]. Emerging research suggests that melatonin plays a crucial role in influencing cell wall structure and composition. For instance, in tomato postharvest ripening, melatonin treatment has been shown to enhance the expression of cell wall-modifying proteins like polygalacturonase and pectinesterase. In this study, Cr treatment resulted in a significant elevation of pectin, HC1, and HC2 levels in both maize leaves and roots. Interestingly, treatment with both H_2S and melatonin further elevated the levels of pectin, HC1, and HC2 under Cr stress conditions. The capacity of the cell wall to bind heavy metals is contingent upon the level of pectin methylation, a process regulated by PME [42,43]. Research has verified that there is a negative correlation between the accumulation of heavy metals in the cell wall and the extent of pectin methylation [44,45]. Furthermore, treatment with H_2S and melatonin has been shown to enhance the activity of PME, facilitating pectin demethylation and increasing the number of Cr-binding sites within the cell wall. This was corroborated through experimentation involving H_2S scavengers and inhibitors of melatonin synthesis (HT and *p*-CPA). The presence of HT had a significant inhibitory effect on the elevation of polysaccharide content in the cell wall caused by melatonin, while the increase in polysaccharide content in the cell wall induced by NaHS was unaffected by *p*-CPA.

Cr-induced stress can lead to the accumulation of ROS and can disturb the equilibrium between ROS generation and breakdown in plants. Prior research has shown that H_2S and melatonin, acting as signaling molecules, mitigate ROS buildup and bolster antioxidant oxidase function in response to heavy metal stress [46,47]. Several researches have substantiated the function of melatonin in enhancing the activities of antioxidant enzymes, particularly under conditions of heavy metal stress [12]. The present investigation reveals that Cr stress notably triggers the buildup of H_2O_2 and $O_2^{\bullet -}$ free radicals, resulting in oxidative damage to the leaves and roots of maize plants. The utilization of NaHS and melatonin resulted in a notable decrease in ROS accumulation in maize seedlings subjected to Cr stress. Furthermore, the application of NaHS and melatonin led to an increase in the activity of SOD, POD, and CAT enzymes under Cr stress conditions. This enhancement in enzyme activity contributed to the preservation of the cell membrane.

ASA and GSH serve as regulatory molecules in the REDOX signal transduction pathway in plant cells during abiotic stress [48,49]. The interplay between these molecules is crucial for maintaining optimal REDOX states, as plants can modulate their REDOX balance by controlling the synthesis and regeneration of ASA and GSH [50]. The ASA–GSH cycle illustrates the interconnected relationship between ASA and GSH, wherein ASA is transformed into the unstable mono dehydroascorbic acid (MDHA) free radical, leading to the production of DHA. Subsequently, DHA is converted back to ASA with the assistance of GSH as an electron donor in the reduction process [51]. Prior research has indicated that maintaining elevated ratios of GSH/GSSG and/or AsA/DHA may play a crucial role in effectively mitigating the accumulation of reactive oxygen species induced by abiotic stress [52]. Nahar et al. [53] proposed that upholding high AsA/DHA and GSH/GSSG ratios is essential to facilitate the functionality of AsA and GSH within the ASA–GSH cycle, and other physiological pathways in the presence of heavy metal stress. The results of this study demonstrate that Cr stress induces an increase in the levels of ASA and GSH, as well as the ratios of ASA to DHA and GSH to GSSG in both the roots and leaves of maize plants. Furthermore, the application of NaHS and melatonin to seedlings resulted in a further enhancement of ASA and GSH content, particularly in the ratios of AsA/DHA and GSH/GSSG, when exposed to Cr stress. These findings suggest that Cr stress disrupts the REDOX balance in plants, leading to an accumulation of DHA and GSSG. NaHS can regulate the REDOX homeostasis of ASA and GSH by increasing the ratios of AsA/DHA and GSH/GSSG, thereby contributing to the tolerance of Cr induced by melatonin. In order to further understand the significance of H_2S and melatonin in clearing ROS under Cr stress, we conducted experiments using H_2S scavengers and melatonin synthesis inhibitors (HT and *p*-CPA). The addition of HT significantly hindered the enhancement of melatonin's antioxidant capabilities, while the addition of *p*-CPA did not impact the ROS clearance capacity induced by NaHS. Thus, it is hypothesized that the efficacy of melatonin may be contingent upon H_2S signaling. While the activation of H_2S and melatonin anti-oxidation mechanisms has been confirmed under abiotic stress conditions, the interplay between H_2S and melatonin in the regulation of ROS metabolism under Cr stress remains poorly understood. Our investigation revealed that H_2S plays a crucial role in mediating the tolerance of maize seedlings to Cr stress induced by melatonin, particularly in mitigating oxidative stress. Consequently, it is suggested that H_2S may serve as a downstream signaling molecule involved in the defense mechanisms of maize seedlings against Cr stress triggered by melatonin.

4. Materials and Methods

4.1. Plant Growth and Experimental Design

The maize seeds (ZD958; Cr-sensitive) were disinfected with a 1% solution of sodium hypochlorite for 10 min, followed by three washes with distilled water. Subsequently, the seeds were placed on double filter paper and incubated at 25 °C in darkness for a period of 3 days. Subsequently, the seedlings were transferred to a plastic container containing 5 L of Hoagland nutrient solution [54], with a pH of 5.8. Hoagland nutrient solution includes KNO_3 (6 mM), $NH_4H_2PO_4$ (1 mM), $MgSO_4{\cdot}7H_2O$ (2 mM), $Ca(NO_3)_2{\cdot}4H_2O$ (4 mM), EDTA-Fe (53.7 μM), $CuSO_4{\cdot}5H_2O$ (0.3 mM), $ZnSO_4{\cdot}7H_2O$ (0.8 μM), $(NH_4)_6Mo_7O_{24}{\cdot}4H_2O$ (0.2 μM), H_3BO_3 (46.3 μM), and $MnCl_2{\cdot}4H_2O$ (9.1 μM). To investigate the effect of melatonin and H_2S on Cr detoxification in maize, 14-day-old seedlings with uniform growth were subjected to foliar spraying with NaHS (50 μM), melatonin (50 μM), hydroxylamine (0.15 mM, a H_2S synthesis inhibitor, HT), and *p*-chlorophenylalanine (100 μM, a melatonin synthesis inhibitor, *p*-CPA), 12 h before the Cr treatment. The concentration of melatonin (50 μM melatonin) used in this study was determined based on findings from our previous research [55]. Cr stress was induced by supplementing the nutrient solution with 100 μM $K_2Cr_2O_7$ [56]. This study comprises eight treatments, including control, 100 μM $K_2Cr_2O_7$, 100 μM $K_2Cr_2O_7$ + 50 μM NaHS, 100 μM $K_2Cr_2O_7$ + 50 μM melatonin, 100 μM $K_2Cr_2O_7$ + 0.15 mM HT, 100 μM $K_2Cr_2O_7$ + 100 μM *p*-CPA, 100 μM $K_2Cr_2O_7$ + 0.15 mM HT + 50 μM melatonin, and 100 μM $K_2Cr_2O_7$ + 100 μM *p*-CPA + 50 μM NaHS. The incubator conditions are as follows: temperature set at 28 °C during the daytime and 23 °C at night, photoperiod of 10 h light and 14 h dark, light intensity of 600 μmol m^{-2} s^{-1}, and relative humidity ranging from 45 to 55%. Each treatment consisted of 10 pots, each containing 12 seedlings. After 7 days of Cr treatment, samples of leaves and roots were collected for subsequent analysis.

4.2. Measurement of Endogenous H_2S Content

The H_2S content was determined using the methodology described Tian et al. [57]. Root and leaf samples weighing 0.2 g were homogenized in 5 mL of phosphate buffer (50 mM, pH 6.8) containing 0.2 M ASA, 0.1 M EDTA, and 0.5 mL of 1 M HCl. The released H_2S was captured using 1% (*w*/*v*) zinc acetate. Subsequently, 0.3 mL of dimethyl p-phenylenediamine (5 mM) dissolved in 3.5 mM H_2SO_4 was added, followed by 0.3 mL of ammonium ferric sulfate (50 mM). After a 15 min reaction period, the absorbance at 667 nm was measured. The standard curve was generated using different concentrations of Na_2S. Each treatment was evaluated using three biological replicates.

4.3. Measurement of Endogenous Melatonin Content

The content of endogenous melatonin was quantified using the methodology described by Chen et al. [58], with the utilization of an ultra-high performance liquid chromatograph (Nexera LC-30AD, Shimadzu, Japan) and mass spectrometer (QTRAP 5500, AB SCIEX, Toronto, Canada). Fresh plant samples weighing 0.5 g were pulverized with liquid nitrogen, followed by the addition of 1 mL of pre-cooled methanol/acetonitrile/water (2:2:1, *v*/*v*/*v*) solution and subsequent vortex mixing. The resulting sample was sonicated in an ice bath for 60 min, then incubated at −20 °C for 1 h to precipitate proteins. Centrifugation was performed at 12,000× g rpm at 4 °C for 20 min, followed by the collection of the supernatant and subsequent vacuum drying. The dried sample was then re-suspended in a 1:1 (v/v) mixture of methanol and water (100 μL) for precipitation, and centrifuged again at 12,000× g rpm and 4 °C for 15 min. The resulting supernatant was collected. Following detection, the peak area of each sample was utilized as the horizontal coordinate to determine the melatonin concentration in the sample solution using a standard curve. Melatonin standard samples were procured from Sigma-Aldrich (Waltham, WA, USA), with three biological replicates measured for each treatment.

4.4. Measurement of Plant Dry Weight and Cr Content

In order to determine biomass, shoots and roots were collected and subsequently dried at 70 °C for 72 h before being weighed. Cr content in the roots and leaves (0.5 g) was analyzed using an inductively coupled plasma mass spectrometer (ICP-MS) (Agilent 7650A, Agilent Technologies, Santa Clara, CA, USA), following the methodology described by Sun et al. [22]. Each treatment was evaluated using three biological replicates.

4.5. Determination of Uronic Acid Content and Pectin Methylase Activity

Fresh tissues (roots and leaves) were extracted using 75% ethanol, and sequentially rinsed with acetone and methanol chloroform (1:1, v/v), followed by removal of the supernatant and freeze-drying of the residue. Subsequently, the pectin, HC1, and HC2 components in the resulting dried cell wall material were fractionated and isolated [59]. The quantification of glucuronic acid in each fraction was determined by utilizing a calibration standard curve established with a known concentration of galacturonic acid (GalA).

The PME activity was measured using the methodology described by Zhan et al. [39]. Fresh root and leaf samples weighing 0.2 g were collected and subsequently ground under liquid nitrogen freezing conditions. Subsequently, 5 mL of extraction solution consisting of 50 mM potassium phosphate buffer, 1 mM EDTA, and 1% PVP-30 was added to the samples. Following homogenization, the sample was centrifuged at 10,000× g for 20 min at 4 °C, yielding the pectin methylase extract in the supernatant. An aliquot of 8 μL of the pectin methylase extract was then mixed with 4 mL of substrate solution, and the absorbance of the resulting solution was measured at 525 nm. The concentration of H^+ in the solution was determined to assess the activity of pectin methyl esterase. Each treatment was evaluated using three biological replicates.

4.6. Assays of Reactive Oxygen Species

The concentration of H_2O_2 was quantified using the method outlined by Loreto and Velikova [60]. Then, 0.2 g of fresh samples were put into powder under liquid nitrogen freezing conditions, and 2 mL of 0.1% (w/v) trichloroacetic acid (TCA) was added. The sample was then mixed with a buffer containing potassium phosphate (10 mM) and potassium iodide (KI) (1 M). Absorption at 415 nm was measured using a spectrophotometer (UV-2550; Shimadzu, Kyoto, Japan) to determine the H_2O_2 content. The $O_2^{\bullet-}$ content was determined following the method described by Jahan et al. [61]. A fresh root sample weighing 0.2 g was pulverized into powder under cryogenic conditions using liquid nitrogen and subsequently combined with 2 mL of a phosphate buffer solution (50 mM) with a pH of 7.8. Following homogenization, the sample underwent centrifugation at 12,000× g for 20 min at 4 °C. The absorbance of the solution was measured at 530 nm. The MDA content was quantified using the method outlined by Sun et al. [22], specifically the thiobarbituric acid method. Fresh samples weighing 0.2 g were treated with 5 mL of a 0.1% TCA extraction solution. Following homogenization, the samples were centrifuged for 5 min at 4 °C at 12,000× g. The resulting sample solution was then combined with 4 mL of a 20% TCA solution containing 0.5% thiobarbituric acid (TBA), and heated in a water bath at 90 °C for 30 min. Absorbance readings were taken at 532 nm after the samples had cooled. Each treatment was performed in triplicate.

4.7. Determination of Superoxide Dismutase, Catalase, and Peroxidase Activities

For enzyme extraction, 0.2 g of fresh samples were cryogenically ground into a powder, followed by the addition of a buffer solution containing 2 mM ascorbate, 2.5 mM HEPES, 2% PVP, and 0.2 mM EDTA. The resulting mixture was homogenized and then centrifuged at 12,000× *g* for 30 min at 4 °C. The supernatant obtained from the centrifugation was utilized for the analysis of antioxidant enzyme activity. SOD activity was evaluated through its capacity to impede the photoreduction of nitrogen blue tetrazole (NBT) [62]. Prior to the quantification of enzyme activity, the incubation solution was subjected to incubation in a water bath at 30 °C. The quantity of enzymes necessary to inhibit the NBT photoreduction reaction by 50% was defined as one unit of SOD activity. The CAT activity was quantified through the decomposition of H_2O_2, as described by Hamurcu et al. [63]. The supernatant was combined with an incubation solution consisting of potassium phosphate buffer (25 mM, pH = 7.6), Na_2EDTA (0.1 mM), and hydrogen peroxide (10 mM). The change in absorbance at 240 nm was monitored to determine CAT activity. POD activity was assessed by measuring the guaiacol oxidation following the addition of H_2O_2, following the method described by Kaya et al. [46]. Each treatment was evaluated using three biological replicates.

4.8. Determination of Non-Enzymatic Antioxidant Content

The determination of ASA content was conducted following the methodology described by Campos et al. [64], wherein 0.1 g of fresh sample was combined with a 5% TCA (*m*/*v*) solution. Subsequently, the mixture was homogenized and centrifuged at 10,000× *g* for 15 min at 4 °C. The resulting supernatant was then introduced into an incubation solution composed of 0.4% H_3PO_4 ethanol solution, 98.8% ethanol, 5% TCA (*m*/*v*), 0.5% erythrophenanthroline ethanol solution (*m*/*v*), and 0.03% $FeCl_3$ ethanol solution (*m*/*v*). Absorption values at 534 nm were recorded using a spectrophotometer (UV-2550; Shimadzu, Kyoto, Japan). The content of dehydroascorbic acid (DHA) was determined by measuring the change in absorption value of dithiothreitol (DTT) at 265 nm.

The determination of GSH content was conducted following the protocol outlined by Wu et al. [65]. A fresh sample weighing 0.2 g was mixed with 2 mL of a 5% sulfosalicylic acid solution and homogenized. The resulting extracts were then centrifuged at 4 °C for 20 min at 12,000× *g*. The total GSH content was quantified by monitoring the absorption rate at 412 nm. The concentration of oxidized glutathione (GSSG) was ascertained following the elimination of GSH through derivatization with 2-vinylpyridine. GSH levels were calculated as the disparity between total GSH and GSSG. Each treatment was evaluated using three biological replicates.

4.9. RNA Extraction and qRT-PCR

The qRT-PCR analysis was conducted following the protocol outlined by Catala et al. [66]. Leaf and root samples weighing approximately 0.1 g and stored at −80 °C were utilized for RNA extraction using the Qiagen RNeasy® Plant Mini Kit (Qiagen, Valencia, CA, USA). Subsequently, cDNA was synthesized through reverse transcription using the iScriptTM cDNA Synthesis Kit (Bio-Rad, Hercules, CA, USA). The resulting cDNA was diluted 50-fold, with 2 µL used for qRT-PCR analysis. The primer sequences can be found in Table S1. The maize *Ubi-2* gene (UniProtKB/TrEMBL; ACC: Q42415) served as an internal reference gene for the normalization of experimental data, as described by Ren et al. [12]. The $2^{-\Delta\Delta Ct}$ method was employed for data analysis.

4.10. Statistical Analysis

The data were analyzed using SPSS software (version 22) for normality testing and statistical analysis, with Sigmaplot 12.0 utilized for data visualization. One-way analysis of variance (ANOVA) was conducted within the general linear model to assess relevant indicators, and post hoc comparisons of treatment differences were made using Duncan's tests, with a significance level set at $p < 0.05$.

5. Conclusions

This study indicates that H_2S, acting as a downstream signaling molecule, plays a role in mediating the tolerance of maize to Cr stress induced by melatonin. The application of exogenous melatonin led to an increase in endogenous H_2S levels, resulting in enhanced tolerance to Cr. This enhancement was evidenced by elevated antioxidant enzyme activity, as well as increased levels of cell wall polysaccharides and PME activity. The addition of HT decreased the Cr tolerance induced by melatonin, while the addition of *p*-CPA did not affect the H_2S-induced tolerance to Cr. The findings of our study offer novel perspectives on the role of melatonin in mitigating Cr toxicity. Future research should incorporate molecular biology methodologies and utilize mutant materials to further elucidate the mechanisms and interactions underlying the enhancement of plant stress tolerance by H_2S and melatonin. This study offers strategies for mitigating heavy metal toxicity and accumulation in maize, while also serving as a theoretical framework for the prevention and control of heavy metal pollution.

Supplementary Materials: The following supporting information can be downloaded at: https://www.mdpi.com/article/10.3390/plants13131763/s1. Table S1: The primers used for qRT-PCR.

Author Contributions: Conceptualization, software, writing—original draft preparation, X.Y. (Xiaoxiao Yang); investigation, software, Q.S. and X.W.; methodology, formal analysis, T.Z., K.F. and G.W.; writing—review and editing, funding acquisition, J.Z., X.Y. (Xiangyang Yuan) and J.R.; All authors have read and agreed to the published version of the manuscript.

Funding: This work was supported by the "scientific and technological innovation talents team of Shanxi" Project (202204051002036), the "National Millet and Sorghum Industry Technical System" Project (CARS-06-14.5-A28), the Doctoral Research Start-up Foundation of Shanxi Agriculture University (2023BQ01), and the Fundamental Research Program of Shanxi Province (202303021212107).

Data Availability Statement: The data are contained within the manuscript.

Conflicts of Interest: The authors declare no conflicts of interest.

References

1. Singh, S.; Prasad, S.M. Management of chromium(VI) toxicity by calcium and sulfur in tomato and brinjal: Implication of nitric oxide. *J. Hazard. Mater.* **2019**, *373*, 212–223. [CrossRef] [PubMed]
2. Alwutayd, K.M.; Alghanem, S.M.S.; Alwutayd, R.; Alghamdi, S.A.; Alabdallah, N.M.; Al-Qthanin, R.N.; Sarfraz, W.; Khalid, N.; Naeem, N.; Ali, B.; et al. Mitigating chromium toxicity in rice (*Oryza sativa* L.) via ABA and 6-BAP: Unveiling synergistic benefits on morphophysiological traits and ASA-GSH cycle. *Sci. Total Environ.* **2024**, *908*, 168208. [CrossRef] [PubMed]
3. Wakeel, A.; Xu, M.; Gan, Y.B. Chromium-induced reactive oxygen species accumulation by altering the enzymatic antioxidant system and associated cytotoxic, genotoxic, ultrastructural, and photosynthetic changes in plants. *Int. J. Mol. Sci.* **2020**, *21*, 728. [CrossRef] [PubMed]
4. Fan, W.J.; Feng, Y.X.; Li, Y.H.; Lin, Y.J.; Yu, X.Z. Unraveling genes promoting ROS metabolism in subcellular organelles of *Oryza sativa* in response to trivalent and hexavalent chromium. *Sci. Total Environ.* **2020**, *744*, 140951. [CrossRef]
5. Shi, J.D.; Zhao, D.; Ren, F.T.; Huang, L. Spatiotemporal variation of soil heavy metals in China: The pollution status and risk assessment. *Sci. Total Environ.* **2023**, *871*, 161768. [CrossRef]
6. Li, Y.Y.; Tian, X.Y.; Liang, J.L.; Chen, X.L.; Ye, J.Y.; Liu, Y.S.; Liu, Y.Y.; Wei, Y.M. Remediation of hexavalent chromium in contaminated soil using amorphous iron pyrite: Effect on leachability, bioaccessibility, phytotoxicity and long-term stability. *Environ. Pollut.* **2020**, *264*, 114804. [CrossRef] [PubMed]
7. Wu, X.W.; Song, H.X.; Guan, C.Y.; Zhang, Z.H. Boron alleviates cadmium toxicity in Brassica napus by promoting the chelation of cadmium onto the root cell wall components. *Sci. Total Environ.* **2020**, *728*, 138833. [CrossRef]
8. Sun, C.L.; Lv, T.; Huang, L.; Liu, X.X.; Jin, C.W.; Lin, X.Y. Melatonin ameliorates aluminum toxicity through enhancing aluminum exclusion and reestablishing redox homeostasis in roots of wheat. *J. Pineal Res.* **2020**, *68*, e12642. [CrossRef]
9. Yuan, Y.; Imtiaz, M.; Rizwan, M.; Dai, Z.H.; Hossain, M.M.; Zhang, Y.H.; Huang, H.L.; Tu, S.X. The role and its transcriptome mechanisms of cell wall polysaccharides in vanadium detoxication of rice. *J. Hazard. Mater.* **2022**, *425*, 127966. [CrossRef]
10. Alsahli, A.A.; Bhat, J.A.; Alyemeni, M.N.; Ashraf, M.; Ahmad, P. Hydrogen sulfide (H_2S) mitigates arsenic (As)-induced toxicity in pea (Pisum sativum L.) plants by regulating osmoregulation, antioxidant defense system, ascorbate glutathione cycle and glyoxalase system. *J. Plant Growth Regul.* **2021**, *40*, 2515–2531. [CrossRef]

11. Bhatta, D.; Adhikari, A.; Kang, S.M.; Kwon, E.H.; Jan, R.; Kim, K.M.; Lee, I.J. Hormones and the antioxidant transduction pathway and gene expression, mediated by Serratia marcescens DB1, lessen the lethality of heavy metals (As, Ni, and Cr) in *Oryza sativa* L. *Ecotoxicol. Environ. Saf.* **2023**, *263*, 115377. [CrossRef] [PubMed]
12. Ren, J.H.; Yang, X.X.; Zhang, N.; Feng, L.; Ma, C.Y.; Wang, Y.L.; Yang, Z.P.; Zhao, J. Melatonin alleviates aluminum-induced growth inhibition by modulating carbon and nitrogen metabolism, and reestablishing redox homeostasis in *Zea mays* L. *J. Hazard. Mater.* **2022**, *423*, 127159. [CrossRef] [PubMed]
13. Muhammad, I.; Ahmad, S.; Shen, W.J. Melatonin-mediated molecular responses in plants: Enhancing stress tolerance and mitigating environmental challenges in cereal crop production. *Int. J. Mol. Sci.* **2024**, *25*, 4551. [CrossRef]
14. Gao, T.; Wang, Z.X.; Dong, Y.L.; Cao, J.; Lin, R.T.; Wang, X.T.; Yu, Z.Q.; Chen, Y.X. Role of melatonin in sleep deprivation-induced intestinal barrier dysfunction in mice. *J. Pineal Res.* **2019**, *67*, e12574. [CrossRef] [PubMed]
15. Colombage, R.; Singh, M.B.; Bhalla, P.L. Melatonin and abiotic stress tolerance in crop plants. *Int. J. Mol. Sci.* **2023**, *24*, 7447. [CrossRef] [PubMed]
16. Gao, Y.Y.; Chen, H.M.; Chen, D.Y.; Hao, G.F. Genetic and evolutionary dissection of melatonin response signaling facilitates the regulation of plant growth and stress responses. *J. Pineal Res.* **2023**, *74*, e12850. [CrossRef] [PubMed]
17. Ahammed, G.J.; Li, Z.; Chen, J.Y.; Dong, Y.F.; Qu, K.H.; Guo, T.M.; Wang, F.H.; Liu, A.R.; Chen, S.C.; Li, X. Reactive oxygen species signaling in melatonin-mediated plant stress response. *Plant Physiol. Biochem.* **2024**, *207*, 108398. [CrossRef] [PubMed]
18. Chen, Q.; Arnao, M.B. Phytomelatonin: An emerging new hormone in plants. *J. Exp. Bot.* **2022**, *73*, 5773–5778. [CrossRef]
19. Jan, R.; Asif, S.; Asaf, S.; Lubna; Du, X.X.; Park, J.R.; Nari, K.; Bhatta, D.; Lee, I.J.; Kim, K.M. Melatonin alleviates arsenic (As) toxicity in rice plants via modulating antioxidant defense system and secondary metabolites and reducing oxidative stress. *Environ. Pollut.* **2022**, *318*, 120868. [CrossRef]
20. Yin, Y.Q.; Hu, J.J.; Tian, X.; Yang, Z.F.; Fang, W.M. Nitric oxide mediates melatonin-induced isoflavone accumulation and growth improvement in germinating soybeans under NaCl stress. *J. Plant Physiol.* **2022**, *279*, 153855. [CrossRef]
21. Altaf, M.A.; Hao, Y.Y.; Shu, H.Y.; Mumtaz, M.A.; Cheng, S.H.; Alyemeni, M.N.; Ahmad, P.; Wang, Z. Melatonin enhanced the heavy metal-stress tolerance of pepper by mitigating the oxidative damage and reducing the heavy metal accumulation. *J. Hazard. Mater.* **2023**, *454*, 131468. [CrossRef] [PubMed]
22. Sun, S.S.; Liu, A.R.; Li, Z.; Guo, T.M.; Chen, S.C.; Ahammed, G.J. Anthocyanin synthesis is critical for melatonin-induced chromium stress tolerance in tomato. *J. Hazard. Mater.* **2023**, *453*, 131456. [CrossRef] [PubMed]
23. Arnao, M.B.; Hernández–Ruiz, J. Melatonin: A new plant hormone and / or a plant master regulator? *Trends Plant Sci.* **2019**, *24*, 38–48. [CrossRef] [PubMed]
24. Hilal, B.; Khan, T.A.; Fariduddin, Q. Recent advances and mechanistic interactions of hydrogen sulfide with plant growth regulators in relation to abiotic stress tolerance in plants. *Plant Physiol. Biochem.* **2023**, *196*, 1065–1083. [CrossRef] [PubMed]
25. Thakur, M.; Anand, A. Hydrogen sulfide: An emerging signaling molecule regulating drought stress response in plants. *Physiol. Plant.* **2021**, *172*, 1227–1243. [CrossRef] [PubMed]
26. Saini, N.; Modolo, L.V.; Deswal, R.; Sehrawat, A.; Yadav, N.; Sangwan, N.S. Expanding Roles of Cross-talk between Hydrogen Sulfide and Nitric Oxide under Abiotic Stress in Plants. *Plant Physiol. Biochem.* 2024, *in press*.
27. Ma, Y.L.; Zhang, W.; Niu, J.; Ren, Y.; Zhang, F. Hydrogen sulfide may function downstream of hydrogen peroxide in salt stress-induced stomatal closure in *Vicia faba*. *Funct. Plant Biol.* **2018**, *46*, 136–145. [CrossRef] [PubMed]
28. Li, Z.; Zhu, Y.Q.; He, X.S.; Yong, B.; Peng, Y.; Zhang, X.Q.; Ma, X.; Yan, Y.H.; Huang, L.K.; Nie, G. The hydrogen sulfide, a downstream signaling molecule of hydrogen peroxide and nitric oxide, involves spermidine-regulated transcription factors and antioxidant defense in white clover in response to dehydration. *Environ. Exp. Bot.* **2019**, *161*, 255–264. [CrossRef]
29. Du, X.Z.; Jin, Z.P.; Liu, D.M.; Yang, G.D.; Pei, Y.X. Hydrogen sulfide alleviates the cold stress through MPK4 in *Arabidopsis thaliana*. *Plant Physiol. Biochem.* **2017**, *120*, 112–119. [CrossRef] [PubMed]
30. Yang, X.; Feng, K.; Wang, G.; Zhang, S.; Zhao, J.; Yuan, X.; Ren, J. Titanium dioxide nanoparticles alleviates polystyrene nanoplastics induced growth inhibition by modulating carbon and nitrogen metabolism via melatonin signaling in maize. *J. Nanobiotechnol.* **2024**, *22*, 262. [CrossRef]
31. Kharbech, O.; Sakouhi, L.; Mahjoubi, Y.; Massoud, M.B.; Debez, A.; Zribi, O.T.; Djebali, W.; Chaoui, A.; Mur, L.A.J. Nitric oxide donor, sodium nitroprusside modulates hydrogen sulfide metabolism and cysteine homeostasis to aid the alleviation of chromium toxicity in maize seedlings (*Zea mays* l.). *J. Hazard. Mater.* **2022**, *424*, 127302. [CrossRef]
32. Sun, Y.P.; Ma, C.; Kang, X.; Zhang, L.; Wang, J.; Zheng, S.; Zhang, T.G. Hydrogen sulfide and nitric oxide are involved in melatonin-induced salt tolerance in cucumber. *Plant Physiol. Biochem.* **2021**, *167*, 101–112. [CrossRef] [PubMed]
33. Hancock, J.T. Hydrogen sulfide and environmental stresses. *Environ. Exp. Bot.* **2019**, *161*, 50–56. [CrossRef]
34. Luo, S.L.; Calderón-Urrea, A.; Jihua, Y.U.; Liao, W.B.; Xie, J.M.; Lv, J.; Feng, Z.; Tang, Z.Q. The role of hydrogen sulfide in plant alleviates heavy metal stress. *Plant Soil* **2020**, *449*, 1–10. [CrossRef]
35. Zhou, M.J.; Zhou, H.; Shen, J.; Zhang, Z.R.; Gotor, C.; Romero, L.C.; Yuan, X.X.; Xie, Y.J. H_2S action in plant life cycle. *Plant Growth Regul.* **2021**, *94*, 1–9. [CrossRef]
36. Moustafa-Farag, M.; Elkelish, A.; Dafea, M.; Khan, M.; Arnao, M.B.; Abdelhamid, M.T.; El-Ezz, A.A.; Almoneafy, A.; Mahmoud, A.; Awad, M.; et al. Role of melatonin in plant tolerance to soil stressors: Salinity, pH and heavy metals. *Molecules* **2020**, *25*, 5359. [CrossRef] [PubMed]

37. Bose, S.K.; Howlader, P. Melatonin plays multifunctional role in horticultural crops against environmental stresses: A review. *Environ. Exp. Bot.* **2020**, *176*, 104063. [CrossRef]
38. Guo, H.M.; Xiao, T.Y.; Zhou, H.; Xie, Y.J.; Shen, W.B. Hydrogen sulfide: Aversatile regulator of environmental stress in plants. *Acta Physiol. Plant.* **2016**, *38*, 16. [CrossRef]
39. Zhan, J.; Huang, H.G.; Yu, H.Y.; Zhang, X.Z.; Zheng, Z.C.; Wang, Y.D.; Liu, T.; Li, T.X. The combined effects of Cd and Pb enhanced metal binding by root cell walls of the phytostabilizer *Athyrium wardii* (Hook.). *Environ. Pollut.* **2020**, *258*, 113663. [CrossRef] [PubMed]
40. Jia, H.L.; Wang, X.; Shi, C.; Guo, J.K.; Ma, P.Y.; Ren, X.H.; Wei, T.; Liu, H.X.; Li, J.S. Hydrogen sulfide decreases Cd translocation from root to shoot through increasing Cd accumulation in cell wall and decreasing Cd^{2+} influx in *Isatis indigotica*. *Plant Physiol. Biochem.* **2020**, *155*, 605–612. [CrossRef]
41. Wu, X.W.; Tian, H.; Li, L.; Wang, X.Q. Polyaspartic acid alleviates cadmium toxicity in rapeseed leaves by affecting cadmium translocation and cell wall fixation of cadmium. *Ecotoxicol. Environ. Saf.* **2021**, *224*, 112685. [CrossRef]
42. Wang, L.; Li, R.; Yan, X.X.; Liang, X.F.; Sun, Y.B.; Xu, Y.M. Pivotal role for root cell wall polysaccharides in cultivar-dependent cadmium accumulation in *Brassica chinensis* L. *Ecotoxicol. Environ. Saf.* **2020**, *194*, 110369. [CrossRef] [PubMed]
43. Yang, S.; Yi, K.; Chang, M.M.; Ling, G.Z.; Zhao, Z.K.; Li, X.F. Sequestration of Mn into the cell wall contributes to Mn tolerance in sugarcane (*Saccharum officinarum* L.). *Plant Soil* **2019**, *436*, 475–487. [CrossRef]
44. Li, T.Q.; Tao, Q.; Shohag, M.J.I.; Yang, X.E.; Sparks, D.L.; Liang, Y.C. Root cell wall polysaccharides are involved in cadmium hyperaccumulation in *Sedum alfredii*. *Plant Soil* **2015**, *389*, 387–399. [CrossRef]
45. Cao, Y.Y.; Qi, C.D.; Li, S.T.; Wang, Z.R.; Wang, X.Y.; Wang, J.F.; Ren, S.X.; Li, X.S.; Zhang, N.; Guo, Y.D. Melatonin alleviates copper toxicity via improving copper sequestration and ROS scavenging in cucumber. *Plant Cell Physiol.* **2019**, *60*, 562–574. [CrossRef] [PubMed]
46. Kaya, C.; Ashraf, M.; Alyemeni, M.N.; Ahmad, P. The role of endogenous nitric oxide in salicylic acid-induced up-regulation of ascorbate-glutathione cycle involved in salinity tolerance of pepper (*Capsicum annuum* L.) plants. *Plant Physiol. Biochem.* **2019**, *147*, 10–20. [CrossRef] [PubMed]
47. Xu, L.; Zhang, F.; Tang, M.J.; Wang, Y.; Dong, J.H.; Ying, J.L.; Chen, Y.L.; Hu, B.; Li, C.; Liu, L.W. Melatonin confers cadmium tolerance by modulating critical heavy metal chelators and transporters in radish plants. *J. Pineal Res.* **2020**, *69*, e12659. [CrossRef] [PubMed]
48. Hasanuzzaman, M.; Bhuyan, M.H.M.; Anee, T.I.; Parvin, K.; Nahar, K.; Al Mahmud, J.; Fujita, M. Regulation of ascorbate-glutathione pathway in mitigating oxidative damage in plants under abiotic stress. *Antioxidants* **2019**, *8*, 384. [CrossRef] [PubMed]
49. Kaya, C. Nitrate reductase is required for salicylic acid-induced water stress tolerance of pepper by upraising the AsA-GSH pathway and glyoxalase system. *Physiol. Plant.* **2021**, *172*, 351–370. [CrossRef] [PubMed]
50. Kapoor, D.; Sharma, R.; Handa, N.; Kaur, H.; Rattan, A.; Yadav, P.; Gautam, V.; Kaur, R.; Bhardwaj, R. Redox homeostasis in plants under abiotic stress: Role of electron carriers, energy metabolism mediators and proteinaceous thiols. *Front. Environ. Sci.* **2015**, *3*, 13. [CrossRef]
51. Kaya, C.; Ashraf, M.; Alyemeni, M.N.; Ahmad, P. Responses of nitric oxide and hydrogen sulfide in regulating oxidative defence system in wheat plants grown under cadmium stress. *Physiol. Plant.* **2020**, *168*, 345–360. [CrossRef]
52. Wang, Y.; Ye, X.Y.; Qiu, X.M.; Li, Z.G. Methylglyoxal triggers the heat tolerance in maize seedlings by driving AsA-GSH cycle and reactive oxygen species-/methylglyoxal-scavenging system. *Plant Physiol. Biochem.* **2019**, *138*, 91–99. [CrossRef] [PubMed]
53. Nahar, K.; Hasanuzzaman, M.; Suzuki, T.; Fujita, M. Polyamines-induced aluminum tolerance in mung bean: A study on antioxidant defense and methylglyoxal detoxification systems. *Ecotoxicology* **2017**, *26*, 58–73. [CrossRef] [PubMed]
54. Hoagland, D.R.; Arnon, D.I. The water-culture method for growing plants without soil. *Calif. Agric. Ext. Serv. Circ.* **1950**, *347*, 1–32.
55. Yang, X.X.; Ren, J.H.; Lin, X.Y.; Yang, Z.P.; Deng, X.P.; Ke, Q.B. Melatonin alleviates chromium toxicity in maize by modulation of cell wall polysaccharides biosynthesis, glutathione metabolism, and antioxidant capacity. *Int. J. Mol. Sci.* **2023**, *24*, 3816. [CrossRef] [PubMed]
56. Yang, X.X.; Ren, J.H.; Yang, W.P.; Xue, J.F.; Gao, Z.Q.; Yang, Z.P. Hydrogen sulfide alleviates chromium toxicity by promoting chromium sequestration and re-establishing redox homeostasis in *Zea mays* L. *Environ. Pollut.* **2023**, *332*, 121958. [CrossRef] [PubMed]
57. Tian, B.; Qiao, Z.; Zhang, L.; Li, H.; Pei, Y. Hydrogen sulfide and proline cooperate to alleviate cadmium stress in foxtail millet seedings. *Plant Physiol. Biochem.* **2016**, *109*, 293–299. [CrossRef]
58. Chen, G.F.; Huo, Y.S.; Tan, D.X.; Liang, Z.; Zhang, W.B.; Zhang, Y.K. Melatonin in Chinese medicinal herbs. *Life Sci.* **2003**, *73*, 19–26. [CrossRef]
59. Yang, J.L.; Zhu, X.F.; Peng, Y.X.; Zheng, C.; Li, G.X.; Liu, Y.; Shi, Y.Z.; Zheng, S.J. Cell wall hemicellulose contributes significantly to aluminum adsorption and root growth in *Arabidopsis*. *Plant Physiol.* **2011**, *155*, 1885–1892. [CrossRef]
60. Loreto, F.; Velikova, V. Isoprene produced by leaves protects the photosynthetic apparatus against ozone damage, quenches ozone products, and reduces lipid peroxidation of cellular membranes. *Plant Physiol.* **2001**, *127*, 1781–1787. [CrossRef]

61. Jahan, M.S.; Guo, S.R.; Baloch, A.R.; Sun, J.; Shu, S.; Wang, Y.; Ahammed, G.J.; Kabir, K.; Roy, R. Melatonin alleviates nickel phytotoxicity by improving photosynthesis, secondary metabolism and oxidative stress tolerance in tomato seedlings. *Ecotoxicol. Environ. Saf.* **2020**, *197*, 110593. [CrossRef]
62. Li, J.T.; Qiu, Z.B.; Zhang, X.W.; Wang, L.S. Exogenous hydrogen peroxide can enhance tolerance of wheat seedlings to salt stress. *Acta Physiol. Plant.* **2011**, *33*, 835–842. [CrossRef]
63. Hamurcu, M.; Sekmen, A.H.; Turkan, İ.; Gezgin, S.; Demiral, T.; Bell, R.W. Induced anti-oxidant activity in soybean alleviates oxidative stress under moderate boron toxicity. *Plant Growth Regul.* **2013**, *70*, 217–226. [CrossRef]
64. Campos, C.N.; Ávila, R.G.; de Souza, K.R.D.; Azevedo, L.M.; Alves, J.D. Melatonin reduces oxidative stress and promotes drought tolerance in young *Coffea arabica* L. plants. *Agric. Water Manag.* **2019**, *211*, 37–47. [CrossRef]
65. Wu, X.X.; He, J.; Ding, H.D.; Zhu, Z.W.; Chen, J.L.; Xu, S.; Zha, D.S. Modulation of zinc-induced oxidative damage in *Solanum melongena* by 6-benzylaminopurine involves ascorbate–glutathione cycle metabolism. *Environ. Exp. Bot.* **2015**, *116*, 1–11. [CrossRef]
66. Catala, R.; Lopez-Cobollo, R.; Castellano, M.M.; Angosto, T.; Alonso, J.M.; Ecker, J.R.; Salinas, J. The *Arabidopsis* 14-3-3 protein RARE COLD INDUCIBLE 1A links low-temperature response and ethylene biosynthesis to regulate freezing tolerance and cold acclimation. *Plant Cell* **2014**, *26*, 3326–3342. [CrossRef]

plants

MDPI

Article

Supplemental Silicon and Boron Alleviates Aluminum-Induced Oxidative Damage in Soybean Roots

Shuwei Wang [1], Haijing Cheng [1] and Yunmin Wei [1,2,*]

[1] College of Life Sciences and Oceanography, Shenzhen University, Shenzhen 518060, China; shuweiwang_21@163.com (S.W.); 13644654490@163.com (H.C.)
[2] College of Optoelectronic Engineering, Shenzhen University, Shenzhen 518060, China
* Correspondence: weiym1024@szu.edu.cn; Tel.: +86-133-4021-3922

Abstract: Aluminum (Al) toxicity in acidic soils is a major abiotic stress that negatively impacts plant growth and development. The toxic effects of Al manifest primarily in the root system, leading to inhibited root elongation and functionality, which impairs the above-ground organs of the plant. Recent research has greatly improved our understanding of the applications of small molecule compounds in alleviating Al toxicity. This study aimed to investigate the role of boron (B), silicon (Si), and their combination in alleviating Al toxicity in soybeans. The results revealed that the combined application significantly improved the biomass and length of soybean roots exposed to Al toxicity compared to B and Si treatments alone. Our results also indicated that Al toxicity causes programmed cell death (PCD) in soybean roots, while B, Si, and their combination all alleviated the PCD induced by Al toxicity. The oxidative damage induced by Al toxicity was noticeably alleviated, as evidenced by lower MAD and H_2O_2 accumulation in the soybean roots treated with the B and Si combination. Moreover, B, Si, and combined B and Si significantly enhanced plant antioxidant systems by up-regulating antioxidant enzymes including CAT, POD, APX, and SOD. Overall, supplementation with B, Si, and their combination was found to alleviate oxidative damage and reduce PCD caused by Al toxicity, which may be one of the mechanisms by which they alleviate root growth inhibition due to Al toxicity. Our results suggest that supplementation with B, Si, and their combination may be an effective strategy to improve soybean growth and productivity against Al toxicity.

Keywords: soybean (*Glycine max* L.); silicon; boron; aluminum; oxidative damage; citric acid

Citation: Wang, S.; Cheng, H.; Wei, Y. Supplemental Silicon and Boron Alleviates Aluminum-Induced Oxidative Damage in Soybean Roots. *Plants* **2024**, *13*, 821. https://doi.org/10.3390/plants13060821

Academic Editors: Violetta Katarzyna Macioszek, Iwona Ciereszko and Andrzej K. Kononowicz

Received: 30 January 2024
Revised: 9 March 2024
Accepted: 10 March 2024
Published: 13 March 2024

1. Introduction

Aluminum (Al) is the third most abundant element on Earth and is a widely available metal. It constitutes about 8.2% of the Earth's crust, following only oxygen and silicon in terms of abundance. In neutral or nearly neutral soil, Al is present in the form of insoluble oxides or aluminosilicates, typically exhibiting a low solubility, which poses no toxic threat to plants. However, in acidic soils with a pH below 5.5, Al mainly exists in the form of $Al(OH)^{2+}$, $Al(OH)^{+}$, and $Al(H_2O)^{3+}$, which can be solubilized and released into the soil. Al ions (mainly Al^{3+}) are highly toxic to most agriculturally cultivated plants [1,2]. In acidic soils, high concentrations of aluminum ions can interfere with plant growth and development by damaging root systems. This interference significantly inhibits root elongation, reduces water and mineral nutrient absorption, and disrupts various physiological processes. Additionally, numerous studies have reported that aluminum induces the rapid production and accumulation of reactive oxygen species (ROS), severely impacting plant cell metabolism, and accelerates Al-induced programmed cell death (PCD) of plants [3–5]. Al-induced lipid peroxidation results in the production and accumulation of malondialdehyde (MDA), which disrupts membrane functions. This effect has been reported in various plants, including tomato (*Solanum lycopersicum* L.), rice (*Oryza sativa* L.), soybean (*Glycine max* L.), tobacco (*Nicotiana tabacum* L.), and pea (*Pisum sativum* L.) [6–8]. Despite their deleterious effects, ROS play a dual role in plants, as they also serve as important signaling molecules to activate the expression of antioxidant mechanisms to alleviate Al stress [9–11]. Additionally, some plants stimulate organic acid (e.g., malate, citrate) exudation from their root surface, thereby chelating and immobilizing the phytotoxic Al^{3+} in the rhizosphere [12–15]. For instance, soybeans can relieve Al stress by secreting citric acid [16,17].

Furthermore, recent studies have demonstrated that silicon (Si) supplementation can potentially mitigate Al toxicity in plants [18–21]. The second most abundant element in the Earth's crust is silicon, which is second only to oxygen in content and constitutes about 27.7% of the Earth's mass [22]. In soil solutions, Si exists as silicic acid ($Si[OH]_4$) and is readily absorbed by the roots of higher plants [23]. Although Si is not a macronutrient for most plants, it plays an important role in plant responses to various abiotic stresses, including Al toxicity [24,25]. Si can alleviate Al toxicity in plants via multiple mechanisms, such as reducing the deposition of Al in the cell wall, increasing antioxidants, forming Si–Al complexes, and enhancing the rhizosphere pH [26–29]. The application of Si to plants exposed to Al stress has been shown to improve the growth and yield of various crops, such as rice, maize, and soybean. These results highlight the potential of Si application in sustainable agriculture [27,30].

Boron (B) is beneficial for plant growth and is an indispensable micronutrient [31,32]. Several studies have shown that B plays a major role in mitigating Al toxicity in many plants, such as *Citrus grandis*, peas, rape seedlings, rice, and so on [33–37]. Supplementation with B effectively reduced the accumulation of Al in cell walls and reduced oxidative damage to roots by modulating their cell wall composition and structure [36,37]. Recent research has reported that B supplementation can alleviate the oxidative stress of Al on citrus roots by reducing the hydrogen peroxide content and lipid peroxidation (indicated by MDA content) [35,37].

Soybean is an important cash crop that is widely cultivated worldwide and is used as a food, feed, and oilseed crop. It is a major source of proteins and edible oil [38]. The growth, biomass production, and essential physiological functions of soybean plants are significantly impaired by aluminum toxicity, thereby adversely impacting productivity [39]. Previous studies have identified Si and B as potential alleviators of Al toxicity in soybeans, but the underlying mechanisms remain poorly understood. Therefore, the present study aimed to explain the inhibitory effects of boron, silicon, and their combination on Al-induced root growth and oxidative damage in soybeans. In addition, the interventions were hypothesized to alleviate Al toxicity in soybeans by up-regulating the antioxidant defense system and secreting organic acids.

2. Results

2.1. *Effects of B, Si, and Their Combination on Soybean Root Growth Parameters under Al Stress*

As shown in Figure 1, B and Si alone and their combination (B + Si) exerted no effect on soybean root growth in the absence of Al stress compared to the non-treated plants (CK). However, upon exposure to 50 μM $AlCl_3$, Al toxicity significantly reduced the root biomass and length of soybeans compared to CK. However, the addition of B and Si significantly alleviated the Al toxicity-induced inhibition of root elongation. Compared to the Al treatment, B and Si treatments increased root length by 16.3% and 31.7%, respectively, and root biomass by 18.5% and 22.4%, respectively. Moreover, the Al + B + Si group demonstrated a significant increase in root length of 68.3% compared to the plants only exposed to Al, which effectively recovered to CK levels. Furthermore, the root biomass of B and Si co-treated plants exposed to Al toxicity significantly increased by 37.4% compared to Al-treated soybean seedlings.

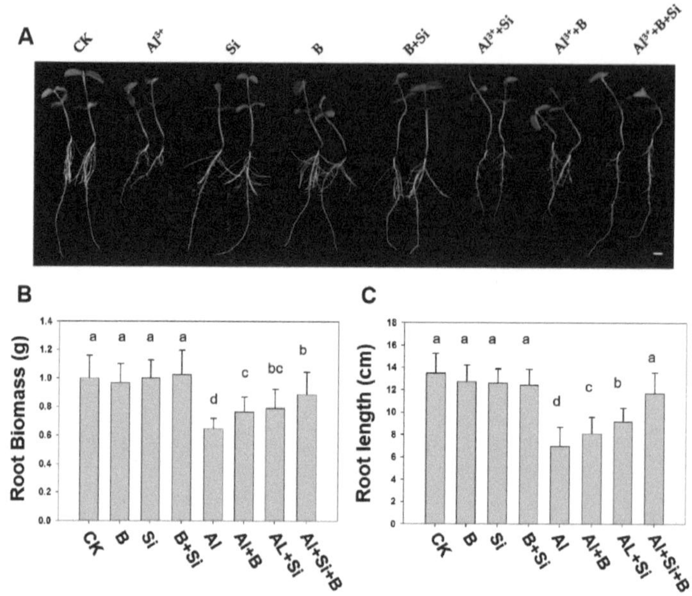

Figure 1. Effects of B, Si, and their combination on soybean root growth parameters under Al stress. (**A**) Symptoms, (**B**) root biomass, (**C**) root length. Here, (1) CK (without Al, B, and Si), (2) + B (16.3 μM H_3BO_3), (3) + Si (1.5 mM Na_2SiO_3), (4) B + Si (16.3μ M H_3BO_3 + 1.5 mM Na_2SiO_3), (5) Al (50 μM $AlCl_3$), (6) Al + B (50 μM $AlCl_3$ + 16.3 μM H_3BO_3), (7) Al + Si (50 μM $AlCl_3$ + 1.5 mM Na_2SiO_3), and (8) Al + B + Si (50 μM $AlCl_3$ + 1.5 mM Na_2SiO_3 + 16.3 μM H_3BO_3). Scale bar, 1 cm. Bars indicate the means of three replicates ± SD. The different letters (a, b, c, d) in each sub-figure represent significant differences at $p < 0.05$.

2.2. Effect of B, Si, and Their Combination on Al Toxicity-Induced PCD and Al Accumulation in Soybean Roots

Treated soybean root tips were stained with Evans blue to investigate the effects of Si, B, and their combination on Al toxicity-induced programmed cell death (PCD) in soybean roots. As shown in Figure 2A, Si and B alone and their combination did not induce PCD in soybean root tips in the absence of Al stress. Upon exposure to 50 μM Al stress, significantly elevated PCD levels were observed in soybean root tips, which were significantly alleviated by B and Si treatment alone and their combination (B + Si). Al stress resulted in the accumulation of Al in root tips, which may contribute to root tip PCD. Therefore, the effects of B, Si, and their combination on Al accumulation in root tips were analyzed using Chrome Azurol S staining and the flame atomic absorption spectrophotometry (FAAS) method. The results revealed that Al stress greatly increased the accumulation of Al in roots, whereas the addition of B and Si alone and their combination reduced Al accumulation by 43.2%, 32.3%, and 28.7%, respectively (Figure 2B,C).

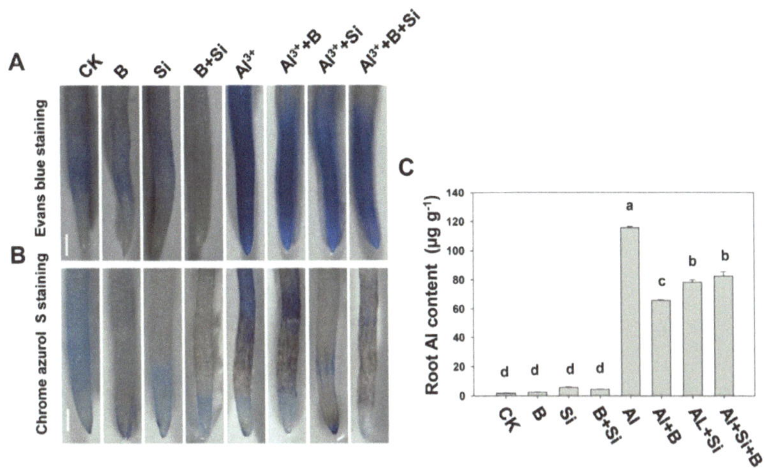

Figure 2. Effect of B, Si, and their combination on Al toxicity-induced PCD and Al^{3+} accumulation in soybean roots. (**A**) Root PCD; (**B**,**C**) root Al content. The bars indicate the means of three replicates ± SD. Scale bar, 1 mm. Different letters (a, b, c, d) in each sub-figure represent significant differences at $p < 0.05$.

2.3. Effects of B, Si, and Their Combination on Lipid Peroxidation and H_2O_2 Content in Soybean Roots under Al Stress

ROS-induced lipid membrane peroxidation and H_2O_2 production induce oxidative damage at the cellular level. The experiment results demonstrated that Al stress rapidly produced MDA and H_2O_2, thereby disrupting cellular structures (Figure 3). A noticeable rise in MDA levels was observed solely in the roots of soybean plants treated with Al^{3+} when compared to plants treated with Si, B, or their combination. The effects of aluminum toxicity were significantly reduced in plants treated solely with silicon or boron compared to plants treated with both silicon and boron. B and Si treatment alone displayed the lowest MDA levels in the roots, showing a reduction of 22.5% and 18.8%, respectively. In contrast, the combination-treated plants exhibited a 5.2% reduction in MDA levels compared to plants exposed to Al^{3+} alone (Figure 3A). Similarly, Al stress significantly elevated the levels of H_2O_2 in roots compared to plants treated with B, Si, and their combination. B, Si, and their combination significantly reduced the production and accumulation levels of H_2O_2 compared to Al-only treated plants, showing a reduction of 11.3%, 8.8%, and 5.6%, respectively (Figure 3B). In summary, B, Si, and their combination significantly decreased H_2O_2 and MDA levels induced by Al toxicity.

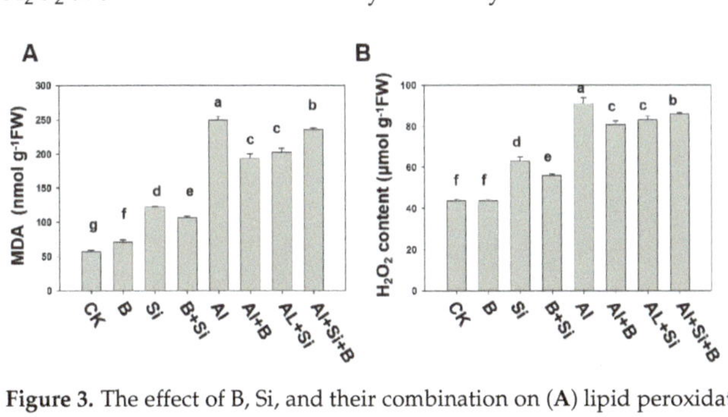

Figure 3. The effect of B, Si, and their combination on (**A**) lipid peroxidation (MDA) and (**B**) H_2O_2 modulation in soybean roots under Al-induced toxicity. The bars represent the means of three replicates ± SD. The different letters (a, b, c, d, e, f, g) in each sub-figure represent significant differences at ($p < 0.05$).

2.4. Effects of B, Si, and Their Combination on Antioxidant Activities in Soybean Roots under Al Stress

The roles of B, Si, and B + Si in mitigating Al toxicity were assessed by analyzing the activity of antioxidant enzymes (APX, CAT, POD, and SOD) in soybean roots under Al stress (Figure 4). In the absence of Al treatment, B, Si, and their combination treatment did not affect CAT activity levels in roots compared to the CK group, as shown in Figure 4B. However, the CAT activities in roots were significantly reduced when plants were exposed to Al stress. The results revealed that B, Si, and their combination significantly enhanced CAT activity by 1.2-, 1.07-, and 1.1-fold, respectively, compared to Al treatment alone. Likewise, SOD activities in roots were significantly increased in all treatments under Al stress compared to the Al treatment alone, showing an increase of 1.4-, 1.3-, and 1.2-fold for B, Si, and their combination, respectively (Figure 4D). Similarly, in terms of POD and APX mediation, Al stress significantly attenuated the activities of both POD and APX compared to the CK group and the B, Si, and B + Si treatment groups without Al (Figure 4A,C). Compared to Al treatment alone, B treatment under Al stress significantly increased root POD activity levels by 1.1-fold, followed by Si and the combination treatment, showing a 1.03- and 1.07-fold increase, respectively. As shown in Figure 4A, the APX activity demonstrated a marked increase of 1.4-fold in roots treated with Al + B. In contrast, Al + Si and Al + B + Si treatments resulted in a 1.3- and 1.2-fold increase, respectively, compared to Al treatment alone. In summary, supplementation with B and Si, as well as their combination, improved the resistance of soybean seedlings to Al stress. This was achieved by increasing the activity of antioxidant enzymes, which in turn reduced the negative effects of oxidative stress caused by Al toxicity in the soybean seedlings.

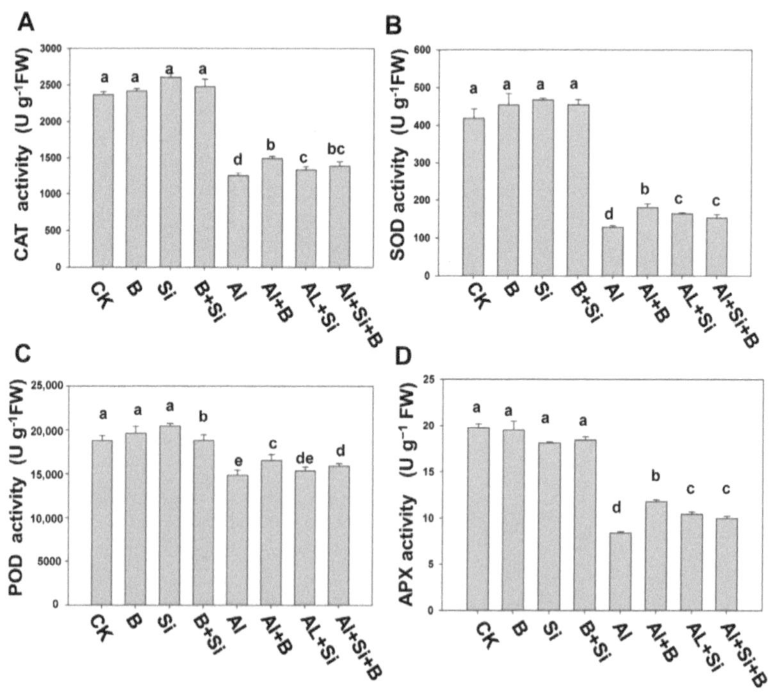

Figure 4. Effects of B, Si, and their combination on antioxidant activities in soybean roots under Al stress. (**A**) APX activity, (**B**) CAT activity, (**C**) POD activity, (**D**) SOD activity. The bars indicate the means of three replicates ± SD. The different letters (a, b, c, d, e) in each sub-figure represent significant differences at $p < 0.05$.

2.5. Effects of B, Si, and Their Combination on the Secretion of Citric Acid in Soybean Roots under Al Stress

Studies have shown that soybeans (*Glycine max*) resist Al stress by secreting citric acid. Our results are consistent with this evidence, indicating a notable rise in the concentration of citric acid in the treatment solution due to Al stress compared to plants that were not subjected to Al treatment (Figure 5A). Interestingly, compared with the control without Al treatment, B, Si, and their combination increased the level of citric acid in the treatment solution when plants were exposed to Al stress. However, B, Si, and their combination treatment showed significantly reduced citric acid levels in the treatment solution under Al stress (Figure 5A). B, Si, and their combination substantially increased citric acid levels compared to Al treatment alone (Figure 5B). Similarly, B, Si, and their combination significantly enhanced the activity of citrate synthase (CS) by 1.3-, 1.1-, and 1.1-fold, respectively, compared to Al treatment alone (Figure 5C).

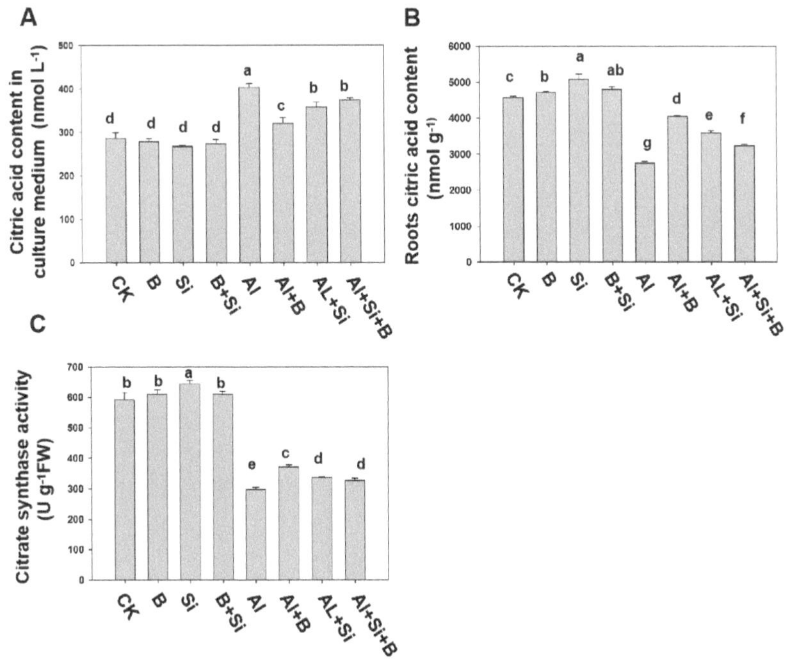

Figure 5. Effects of Si, B, and their combination on the secretion of citric acid in soybean roots under Al stress. (**A**) Citric acid in culture medium, (**B**) citric acid in roots, (**C**) citric synthase (CS) activity in roots. The bars indicate the means of three replicates ± SD. The different letters (a, b, c, d, e, f, g) in each sub-figure represent significant differences at $p < 0.05$.

3. Discussion

Al toxicity severely limits plant growth and productivity by negatively affecting plant physiology and metabolic processes in acidic soils [1,7,40,41]. Numerous studies have demonstrated that Al stress inhibits plant root elongation. This study revealed the inhibited root growth potential and parameters under Al toxicity (50 µM Al) (Figure 1). Several mechanisms have been proposed for plant resistance to aluminum toxicity, and it has become evident that B and Si play an indispensable role in plant tolerance to aluminum toxicity [42]. In our current study, the B + Si combination and B and Si alone significantly reversed the root growth inhibition induced by Al toxicity (Figure 1C), which is consistent with earlier reports in citrus and date palm seedlings [36]. Programmed cell death (PCD) plays an essential role in plant development and stress. Al stress has been found to induce PCD in a variety of plants, such as peanut, pea, and barley [3]. Al-induced PCD of peanut root tip cells is responsible for Al toxicity and shows a direct relationship with the suppression of root elongation caused by Al [43]. Consistent with other studies, our results indicated that Al toxicity causes PCD in soybean roots, while B, Si, and their combination alleviated the PCD induced by Al toxicity (Figure 2A). There is evidence that Al-induced PCD is closely related to ROS bursts [44]. Supplementation with B, Si, and their combination alleviated oxidative damage and thereby reduced PCD caused by Al toxicity, which may be one of the mechanisms by which they alleviate root growth inhibition due to Al toxicity. However, the specific mechanism remains to be further studied in the future.

B and Si supplementation has been hypothesized to enhance plant antioxidant defense systems, thereby mitigating Al toxicity [42,45]. As stated above, MDA and H_2O_2 levels were significantly elevated in soybean roots under Al stress (Figure 3A,B). The accumulation of H_2O_2 may enhance the production of hydroxyl radicals and thus lipid peroxidation (Figure 2A,B), which may explain the inhibitory effect of Al toxicity on root elongation. Moreover, higher root PCD levels and relative root length parallelled higher H_2O_2 production (Figure 1A,C, Figures 2A and 3B), indicating that Al-induced oxidative bursts play an active role in the induction of PCD. Moreover, B, Si, and their combination reduced MDA and H_2O_2 levels and alleviated root growth inhibition under Al stress (Figures 1 and 2), suggesting that B, Si, and their combination mitigates Al toxicity by reducing the accumulation of ROS [36,42,46].

Plants have developed various protective mechanisms, both enzymatic and non-enzymatic, to efficiently eliminate ROS. APX, CAT, POD, and SOD are the four key antioxidant enzymes in the enzymatic detoxification system. Our results indicate that aluminum toxicity significantly inhibited SOD activities, while the addition of B, Si, and their combination alleviated this inhibition (Figure 4D). It can be confirmed that the addition of B, Si, and their combination is related to the reduction in the accumulation of free radicals in cells and the damage to the membrane system. B, Si, and their combination can activate APX, CAT, and POD, which are the antioxidant enzymes responsible for scavenging H_2O_2 under Al stress (Figure 4A–C). Consistent with our findings, Pontigo et al. observed that Si significantly reduced lipid peroxidation induced by Al toxicity, and similarly, B application reduced oxidative damage to plants subjected to Al stress [42,45,46]. In summary, supplementation with B, Si, and their combination mitigated oxidative damage by reducing the accumulation of ROS and MAD and triggering the up-regulation of the antioxidant system (Figures 3 and 4).

Many studies have confirmed a significant correlation between Al toxicity resistance and the level of Al-dependent release of organic acids; in particular, the exudation of citric acid from roots determines the Al resistance of the plant [1,33]. Under Al stress, the treatment solution experienced a noteworthy decline in citric acid levels when B, Si, or a combination of both were introduced, as evidenced by our results (Figure 5A). However, addition of B, Si, and their combination enhanced citric acid levels and CS activity in roots under Al stress (Figure 5B,C). This may suggest that although both B and Si are involved in regulating the metabolic pathway of citric acid synthesis in soybean roots, they do not serve as the primary mechanism for alleviating Al toxicity-induced inhibition of root growth. However, there are still many problems to be solved concerning how B, Si, and their interactions regulate the internal and external tolerance mechanisms of plants subjected to Al toxicity, especially the organic acid pathway.

Based on our results showing the positive roles of Si, B, and their combination in mitigating Al toxicity in soybeans, adding B and Si to the soil or through foliar sprays may help slow the effects of Al toxicity in soybeans. For areas where Al toxicity in the soil is a serious problem, you can consider adding B and Si amendments to increase the biological activity of the soil, thereby improving the growing environment of plants. However, acidic soil is a complex environment influenced by numerous factors, including plant species and age, necessitating further research to provide a theoretical basis for B and Si to mitigate Al toxicity. In the future, through in-depth research on the mechanism of these trace elements, their application methods and dosages in plant cultivation can be optimized to achieve the best reduction in Al toxicity. More experiments and trials are still needed to verify the molecular mechanisms and actual roles of B and Si in mitigating Al toxicity. Taken together, B and Si have potential applications in mitigating Al toxicity in plants, but further research and practice are needed to determine the best application methods and effect evaluations.

4. Materials and Methods

4.1. Plant Growth and Experimental Design

Soybean seeds underwent sterilization by immersing them in a solution containing 0.5% sodium hypochlorite for a period of 15 min. The seeds were rinsed several times with distilled water, germinated on trays for 3 days, and watered with distilled water daily. After 3 days of seed germination, seedlings of uniform length with a uniform number of roots were selected and transferred to 15 L plastic pots, each with 20 seedlings, containing 25% strength Hoagland solution without Al. After true leaves emerged, seedlings with similar root lengths were transferred to solutions containing 0.5 mM $CaCl_2$ at pH 4.3 and pre-treated for 1 day. The seedlings were then divided into the following eight groups: (1) CK (solutions containing 0.5 mM $CaCl_2$, pH 4.3), (2) B (solutions containing 0.5 mM $CaCl_2$ and 16.3 μM H_3BO_3, pH 4.3), (3) Si (solutions containing 0.5 mM $CaCl_2$ and 1.5 mM Na_2SiO3, pH 4.3), (4) B + Si (solutions containing 0.5 mM $CaCl_2$, 16.3 μM H_3BO_3, and 1.5 mM Na2SiO3, pH 4.3), (5) Al (solutions containing 0.5 mM $CaCl_2$ and 50 μM $AlCl_3$, pH 4.3), (6) Al + B (solutions containing 0.5 mM $CaCl_2$, 50 μM $AlCl_3$, and 16.3 μM H_3BO_3, pH 4.3), (7) Al + Si (solutions containing 0.5 mM $CaCl_2$, 50 μM $AlCl_3$, and 1.5 mM Na_2SiO_3, pH 4.3), and (8) Al + B + Si (solutions containing 0.5 mM $CaCl_2$, 50 μM $AlCl_3$, 16.3 μM H_3BO_3, and 1.5 mM Na_2SiO_3) (Figure 6). All the solutions were prepared in distilled water. The plastic pots were prepared in triplicate and placed in a greenhouse in a randomized arrangement; fresh solutions were replaced every other day. After 7 days of treatment, the lengths and fresh weights of roots were measured, and seedling roots were carefully collected and put in a −80 °C refrigerator after quick-freezing with liquid nitrogen.

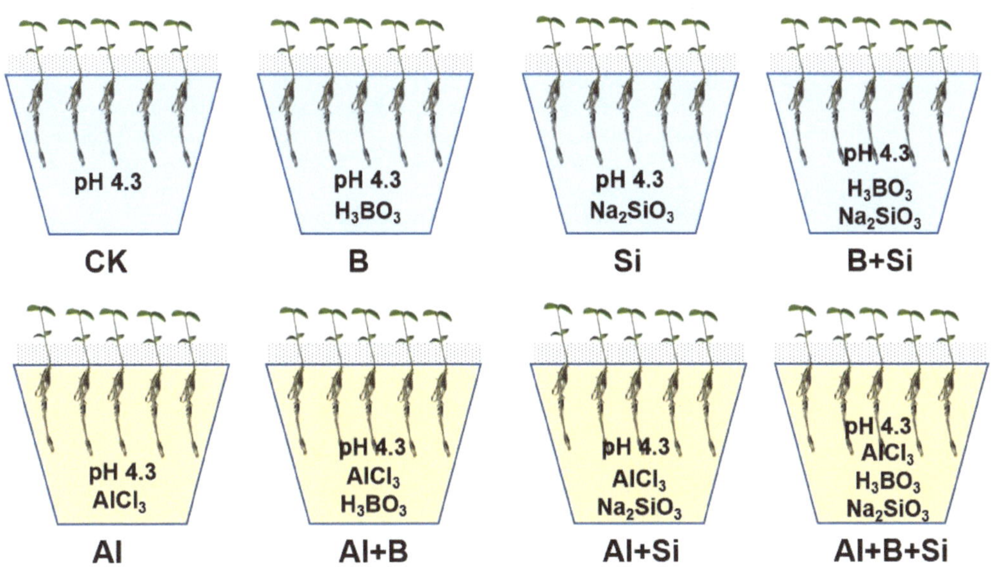

Figure 6. An illustrative experimental diagram explaining the eight different treatments.

4.2. Root Biomass and Length Measurement

In each repetition, five plants were randomly selected, and their roots were separated and dried on paper towels for measurement using an electronic analytical balance. The length of the roots was evaluated with the help of a graduated ruler.

4.3. Determination of Al Content in Root

The Al content in the dried roots was analyzed. The root samples were ashed at 500 °C for 8 h and treated with 2 M HCl. Subsequently, 4 g of sodium hydroxide and 1 g of sodium peroxide was added to the mixture, which was then melted in a muffle furnace at 650 °C for 15 min. The resulting solution was filtered, and the total amount of Al was quantified by FAAS at 324.7 nm [47]. Al content was localized in roots by Chrome Azurol S staining. The roots were fixed in 2.5% (*v*/*v*) glutaraldehyde in 0.1 M sodium phosphate buffer (pH 7.2). Subsequently, the samples were stained with 0.5% Chrome Azurol S for 1 h. A positive reaction was determined by bluish staining. Images were taken using a light microscope (Axio Zoom.V16, Carl Zeiss, Oberkochen, Germany) equipped with software to capture images (AxioVision software Release 4.8, Carl Zeiss Vision).

4.4. Root Cell Death Detection

Al-induced cell death in the roots was assessed following Evans blue staining (0.005%, *w*/*v*). The roots were mixed and incubated with the dye for 10 min (Delphine Arbelet-Bonnin2018). Dead cells appeared with a bluish stain. Images were captured using a light microscope.

4.5. Lipid Peroxidation (MDA) and Hydrogen Peroxide (H_2O_2) Content in Soybean Roots

The extent of lipid peroxidation and the formation of MDA and H_2O_2 content in the roots was measured using an MDA content assay kit (Shanghai Sangon Biotech, Shanghai, China) and a commercially available Hydrogen Peroxide (H_2O_2) Content Assay Kit (Shanghai Sangon Biotech, Shanghai, China). Briefly, approximately 0.1 g of sample was placed into 1 mL of extraction solution and then centrifuged at 4 °C and 12,000 rpm for 10 min. Subsequently, 300 µL of supernatant was added to a reaction solution containing 0.5% thiobarbituric acid (TBA), followed by incubation at 95 °C for 30 min and centrifugation. The MDA content was determined and calculated according to the manufacturer's instructions. Furthermore, 0.1 g of sample was placed into 1 mL of acetone. The mixture was homogenized in an ice bath and then centrifuged. Subsequently, 2.0 mL of a solution containing 20% sulfuric acid (*v*/*v*) and 0.1% titanium tetrachloride (*v*/*v*) was added to the 200 µL supernatant, followed by centrifugation. The supernatants were analyzed at 415 nm using an instrument to obtain the concentration of H_2O_2.

4.6. Determination of Antioxidant Enzyme Activities in Soybean Roots

Commonly analyzed antioxidant enzymes in roots include ascorbate peroxidase (APX), catalase (CAT), peroxidase (POD), and superoxide dismutase (SOD). The following kits were used for detection: APX Activity Assay Kit, CAT Activity Assay Kit, POD Activity Assay Kit, and SOD Activity Assay Kit (Shanghai Sangon Biotech, Shanghai, China). Briefly, 0.1 g of frozen sample was homogenized and extracted in PBS buffer. After centrifugation, the supernatant was collected via pipetting for enzyme activity determination. Subsequently, the activities of these enzymes were measured according to the manufacturer's instructions. The activity of superoxide dismutase (SOD; EC 1.15.1.1) was measured using the total superoxide dismutase assay kit with WST-8. In the coupled reaction system of WST-8 and xanthine oxidase (XO), the SOD enzyme activity in the reaction system was defined as one unit of enzyme activity (U/mL) when the inhibition percentage reached 50%. The peroxidase (POD; EC 1.11.1.7) activity was measured by observing the oxidation of guaiacol after the addition of H_2O_2. One unit (U) of POD activity was defined as an increase of 0.5 in absorbance at 470 nm per minute per gram of root sample in the reaction system. Catalase (CAT; EC 1.11.1.6) activity was determined by monitoring the decrease in absorbance at 240 nm, indicating a decrease in H_2O_2 concentration. One unit (U) of

CAT activity was defined as the degradation of 1 M H_2O_2 per minute per gram of root sample. The activity of the enzyme ascorbate peroxidase (APX; EC 1.11.1.11) was measured by monitoring the reduction in absorbance at 290 nm caused by the oxidation of ascorbic acid by the APX-H_2O_2 complex. The unit (U) of APX activity is defined as the oxidation of 1 M of ascorbate per minute per gram of root sample [47].

4.7. Determination of Citrate Synthase Activity and Citric Acid Content in Soybean Roots and Citric Acid Content in Treatment Solution

The following kits were used for detection: Citrate Synthase (CS) Activity Assay Kit (Shanghai Sangon Biotech, Shanghai, China), and Citric Acid (CA) Content Assay Kit (Shanghai Sangon Biotech, Shanghai, China). The activities of CS enzymes and the citric acid content, as well as the citric acid content in the treatment solution, were measured following the manufacturer's instructions.

4.8. Statistical Analysis

Microsoft Office Excel 2019 and SigmaPlot 15.0 software were used to create all graphs and data analyses, and Photoshop CX6 was adopted for the drawing. One-way ANOVA and Duncan's tests ($p < 0.05$) were performed to detect significant differences among the various treatments, denoted by distinct lower-case letters.

5. Conclusions

This study aimed to investigate the roles of B, Si, and their combination in alleviating Al toxicity in soybeans. The results revealed that the combined application significantly improved the growth and biomass of soybean roots exposed to Al toxicity compared to B and Si treatments alone. Our results indicated that Al toxicity causes PCD in soybean roots, while B, Si, and their combination all alleviated the PCD induced by Al toxicity. The administration of a combination of B and Si to soybeans resulted in a notable decrease in Al-induced oxidative stress, which was supported by the decreased levels of MAD and H_2O_2 production. Moreover, B, Si, and their combination significantly enhanced the plants' antioxidant systems by up-regulating antioxidant enzyme activities (APX, CAT, POD, and SOD) compared to plants under Al stress alone. Overall, supplementation with B, Si, and their combination alleviated oxidative damage and reduced PCD induced by Al toxicity, which may be one of the mechanisms involved in alleviate root growth inhibition due to Al toxicity. However, the exact mechanisms underlying the regulatory roles of B and Si in the mechanisms of internal and external tolerance against Al toxicity in plants remain incompletely understood, and there are still many issues to be explored in future research: (1) How do Si, B and their complexes mitigate Al toxicity at the cellular and molecular levels in soybeans? (2) How do Si and B complexes interact to better mitigate the effects of Al toxicity on soybeans?

Author Contributions: Project administration and supervision, Y.W.; funding acquisition, Y.W.; investigation, H.C.; data curation, H.C. and S.W.; writing—original draft, S.W.; writing—review and editing, Y.W. All authors have read and agreed to the published version of the manuscript.

Funding: This research received no external funding.

Data Availability Statement: Data is contained within the article.

Conflicts of Interest: The authors declare no conflict of interest.

References

1. Kochian, L.V.; Piñeros, M.A.; Liu, J.; Magalhaes, J.V. Plant adaptation to acid soils: The molecular basis for crop aluminum resistance. *Annu. Rev. Plant Biol.* **2015**, *66*, 571–598. [CrossRef]
2. Bojórquez-Quintal, E.; Escalante-Magaña, C.; Echevarría-Machado, I.; Martínez-Estévez, M. Aluminum, a friend or foe of higher plants in acid soils. *Front. Plant Sci.* **2017**, *8*, 1767. [CrossRef]

3. Huang, W.; Yang, X.; Yao, S.; LwinOo, T.; He, H.; Wang, A.; Li, C.; He, L. Reactive oxygen species burst induced by aluminum stress triggers mitochondria-dependent programmed cell death in peanut root tip cells. *Plant Physiol. Biochem.* **2014**, *82*, 76–84. [CrossRef]
4. Ofoe, R.; Gunupuru, L.R.; Wang-Pruski, G.; Fofana, B.; Thomas, R.H. Abbey, Lord Seed priming with pyroligneous acid mitigates aluminum stress, and promotes tomato seed germination and seedling growth. *Plant Stress* **2022**, *4*, 100083. [CrossRef]
5. Guo, P.; Qi, Y.P.; Cai, Y.T.; Yang, T.Y.; Yang, L.T.; Huang, Z.R.; Chen, L.S. Aluminum effects on photosynthesis, reactive oxygen species and methylglyoxal detoxification in two Citrus species differing in aluminum tolerance. *Tree Physiol.* **2018**, *38*, 1548–1565. [CrossRef]
6. Ofoe, R.; Thomas, R.H.; Asiedu, S.K.; Wang-Pruski, G.; Fofana, B.; Abbey, L. Aluminum in plant: Benefits, toxicity and tolerance mechanisms. *Front. Plant Sci.* **2023**, *13*, 1085998. [CrossRef]
7. Yamamoto, Y. Aluminum toxicity in plant cells: Mechanisms of cell death and inhibition of cell elongation. *Soil Sci. Plant Nutr.* **2019**, *65*, 41–55. [CrossRef]
8. Chen, D.; Chen, D.; Xue, R.; Long, J.; Lin, X.; Lin, Y.; Jia, L.; Zeng, R.; Song, Y. Effects of boron, silicon and their interactions on cadmium accumulation and toxicity in rice plants. *J. Hazard. Mater.* **2019**, *367*, 447–455. [CrossRef]
9. Chauhan, D.K.; Yadav, V.; Vaculík, M.; Gassmann, W.; Pike, S.; Arif, N.; Singh, V.P.; Deshmukh, R.; Sahi, S.; Tripathi, D.K. Aluminum toxicity and aluminum stress-induced physiological tolerance responses in higher plants. *Crit. Rev. Biotechnol.* **2021**, *41*, 715–730. [CrossRef]
10. Liu, H.; Zhu, R.; Shu, K.; Lv, W.; Wang, S.; Wang, C. Aluminum stress signaling, response, and adaptive mechanisms in plants. *Plant Signal. Behav.* **2022**, *17*, 2057060. [CrossRef]
11. Sachdev, S.; Ansari, S.A.; Ansari, M.I.; Fujita, M.; Hasanuzzaman, M. Abiotic stress and reactive oxygen species: Generation, signaling, and defense mechanisms. *Antioxidants* **2021**, *10*, 277. [CrossRef]
12. Du, H.; Ryan, P.R.; Liu, C.; Li, H.; Hu, W.; Yan, W.; Huang, Y.; He, W.; Luo, B.; Zhang, X.; et al. *ZmMATE6* from maize encodes a citrate transporter that enhances aluminum tolerance in transgenic *Arabidopsis thaliana*. *Plant Sci.* **2021**, *311*, 111016. [CrossRef]
13. Melo, J.O.; Martins, L.G.C.; Barros, B.A.; Pimenta, M.R.; Lana, U.G.P.; Duarte, C.E.M.; Pastina, M.M.; Guimaraes, C.T.; Schaffert, R.E.; Kochian, L.V.; et al. Repeat variants for the SbMATE transporter protect sorghum roots from aluminum toxicity by transcriptional interplay in cis and trans. *Proc. Natl. Acad. Sci. USA* **2019**, *116*, 313–318. [CrossRef]
14. Ligaba, A.; Katsuhara, M.; Ryan, P.R.; Shibasaka, M.; Matsumoto, H. The *BnALMT1* and *BnALMT2* genes from rape encode aluminum-activated malate transporters that enhance the aluminum resistance of plant cells. *Plant Physiol.* **2006**, *142*, 1294–1303. [CrossRef]
15. Zhu, X.; Wang, P.; Bai, Z.; Herde, M.; Ma, Y.; Li, N.; Liu, S.; Huang, C.F.; Cui, R.; Ma, H.; et al. Calmodulin-like protein CML24 interacts with CAMTA2 and WRKY46 to regulate ALMT1-dependent Al resistance in *Arabidopsis thaliana*. *New Phytol.* **2022**, *233*, 2471–2487. [CrossRef]
16. Wang, Z.; Liu, Y.; Cui, W.; Gong, L.; He, Y.; Zhang, Q.; Meng, X.; Yang, Z.; You, J. Characterization of GmMATE13 in its contribution of citrate efflux and aluminum resistance in soybeans. *Front. Plant Sci.* **2022**, *13*, 1027560. [CrossRef]
17. Wen, Z.; Yang, M.; Han, H.; Fazal, A.; Liao, Y.; Ren, R.; Yin, T.; Qi, J.; Sun, S.; Lu, G.; et al. Mycorrhizae Enhance Soybean Plant Growth and Aluminum Stress Tolerance by Shaping the Microbiome Assembly in an Acidic Soil. *Microbiol. Spectr.* **2023**, *11*, e0331022. [CrossRef]
18. Zhang, W.; Xie, Z.; Lang, D.; Cui, J.; Zhang, X. Beneficial effects of silicon on abiotic stress tolerance in legumes. *J. Plant Nutr.* **2017**, *40*, 2224–2236. [CrossRef]
19. Zhu, Y.X.; Gong, H.J.; Yin, J.L. Role of silicon in mediating salt tolerance in plants: A Review. *Plants* **2019**, *8*, 147. [CrossRef]
20. Bhat, J.A.; Shivaraj, S.M.; Singh, P.; Navadagi, D.B.; Tripathi, D.K.; Dash, P.K.; Solanke, A.U.; Sonah, H.; Deshmukh, R. Role of silicon in mitigation of heavy metal stresses in crop plants. *Plants* **2019**, *8*, 71. [CrossRef]
21. Dorneles, A.O.S.; Pereira, A.S.; Sasso, V.M.; Possebom, G.; Tarouco, C.P.; Schorr, M.R.W.; Rossato, L.; Ferreira, P.A.A.; Tabaldi, L.A. Aluminum stress tolerance in potato genotypes grown with silicon. *Bragantia* **2019**, *78*, 12–25. [CrossRef]
22. Souri, Z.; Khanna, K.; Karimi, N.; Ahmad, P. Silicon and Plants: Current Knowledge and Future Prospects. *J. Plant Growth Regul.* **2021**, *40*, 906–925. [CrossRef]
23. Etesami, H.; Jeong, B.R. Silicon (Si): Review and future prospects on the action mechanisms in alleviating biotic and abiotic stresses in plants. *Ecotoxicol. Environ. Saf.* **2018**, *147*, 881–896. [CrossRef]
24. Pavlovic, J.; Kostic, L.; Bosnic, P.; Kirkby, E.A.; Nikolic, M. Interactions of Silicon with Essential and Beneficial Elements in Plants. *Front. Plant Sci.* **2021**, *12*, 697592. [CrossRef]
25. Cheraghi, M.; Motesharezadeh, B.; Mousavi, S.M.; Ma, Q.; Ahmadabadi, Z. Silicon (Si): A Regulator Nutrient for Optimum Growth of Wheat Under Salinity and Drought Stresses—A Review. *J. Plant Growth Regul.* **2023**, *42*, 5354–5378. [CrossRef]
26. Xiao, Z.; Ye, M.; Gao, Z.; Jiang, Y.; Zhang, X.; Nikolic, N.; Liang, Y. Silicon Reduces Aluminum-Induced Suberization by Inhibiting the Uptake and Transport of Aluminum in Rice Roots and Consequently Promotes Root Growth. *Plant Cell Physiol.* **2022**, *63*, 340–352. [CrossRef]
27. Xiao, Z.; Yan, G.; Ye, M.; Liang, Y. Silicon relieves aluminum-induced inhibition of cell elongation in rice root apex by reducing the deposition of aluminum in the cell wall. *Plant Soil* **2021**, *462*, 189–205. [CrossRef]
28. Vega, I.; Nikolic, M.; Pontigo, S.; Godoy, K.; de La Luz Mora, M.; Cartes, P. Silicon improves the production of high antioxidant or structural phenolic compounds in barley cultivars under aluminum stress. *Agronomy* **2019**, *9*, 388. [CrossRef]

29. Kopittke, P.M.; Gianoncelli, A.; Kourousias, G.; Green, K.; McKenna, B.A. Alleviation of Al toxicity by Si Is associated with the formation of Al–Si complexes in root tissues of sorghum. *Front. Plant Sci.* **2017**, *8*, 2189. [CrossRef]
30. de Sousa, A.; Saleh, A.M.; Habeeb, T.H.; Hassan, Y.M.; Zrieq, R.; Wadaan, M.A.M.; Hozzein, W.N.; Selim, S.; Matos, M.; AbdElgawad, H. Silicon dioxide nanoparticles ameliorate the phytotoxic hazards of aluminum in maize grown on acidic soil. *Sci. Total Environ.* **2019**, *693*, 133636. [CrossRef]
31. Kohli, S.K.; Kaur, H.; Khanna, K.; Handa, N.; Bhardwaj, R.; Rinklebe, J.; Ahmad, P. Boron in plants: Uptake, deficiency and biological potential. *Plant Growth Regul.* **2023**, *100*, 267–282. [CrossRef]
32. Yan, L.; Li, S.; Cheng, J.; Zhang, Y.; Jiang, C. Boron-mediated lignin metabolism in response to aluminum toxicity in citrus (*Poncirus trifoliata* (L.) Raf.) root. *Plant Physiol. Biochem.* **2022**, *185*, 1–12. [CrossRef]
33. Riaz, M.; Yan, L.; Wu, X.; Hussain, S.; Aziz, O.; Jiang, C. Mechanisms of organic acids and boron induced tolerance of aluminum toxicity: A review. *Ecotoxicol. Environ. Saf.* **2018**, *165*, 25–35. [CrossRef]
34. Yan, L.; Riaz, M.; Wu, X.; Du, C.; Liu, Y.; Jiang, C. Ameliorative effects of boron on aluminum induced variations of cell wall cellulose and pectin components in trifoliate orange (*Poncirus trifoliate* (L.) Raf.) rootstock. *Environ. Pollut.* **2018**, *240*, 764–774. [CrossRef]
35. Yan, L.; Riaz, M.; Du, C.; Liu, Y.; Zeng, Y.; Jiang, C. Ameliorative role of boron to toxicity of aluminum in trifoliate orange roots. *Ecotoxicol. Environ. Saf.* **2019**, *179*, 212–221. [CrossRef]
36. Yan, L.; Riaz, M.; Liu, J.; Liu, Y.; Zeng, Y.; Jiang, C. Boron reduces aluminum deposition in alkali-soluble pectin and cytoplasm to release aluminum toxicity. *J. Hazard. Mater.* **2021**, *401*, 123388. [CrossRef]
37. Zhu, C.Q.; Cao, X.C.; Zhu, L.F.; Hu, W.J.; Hu, A.Y.; Abliz, B.; Bai, Z.G.; Huang, J.; Liang, Q.D.; Sajid, H.; et al. Boron reduces cell wall aluminum content in rice (*Oryza sativa*) roots by decreasing H 2 O 2 accumulation. *Plant Physiol. Biochem.* **2019**, *138*, 80–90. [CrossRef]
38. Liu, S.; Zhang, M.; Feng, F.; Tian, Z. Toward a "Green Revolution" for Soybean. *Mol. Plant* **2020**, *13*, 688–697. [CrossRef]
39. Wang, H.; Yin, X.; Du, D.; Liang, Z.; Han, Z.; Nian, H.; Ma, Q. *GsMYB7* encoding a R2R3-type MYB transcription factor enhances the tolerance to aluminum stress in soybean (*Glycine max* L.). *BMC Genom.* **2022**, *23*, 529. [CrossRef]
40. Sade, H.; Meriga, B.; Surapu, V.; Gadi, J.; Sunita, M.S.L.; Suravajhala, P.; Kavi Kishor, P.B. Toxicity and tolerance of aluminum in plants: Tailoring plants to suit to acid soils. *Biometals* **2016**, *29*, 187–210. [CrossRef]
41. He, H.; Li, Y.; He, L.F. Aluminum toxicity and tolerance in Solanaceae plants. *S. Afr. J. Bot.* **2019**, *123*, 23–29. [CrossRef]
42. Bilal, S.; Khan, A.; Imran, M.; Khan, A.L.; Asaf, S.; Al-Rawahi, A.; Al-Azri, M.S.A.; Al-Harrasi, A.; Lee, I.J. Silicon-and Boron-Induced Physio-Biochemical Alteration and Organic Acid Regulation Mitigates Aluminum Phytotoxicity in Date Palm Seedlings. *Antioxidants* **2022**, *11*, 1063. [CrossRef]
43. He, H.; Huang, W.; Oo, T.L.; Gu, M.; He, L.F. Nitric oxide inhibits aluminum-induced programmed cell death in peanut (*Arachis hypoganea* L.) root tips. *J. Hazard. Mater.* **2017**, *333*, 285–292. [CrossRef]
44. Huang, J.; Han, R.; Ji, F.; Yu, Y.; Wang, R.; Hai, Z.; Liang, W.; Wang, H. Glucose-6-phosphate dehydrogenase and abscisic acid mediate programmed cell death induced by aluminum toxicity in soybean root tips. *J. Hazard. Mater.* **2022**, *425*, 127964. [CrossRef]
45. Pontigo, S.; Godoy, K.; Jiménez, H.; Gutiérrez-Moraga, A.; Mora, M.D.L.L.; Cartes, P. Silicon-mediated alleviation of aluminum toxicity by modulation of al/si uptake and antioxidant performance in ryegrass plants. *Front. Plant Sci.* **2017**, *8*, 642. [CrossRef]
46. Yan, L.; Li, S.; Cheng, J.; Liu, Y.; Liu, J.; Jiang, C. Boron contributes to excessive aluminum tolerance in trifoliate orange (*Poncirus trifoliata* (L.) Raf.) by inhibiting cell wall deposition and promoting vacuole compartmentation. *J. Hazard. Mater.* **2022**, *437*, 129275. [CrossRef]
47. Liu, C.; Cheng, H.; Wang, S.; Yu, D.; Wei, Y. Physiological and Transcriptomic Analysis Reveals That Melatonin Alleviates Aluminum Toxicity in Alfalfa (*Medicago sativa* L.). *Int. J. Mol. Sci.* **2023**, *24*, 17221. [CrossRef]

plants

MDPI

Review

Insights into Plant Sensory Mechanisms under Abiotic Stresses

Songsong Jin [1,2], Mengting Wei [1], Yunmin Wei [1,2] and Zhonghao Jiang [1,*]

[1] College of Life Sciences and Oceanography, Shenzhen University, Shenzhen 518060, China; jsssnd@szu.edu.cn (S.J.); 2100252009@email.szu.edu.cn (M.W.); weiym1024@163.com (Y.W.)
[2] College of Physics and Optoelectronic Engineering, Shenzhen University, Shenzhen 518060, China
* Correspondence: zh-jiang@szu.edu.cn; Tel.: +86-188-1066-8122

Abstract: As sessile organisms, plants cannot survive in harmful environments, such as those characterized by drought, flood, heat, cold, nutrient deficiency, and salt or toxic metal stress. These stressors impair plant growth and development, leading to decreased crop productivity. To induce an appropriate response to abiotic stresses, plants must sense the pertinent stressor at an early stage to initiate precise signal transduction. Here, we provide an overview of recent progress in our understanding of the molecular mechanisms underlying plant abiotic stress sensing. Numerous biomolecules have been found to participate in the process of abiotic stress sensing and function as abiotic stress sensors in plants. Based on their molecular structure, these biomolecules can be divided into four groups: Ca^{2+}-permeable channels, receptor-like kinases (RLKs), sphingolipids, and other proteins. This improved knowledge can be used to identify key molecular targets for engineering stress-resilient crops in the field.

Keywords: abiotic stresses; crop productivity; sensors; Ca^{2+}-permeable channels; RLKs; sphingolipids

Citation: Jin, S.; Wei, M.; Wei, Y.; Jiang, Z. Insights into Plant Sensory Mechanisms under Abiotic Stresses. *Plants* **2024**, *13*, 1907. https://doi.org/10.3390/plants13141907

Academic Editors: Violetta Katarzyna Macioszek, Iwona Ciereszko and Andrzej K. Kononowicz

Received: 22 May 2024
Revised: 8 July 2024
Accepted: 8 July 2024
Published: 10 July 2024

1. Introduction

As sessile organisms, plants are unable to move to favorable environments and must trigger numerous responses to survive when they are challenged by adverse environmental conditions, such as drought, flooding, heat, cold, nutrient deficiency, and salt or toxic metal stress. These abiotic stresses impair plant development and reproduction, leading to decreased crop productivity [1–4]. Therefore, understanding the molecular basis of how plants perceive and respond to abiotic stresses is critical for food security worldwide.

Relative to the identification of downstream cellular signaling pathways and physiological responses to these abiotic stresses, research on plant sensory mechanisms has long been lacking. In nature, multiple abiotic stresses often occur simultaneously, such as flood stress, which simultaneously imposes hypoxia, photosynthesis reduction, and mechanical stress [5,6]. Furthermore, even an individual abiotic stress can trigger more than one stress; for example, salt stress can induce three stresses: osmotic stress, ionic stress, and secondary stress (such as oxidative stress) [7,8]. This complexity of environmental abiotic stresses makes it difficult to identify abiotic stress sensors in plants. As a pioneer, the primary abiotic stress sensors should detect the occurrence of the pertinent stressor at an early stage and convert external stimuli to cellular signals. It is difficult to demonstrate that a biomolecule directly senses stress. The identification of most putative stress sensors is based on indirect approaches. For instance, impairing the function of sensors is expected to affect the levels of second messengers such as calcium (Ca^{2+}), reactive oxygen species (ROS), nitric oxide (NO), and phospholipids [8–10]. To identify abiotic stress sensors, an early readout of the signaling pathway should be chosen to avoid complications from downstream signaling interactions and integration. Rapid changes in the intracellular concentration of free Ca^{2+} constitute one of the earliest signaling events in plants in response to external stressors [11,12]. A transient increase in cytosolic Ca^{2+} concentrations ($[Ca^{2+}]$) is a common theme of abiotic stress sensing and is detected by bioluminescence-based aequorin technology, in which the aequorin system is used to detect Ca^{2+} signaling induced by abiotic stimuli [11,12]. This spatiotemporally defined increase in the cytoplasmic Ca^{2+} concentration is caused by the combined effects of Ca^{2+} influx and efflux mechanisms, which depend on numerous plasma-membrane-localized Ca^{2+}-permeable channels or transporters [9,13–15]. In addition, different abiotic stresses activate different Ca^{2+} signature, including tissue-specific and stress-specific differences in Ca^{2+} peak amplitude, oscillation pattern, and Ca^{2+} wave propagation in cells [8–10].

Abiotic stress signals can simultaneously impact all parts of cells and are perceived at various locations of the cells [16]. Biological membrane-anchored proteins, such as Ca^{2+}-permeable channels and receptor-like kinases (RLKs), are regarded as candidates for abiotic stress sensors. For example, plasma-membrane-located cyclic-nucleotide-gated calcium channels (CNGCs) can monitor the fluidity of cellular membranes and sense extreme temperature stress [17–19]. The hydrogen-peroxide-induced Ca^{2+} increases 1 (HPCA1), as a leucine-rich repeat receptor kinase, is a representative of RLKs and functions in sensing oxidative stress [10]. Membrane lipids interact with membrane proteins and regulate their fuctions. Sphingolipids are a lipid composition of plasma membrane and modulate cellular signal transduction events [20]. Recently, glycosylinositol phosphoceramides (GIPCs) have been identified as a salt stress sensor [8]. In addition, non-membrane proteins located in the nucleus and cytosol are also found to participate in sensing abiotic stresses and are dependent on protein conformation changes, such as EARLY FLOWERING 3 (ELF3) [21] and phytochrome B (phyB) [22,23].

In this review, we summarize and discuss recent studies in the field of plant sensing of abiotic stresses. Numerous biomolecules have been found to participate in abiotic stress sensing and function as abiotic stress sensors in plants. Based on their molecular structure, these biomolecules can be divided into four groups: Ca^{2+}-permeable channels, RLKs, sphingolipids, and other proteins. Here, we focus on initial abiotic stress signal perception and do not elaborate on the roles of the above-mentioned biomolecules in other bioprocesses, such as plant growth and development and response to biotic stress, which have been discussed in detail in several recent reviews [16,24–28]. The analysis of the above-mentioned biomolecules based on molecular structure provides a new research direction for uncovering more abiotic stress sensors in plants. Understanding the mechanisms by which plants sense stressful environments will provide information for genetically engineering abiotic stress-tolerant crops to meet the demand for increased food production for an increasing world population.

2. Classification of Biomolecules Participating in Abiotic Stress Sensing

2.1. Ca^{2+}-Permeable Channels

Changes in the cytosolic Ca^{2+} concentration are regulated by Ca^{2+}-permeable channels. There are more than 40 genes encoding putative Ca^{2+} channels in the Arabidopsis genome and more than 20 genes encoding putative Ca^{2+} channels in the rice genome, some of which may function in abiotic stress sensing [14,29,30].

2.1.1. Hyperosmolarity-Gated Calcium-Permeable Channel Family of Proteins (OSCAs)

In Arabidopsis, the plasma-membrane-localized OSCA1 was identified as the first putative osmosensor using a Ca^{2+} imaging-based forward genetic screen (Table 1) [9]. The *osca1* mutant was isolated from EMS-mutagenized aequorin-expressing Arabidopsis seedlings, demonstrating decreased Ca^{2+} accumulation in guard cells and root cells when exposed to sorbitol but not in response to H_2O_2 or ABA. Under osmotic stress conditions, reduced primary root length and leaf area and attenuated transpiration were observed in *osca1* seedlings [9]. *AtOSCA1* is a founding member of a gene family that includes at least 15 ion channels with similar structural features in Arabidopsis, and the OSCA family may mediate the sensing of hyperosmotic conditions [9,31]. Recently, OSCA2.1 and OSCA2.2 were identified as essential hypo-osmosensors in Arabidopsis. The double-loss-function mutant *osca2.1/osca2.2* defected in pollen germination and was sensitive to hypo-osmolarity. OSCA2.1 and OSCA2.2 perceived extracellular water status and converted it into Ca^{2+} spiking in pollen [32]. In rice, a genome-wide survey found 11 genes in the entire *OSCA* family, containing a conserved DUF221 domain [31]. In maize, 12 *ZmOSCAs* were identified from the genome database of maize [33]. The expression profiles of *OSCAs* are variant in different tissues and under diverse abiotic stresses [33,34]. Subsequently, the cryo-electron microscopy (cryo-EM) structures and functions of the OSCA1 homologs CALCIUM PERMEABLE STRESS-GATED CATION CHANNEL1 (AtCSC1A/AtOSCA1.2), AtOSCA1.1, AtOSCA3.1, and OsOSCA1.2 were analyzed, providing a model of how they could mediate hyperosmolality sensing and transport pathway gating. They share similar protein fold and topology. OSCA channels are dimeric architectures containing 11 transmembrane (TM) helices, associated extracellular loops, intracellular loops, and an intracellular soluble domain [31,35–38]. However, the gating mechanisms of OSCAs remain poorly understood and need to be determined.

Table 1. List of discussed Ca^{2+}-permeable channels involved in plant sensing of abiotic stresses.

Groups	Type	Species	Biomolecule Names	Functions	References
I: Ca^{2+}-permeable channels	Hyperosmolarity-gated calcium-permeable channel family of proteins (OSCAs)	*Arabidopsis thaliana*	AtOSCA1	Hyper-osmosensors; Regulate primary root length, leaf area, and transpiration	[9]
			AtOSCA2.1 AtOSCA2.2	Hypo-osmosensors; Regulate pollen germination	[32]
	Cyclic-nucleotide-gated calcium channels (CNGCs)	*Oryza sativa*	OsCNGC9	Tolerance to chilling shock	[15]
			OsCNGC14	Tolerance to extreme temperatures; Regulate H_2O_2 accumulation	[17]
			OsCNGC16		
		Arabidopsis thaliana	AtCNGC2	Tolerance to extreme temperatures	[17,18]
			AtCNGC4		
			AtCNGC6	Tolerance to heat stress	[19]
	ANNEXIN proteins (ANNs)	*Arabidopsis thaliana*	AtANN1	Tolerance to extreme temperatures and salt stress	[39–41]
			AtANN4		[39,40,42]
	Glutamate receptor-like proteins (GLRs)	*Arabidopsis thaliana*	AtGLR3.3	Regulate membrane depolarization	[43]
			AtGLR3.4	Tolerance to touch and cold stress	[44]
	Mid1-complementing activity proteins (MCAs)	*Arabidopsis thaliana*	AtMCA1	Tolerance to mechanical stress	[13,45–47]
			AtMCA2		[45,46]

2.1.2. CNGCs

The fluidity of cellular membranes is related to Ca^{2+} influx and influenced by extreme temperatures, such as decreases caused by cold stress or increases caused by heat stress, which may be sensed by plasma membrane CNGCs (Table 1) [48]. CNGCs have been reported to play important roles in plant thermal sensing and acquired thermotolerance. In rice, *OsCNGC9* confers enhanced chilling tolerance by mediating cold-induced Ca^{2+} influx. The *cds1* (*cell death and susceptible to blast 1*), a loss-of-function mutant of *OsCNGC9*, is more sensitive to chilling shock. OsCNGC9 is phosphorylated and activated by OsSAPK8, a homolog of AtOST1, to trigger cytoplasmic calcium elevation. In addition, OsDREB1A positively regulates transcriptional expression of *OsCNGC9* [15]. OsCNGC14 and OsCNGC16 play vital roles in response to heat stimulation as well as low-temperature stress by impacting cytosolic calcium increase in rice. Their loss-of-function mutants, *cngc14* and *cngc16*, are generated by genome editing and have more withered and yellower leaves, lower survival rates, higher H_2O_2 accumulation, more cell damage, and stronger defect in calcium signal compared to the wild-type *cv Nipponbare* plants under a heat treatment or chilling treatment [17]. Furthermore, their homologs, AtCNGC2 and AtCNGC4, in Arabidopsis lead to chilling tolerance and acquired thermotolerance by regulating cytosolic Ca^{2+} [17,18]. Two loss-of-function mutants of *AtCNGC2* and *AtCNGC4*, *cngc2* and *cngc4*, display reduced hypocotyl elongation, rosette growth, and fresh weight under chilling conditions [17]. Disruption of *AtCNGC2*, *AtCNGC4*, and *AtCNGC6* results in a hyper-thermosensitive phenotype, such as a defect in heat-induced increases in cytosolic Ca^{2+}, reduced expression of heat shock protein (HSP) genes and abolished thermotolerance [18,19]. CNGCs are reported as putative abiotic stress sensors based on their direct regulation to cytosolic Ca^{2+} concentration in cell, but their activation mechanisms remain to be investigated.

2.1.3. ANNEXIN Proteins (ANNs)

ANNs act as Ca^{2+}-permeable transporters and regulate stress-induced cytosolic free Ca^{2+} ($[Ca^{2+}]_{cyt}$) elevations in plant response to abiotic stress (Table 1). In Arabidopsis, MYB30, a R2R3-MYB transcription factor, regulates negatively $[Ca^{2+}]_{cyt}$ in response to oxidation stress and heat stresses, which depend on the function of ANN proteins [39]. Transcriptional expression of *ANN1* and *ANN4* is repressed when MYB30 binds to their promoters. The single mutant *myb30* is sensitive to MV and heat treatment. However, the triple mutant *myb30 ann1 ann4* displays an attenuated phenotype compared with *myb30* under MV and heat treatment. In addition, the application of $LaCl_3$, a calcium channel blocker, can suppress the MV and heat sensitivity of *myb30* [39]. AtANN1 and AtANN4 are also found to mediate cold-induced Ca^{2+} influx and confer enhanced freezing tolerance in Arabidopsis. The loss of function of AtANN1 and AtANN4 exhibits reduced freezing tolerance. The single mutant *atann1*, *atann4-1* and double mutant *atann1 atann4-1* display lower survival rates than the wide type [40,42]. AtANN1 is phosphorylated by the OST1/SnRK2.6 kinase and acts downstream of OST1 in responses to freezing shock. The cascade linking OST1-AtANN1 triggers cold-induced $[Ca^{2+}]_{cyt}$ elevation and activates the cold response to acclimate to freezing conditions [40]. Under salt stress conditions, root epidermal net Na^+ influx in *atann1* is significantly higher than the wild type, while transient $[Ca^{2+}]_{cyt}$ increase is significantly lower in *atann1* than the wild type. This phenomenon shows that AtANN1 restricts Na^+ Influx and positively regulates NaCl-induced Ca^{2+} influx to improve salt stress tolerance [41]. Moreover, AtANN4 plays an important role in plant responses to salt stress by a negative feedback regulatory loop. AtANN4 interacts with the SOS2-SCaBP8 complex to increase salt-induced Ca^{2+} influx and, then, initiates a specific salt-induced calcium signal [42]. ANNs are found to play multifaceted roles in plants; however, knowledge about their functions is still in its infancy. Further research is needed to exploit the potential abiotic stress sensing mechanisms mediated by ANNs in plants.

2.1.4. Glutamate Receptor-like Proteins (GLRs)

Plant GLRs are ligand-gated ion channels and act as Ca^{2+}-permeable channels to mediate Ca^{2+} signaling (Table 1) [43,49]. GLRs can perceive environmental stress signals and convert them into specific stress-induced Ca^{2+} elevations that propagate to distant organs, to initiate stress defense responses in the whole plant [50]. In Arabidopsis, there are 20 members in the GLR family [51]. The application of glutamate (Glu) and glycine (Gly) triggers a very large and fast transient spike in $[Ca^{2+}]_{cyt}$ accompanied membrane depolarization, suggesting that Glu and Gly participate to control the ligand-mediated gating of calcium in plants [52,53]. The membrane depolarization and associated rise in cytosolic Ca^{2+} triggered by Glu are abolished in *glr3.3-1* and *glr3.3-2*, two loss-of-function mutants, indicating GLR3.3 mediates Glu-triggered Ca^{2+} influx [43]. Additionally, AtGLR3.4 participates in response to abiotic stress stimuli, such as touch, osmotic stress, or cold stress [44]. GLRs are emerging as a novel signaling molecule involved in plant sensory mechanisms under abiotic stresses. Additional research is needed to explore the specific functions and activation mechanisms of GLRs in plant responses to abiotic stresses.

2.1.5. Mid1-Complementing Activity Proteins (MCAs)

MCAs, which are Ca^{2+}-permeable mechanosensitive ion channels localized to the plasma membrane, have been identified as mechanosensitive sensors (Table 1). Arabidopsis transgenic lines MCA1ox (overexpressed *AtMCA1* cDNA) roots accumulate a greater extent of Ca^{2+} than wide-type and *mca1-null* (a T-DNA insertion mutant of *AtMCA1*) roots, suggesting that *AtMCA1* promotes Ca^{2+} uptake in roots. Under hypo-osmotic stress or treatment of the anionic amphipath trinitrophenol (TNP), generating membrane distortion, MCA1ox seedlings show greater $[Ca^{2+}]_{cyt}$ increase compared with wide-type and *mca1-null* seedlings. Growing on a lower (harder) medium containing 1.6% agar covered with an upper (softer) medium containing 0.8% agar, only the primary roots of *mca1-null* seedlings cannot penetrate a harder agar medium from a softer one. These studies indicate that MCA1 promotes Ca^{2+} influx upon plasma membrane distortion, which leads to mechanosensing and soil hardness sensing [13]. MCA2 has been identified as a Ca^{2+}-permeable mechanosensitive channel and directly activated by sensing membrane tension. MCA1 (1-173) and MCA2 (1-173), the N-terminal 173 residues of MCA1 and MCA2, are able to mediate Ca^{2+} influx and maintain mechanosensitivity [45]. In addition, MCA1 and MCA2 are involved in the process of sensing gravity signals in plants. Hypergravity stress induces the expression of *MCA1* and *MCA2*. The degree of hypergravity suppressing the elongation growth of hypocotyls in *mca-null* mutants is lower than that in the wide type. The overexpression of *MCAs* leads the plant to become sensitive to increased gravity [46]. The gravistimulation-induced very slow Ca^{2+} increase is defective in *mca1-null* mutants [47]. However, the direct involvement of MCAs in plant sensory mechanisms under abiotic stresses is still unclear and needs to be investigated.

2.2. RLKs

In Arabidopsis, the largest protein family, RLKs, contains 610 members, including 417 receptor kinases; the other 193 members lack the signature signal sequence and/or transmembrane sequence [54,55]. Multiple lines of evidence suggest that plant RLKs play a vital role in perceiving external signals under abiotic stress. Several transmembrane RLKs have been reported to play vital roles in abiotic stress sensing, including *Catharanthus roseus* receptor-like kinase 1-like (*Cr*RLK1L) family protein FERONIA (FER), THESEUS1 (THE1), HERCULES1 and 2 (HERK1 and 2), MALE DISCOVERER1-INTERACTING RECEPTOR LIKE KINASE 2/LEUCINE-RICH REPEAT KINASE FAMILY PROTEIN INDUCED BY SALT STRESS (MIK2/LRR-KISS), HPCA1, root meristem growth factor receptors and plant elicitor peptide receptors (RGFRs and PEPRs), and the aluminum ion sensor Al Resistance1 (ALR1) [7,10,56–64] (Table 2).

Table 2. List of discussed RLKs involved in plant sensing of abiotic stresses.

Groups	Type	Species	Biomolecule Names	Functions	References
II: RLKs	Cell wall integrity sensors	*Arabidopsis thaliana*	FERONIA (FER)	Tolerance to salt stress and metal ion stresses; Maintain cell wall integrity	[7,56,60,61]
			THESEUS1 (THE1)		[57–61]
			HERCULES1 (HERK1)		[60,61]
			HERCULES2 (HERK2)		
			MIK2/LRR-KISS	Tolerance to salt stress; Maintain cell wall integrity	[62]

Table 2. *Cont.*

Groups	Type	Species	Biomolecule Names	Functions	References
II: RLKs	Hydrogen-peroxide-induced Ca^{2+} increases 1 (HPCA1)	*Arabidopsis thaliana*	HPCA1	H_2O_2 sensor; Tolerance to oxidative stress; Regulates stomatal movement	[10]
	Root meristem growth factor receptors and plant elicitor peptide receptors (RGFRs and PEPRs)	*Arabidopsis thaliana*	RGFR1	Sense extracellular pH in plants; Promote plant immunity	[63]
			RGFR4		
			PEPR1		
			PEPR2		
	Aluminum ion sensor Al Resistance1 (ALR1)	*Arabidopsis thaliana*	ALR1	Reduces Al toxicity; Regulates ROS generation	[64]

MIK2/LRR-KISS: male discoverer1-interacting receptor like kinase 2/leucine-rich repeat kinase family protein induced by salt stress.

2.2.1. Cell Wall Integrity Sensors

Salt stress and drought stress both induce hyperosmotic stress in plant cells, leading to the loss of turgor pressure, plasmolysis, and detachment of the plasma membrane from the cell wall. The integrity of the cell wall can be monitored by plasma-membrane-located RLKs. *Cr*RLK1Ls, wall-associated kinases (WAKs), and LRR-RKs are putative sensors of cell wall integrity. Here, FER, THE1, HERK1 and 2, and MIK2 are selected to display their important roles in abiotic stress sensing (Table 2).

FER is the most intensively studied member of the CrRLK1L family. In Arabidopsis, the plasma-membrane-located receptor-like kinase FER, which senses salt-induced cell wall changes, interacts with pectin in the cell wall and elicits salt-induced Ca^{2+} transients to maintain cell wall integrity under salt stress [7]. Loss-of-function *fer* mutants display a significantly lower root growth rate within 24 h treated by salt stress and are unable to fully recover their growth rate after escaping salt stress. Root growth defects in *fer* mutants can only be observed under salt stress but not under hyperosmotic stress, indicating that FER participates in plant responses to sodium ion stress, rather than the associated osmotic stress [7]. Three cell wall leucine-rich repeat extensins, LRX3/4/5, have been found to positively regulate salt tolerance in plants. Their loss-of-function double mutant *lrx34* and triple mutant *lrx345* are hypersensitive to salt stress. Retarded growth is observed in *lrx34* and *lrx345*. These phenotypes are similar with *fer* mutans. Coimmunoprecipitation (Co-IP) assays and an in vitro pull-down assay show that two secretory peptides RALF22/23 interact with LRX3/4/5 and FER, respectively. The overexpression of *RALF22* or *RALF23* leads to retarded growth, increased accumulation of anthocyanin, and hypersensitivity to NaCl in the plant, which is similar to *lrx345* and *fer* mutants. RALF peptides induce the internalization of FER and act as a negative regulator of FER function in salt tolerance [56]. The LRX–RALF–FER module functions in sensing high-salinity-induced cell wall disruptions [7,56].

The cell-wall sensing receptor kinase THE1 plays a vital role in the process of cell elongation and not cell division in the hypocotyl [57]. Mutation of THE1 does not affect seedling growth in the background of wild-type plants but attenuates growth inhibition and ectopic lignification in the background of mutants with the mutation of cellulose synthase CESA6, indicating that THE1 mediates growth inhibition led by defective cellulose synthesis in plants [58]. THE1 is reported as a pH-dependent receptor for RALF34 by micro-scale thermophoresis (MST) assay. The RALF34-THE1 signaling module fine-tunes lateral root initiation in a manner dependent on FER [59].

To investigate whether salt application disturbs other CrRLK1Ls, apart from FER, Gigli-Bisceglia et al. collected six available cell-wall-integrity-sensing-associated mutants from the CrRLK1L protein family, including *the1-1*, *the1-4*, *herk1*, *fer-4*, *the1-1 fer-4*, and *herk1 the1-4*. After 10 days of growth in medium containing 150 mM NaCl, only *the1-4*, *herk1 the1-4*, *fer-4*, and *the1-1 fer-4* displayed significant salt-induced cotyledon bleaching. The single mutant *fer-4* and double mutant *the1-1 fer-4* showed similar rates of cotyledon bleaching. The double mutant *herk1 the1-4* showed enhanced cotyledon bleaching compared with that in the single mutant *the1-4*. These results suggested that FER alone or HERK1/THE1 positively regulates salt tolerance and cell-wall-integrity-dependent salt-sensing mechanisms are complex [60].

HERK1 and 2, THE1, and FER, which act as cell wall integrity sensors, differentially regulate growth adaptation triggered by metal ion stresses [61]. Based on a hypocotyl elongation assay, loss-of-function mutants, *herk1*, *herk2.1*, and *herk2.2* displayed enhanced hypocotyl elongation in response to Cd, Ni, Zn, and Pb. The loss-of-function mutant *the1-6* showed Ni-specific promotion of hypocotyl elongation in the dark. The loss-of-function mutant *fer-4* displayed strong hypocotyl elongation in response to Cd, Cu, Pb, and Zn. Based on a root growth assay, *herk1*, *herk2.1*, and *herk2.2* displayed inhibited root growth in response to Cd, Cu, and Ni. The mutant *the1-6* showed attenuated Cd-specific root growth. The mutant *fer-4* displayed inhibited root growth in response to Ni [61]. These studies indicate the functional diversity of CrRLK1L family proteins in plant response to metal stress.

As a primary component in cell-wall-integrity sensing, MIK2 links cell-wall-integrity sensing to plant development and environmental acclimation. The structure of MIK2 protein contains three domains: an extracellular domain consisting of 24 LRRs, a single-pass transmembrane domain, and an intracellular kinase domain [62]. When grown vertically on MS medium, *mik2-1* and *mik2-2*, two loss-of-function mutants of *MIK2*, showedleft-ward root skewing, while *the1-1* and *the1-4* did not. In the background of the *the1-1* mutant, this effect of *mik2-1* was abolished. In addition, two cellulose biosynthesis inhibitors isoxaben (ISX) and 2,6-di-chlorobenzonitrile (DCB) impaired left-ward root skewing in *mik2-1*. These phenotypes suggest that MIK2 regulates root angle by a THE1- and cellulose synthase-dependent manner. Furthermore, *mik2-1* and *mik2-2* are sensitive to salt stress, while *the1-1*, *the1-4*, and *mik2-1 the1-1* did not, indicating that MIK2 improves salt tolerance in a THE1-dependent manner [62].

2.2.2. HPCA1

As an important second messenger, ROS also plays a key role in abiotic stress sensing. The major forms of ROS include hydrogen peroxide (H_2O_2), superoxide (O_2^-), singlet oxygen ($^1O_2$), and the hydroxyl radical (HO) based on their properties and chemical reactivity [65]. Recently, considerable attention has been given to H_2O_2 because of its prominent role in the regulation of biological activity in cells during the lifecycle of plants [66–68]. GUARD CELL HYDROGEN PEROXIDE-RESISTANT1 (GHR1) can monitor H_2O_2 signaling and mediate ABA- and H_2O_2-regulated stomatal movement in Arabidopsis [69]. However, the mechanism underlying the initial sensing of H_2O_2 remains poorly understood. In Arabidopsis, *HPCA1* is a novel extracellular H_2O_2 sensor encoding a leucine-rich repeat receptor kinase located in the plasma membrane (Table 2) [10]. The *hpca1* mutant is defective in extracellular H_2O_2-induced Ca^{2+} influx and ABA signaling pathways in guard cells, leading to reduced stomatal closure. H_2O_2 activates HPCA1 via covalent modification of extracellular cysteine residues in the extracellular domain of HPCA1, leading to autophosphorylation of HPCA1 and subsequent activation of plasma membrane-localized Ca^{2+} channels [10]. However, HPCA1-gated Ca^{2+} channels in plants remain to be identified.

2.2.3. RGFRs and PEPRs

Extracellular pH plays an important role in regulating various biological processes in plants, such as nutrient uptake, cell-to-cell communication, and plant–microbe interactions [70]. The identification of the plant cell surface peptide-receptor complexes, including RGF1-RGFRs and Pep1-PEPRs, represents a significant breakthrough in understanding how plants sense extracellular pH (Table 2) [63]. RGFRs and PEPRs belong to the leucine-rich repeat receptor kinase family. The acidic extracellular pH in the root apical meristem (RAM) region is alkalinized by pattern-triggered immunity (PTI). The interaction between

RGF1 and its receptors (RGFRs), which regulate RAM growth, is acid-dependent and inhibited by extracellular alkalinization through the pH sensor sulfotyrosine, while the binding of plant elicitor peptides (Peps) to its receptors (PEPRs) is alkaline-dependent and promoted by extracellular alkalinization through the pH sensor Glu/Asp, which promotes immunity [63]. However, whether plant cell-surface peptide-receptor complexes sense abiotic-stress-triggered extracellular alkalization needs to be elucidated in further studies.

2.2.4. ALR1

Recently, another LRR receptor-like kinase, ALR1, was identified as a plant aluminum (Al) ion sensor (Table 2) [64]. As a highly phytotoxic ion, Al at very low micromolar concentrations can cause cellular damage and inhibit root growth, leading to a severe reduction in crop production [71]. To reduce Al toxicity, ALR1 specifically binds to Al ions through the intracellular cytoplasmic domain, recruits its coreceptor kinase BAK1, and promotes ALR1-dependent phosphorylation of the NADPH oxidase RbohD, thereby increasing ROS generation. In turn, ROS oxidatively modifies the RAE1 F-box protein. Subsequently, the RAE1-dependent proteolysis of STOP1 is inhibited to activate organic acid anion secretion to detoxify Al. A functional analysis of ALR has provided novel insights into ion-sensing mechanisms in living organisms [64].

Moreover, there are several plant RLKs reported to sense temperature changes and function as an indispensable component in response to extreme temperature stress. A novel calcium/calmodulin-regulated receptor-like cytoplasmic kinase CRLK1 and its paralog CRLK2 positively regulate chilling and freezing tolerance [72,73]. The plasma-membrane-localized RLCK, cold-responsive protein kinase 1 (CRPK1), interacts with and phosphorylates 14-3-3 proteins to reduce freezing tolerance in plants [74]. Shen et al. found that the receptor-like kinase ERECTA (ER) performs a vital role in conferring thermotolerance in *Arabidopsis thaliana*, rice, and tomato [75]. Thermo-Sensitive Genic Male Sterile 10 (TMS10) and its close homolog TMS10-Like (TMS10L), which encode two rice leucine-rich repeat–receptor-like kinases, mediate tapetal degeneration and male fertility by buffering environmental temperature changes [76].

2.3. Sphingolipids

As one of three main classes of lipids in eukaryotic plasma membranes, sphingolipids are required for preserving normal cellular functions [77]. In plants, sphingolipids are divided into four classes: free long-chain bases (LCBs), ceramides, glycosylceramides, and glycosylinositol phosphoceramides (GIPCs). GIPCs are the major class of plant sphingolipids (64% of total sphingolipids) and represent ~25% of the plasma membrane lipids in Arabidopsis leaves and ~40% in tobacco leaves [78,79]. GIPC has been proposed to be a bioactive molecule involved in cell wall anchoring, cell surface recognition, and lipid-mediated protein anchoring [77,80,81]. To date, there is little information about the precise roles of GIPCs in plants.

A recent study confirmed the possibility of sphingolipid function in abiotic stress sensing in plants. In Arabidopsis, Jiang and colleagues reported that plant cell-surface GIPCs function as a salt stress sensor (Table 3) [8]. Based on screening for salt-stress-induced Ca^{2+} transients, *monocation-induced* $[Ca^{2+}]_i$ *increase 1* (*MOCA1*), which encodes an inositol phosphorylceramide glucuronosyltransferase (IPUT1), was identified to be required for $[Ca^{2+}]_i$ increase. IPUT1 resides on plasma membranes and ER membranes and catalyzes the biosynthesis of the sphingolipid glycosyl inositol phosphorylceramide (GIPC). The *moca1* mutant plant exhibits reduced Ca^{2+} spikes initiated by monovalent cations (Na^+ as well as K^+ and Li^+) but exhibits no change in the Ca^{2+} spikes initiated by oxidative stress (caused by high concentrations of H_2O_2), cold stress, osmotic stress (caused by high concentrations of sorbitol), or multivalent cations. Interestingly, the growth of *moca1* seedlings is inhibited by Na^+ but not by K^+ or Li^+. Isothermal titration calorimetry (ITC) analyses revealed that Na^+ ions bind to GIPCs, which induces depolarization of the cell membrane. MOCA1-dependent GIPC senses changes in extracellular Na^+ concentrations and leads to salt-dependent intracellular Ca^{2+} spikes via unknown Ca^{2+} transporter(s) [8]. The Ca^{2+}-permeable transporters AtANN1 and AtANN4 may be candidate GIPC-gated Ca^{2+} transporters whose mutation leads to a disrupted salt-stress-induced Ca^{2+} signature [41,42].

Table 3. List of discussed sphingolipids and "other proteins" involved in plant sensing of abiotic stresses.

Groups	Type	Species	Biomolecule Names	Functions	References
III: Sphingolipids	Glycosylinositol phosphoceramides (GIPCs)	*Arabidopsis thaliana*	GIPCs	Salt stress sensor	[8]
IV: Other proteins	Extreme temperature stress sensors	*Arabidopsis thaliana*	Chilling tolerance divergence 1 (COLD1)	Tolerance to chilling shock	[3]
			EARLY FLOWERING 3 (ELF3)	Tolerance to heat stress	[21]
			THERMO-WITH ABA-RESPONSE 1 (TWA1)	Tolerance to heat stress	[82]
			Heat shock proteins (HSPs)	Tolerance to heat stress	[83,84]
			Phytochrome B (phyB)	Tolerance to extreme temperatures	[22,23,85]
	Mechanosensitive sensors	*Arabidopsis thaliana*	AtMSL8	Tolerance to mechanical stress	[86]
			AtTPK1		[87]
		Hordeum vulgare	HvTPK1		
		Oryza sativa	OsTPKa		
	Hypoxia stress sensors	*Arabidopsis thaliana*	Ethylene Response Factor (ERF) Group VII transcription factors (ERFVIIs)	Tolerance to hypoxia stress	[88–90]
			PLANT CYS OXIDASE 1 (PCO1)		[91]
			PCO2		

2.4. Other Proteins

In addition to Ca^{2+}-permeable channels, RLKs, and sphingolipids, other proteins are involved in abiotic stress sensing. Because of their diverse protein structures, they are assigned to a fourth group, named "Other proteins", which are involved in sensing extreme temperature stress, mechanical stress, and hypoxia stress.

2.4.1. Extreme Temperature Stress Sensors

With respect to extreme temperature stress, some proteins sense only cold stress or heat stress, such as chilling tolerance divergence 1 (COLD1), ELF3, THERMO-WITH ABA-RESPONSE 1 (TWA1), and heat shock proteins (HSPs), while some proteins can sense cold stress and heat stress simultaneously, such as phyB [3,21–23,82–85] (Table 3). In rice, the transmembrane protein COLD1 has been identified as a potential cold sensor that interacts with G-protein α subunit 1 (RGA1) to activate Ca^{2+} channels and enhance rice cold tolerance [3]. As a thermosensor in Arabidopsis, the nuclear protein ELF3 negatively regulates elevated temperature tolerance. ELF3 contains a polyglutamine (polyQ) repeat embedded within a predicted prion domain (PrD). ELF3-GFP forms speckles within minutes in response to higher temperatures in a PrD-dependent manner. In vitro, the ELF3 PrD reversibly formed liquid droplets in response to temperature, reflecting a direct biophysical response conferred by the PrD. The ability of temperature to rapidly shift ELF3 between active and inactive states occurs via phase transition [21]. Recently, TWA1 has been found to function as a temperature sensor that is required for basal and acquired thermotolerance in Arabidopsis [82]. TWA1 is a transcriptional co-regulator. At elevated temperatures, TWA1 accumulates in nuclear subdomains, changes its conformation, and physically interacts with JASMONATE-ASSOCIATED MYC-LIKE (JAM) transcription factors and TOPLESS (TPL) and TOPLESS-RELATED (TPR) proteins, triggering transcriptional upregulation of the heat shock transcription factor A2 (HSFA2) and HSPs [82]. HSPs have been reported to participate in high-temperature perception. Heat stress leads to protein denaturation and misfolded proteins, causing protein aggregation, which is sensed by HSPs. When HSPs bind to aggregated proteins, heat shock factor (HSF) transcription factors are released to activate heat stress responses [83,84]. In addition, the canonical photoreceptor phyB, which is also implicated as a temperature sensor, plays an important role in plant responses to both warm and cold temperatures through its temperature-dependent reversion from the active Pfr state (a far-red-light-absorbing form) to the inactive Pr state (a red-light-absorbing form) [85]. Under cold stress, the stabilization of phyB is induced via the accumulation of the key transcription factor C-REPEAT BINDING FACTOR (CBF), which interacts with PHYTOCHROME-INTERACTING FACTOR 3 (PIF3) to attenuate the mutually assured destruction of PIF3–phyB. Cold-stabilized phyB enhances freezing tolerance in Arabidopsis [22]. Under elevated temperatures, the phyB Pfr-to-Pr reversion is facilitated to release the inhibition of the Pfr dimer on PIF4 and PIF7 and subsequently activate downstream responses [23].

2.4.2. Mechanosensitive Sensors

Mechanosensitive ion channels of small conductance (MscS)-like proteins (MSLs) and two-pore potassium (TPK) family proteins have been identified as mechanosensitive sensors [86,87] (Table 3). In Arabidopsis, MscS-like 8 (AtMSL8) has been identified as a sensor of hypo-osmotic-stress-induced membrane tension in pollen [86]. AtMSL8 is a pollen-specific, membrane-tension-gated ion channel and decreases the survival rates of pollen grains exposed to the hypo-osmotic shock of rehydration. The hypo-osmolarity induced increases membrane tension and leads to the opening of MSL8, allowing ion efflux, which protects the cells from internal osmotic pressure [86]. Plant TPKs play an important role in vacuolar K^+ homeostasis and are regulated by Ca^{2+} and 14-3-3 proteins. In Arabidopsis, rice, and barley, vacuolar TPKs can act as intracellular osmosensors via the detection of small perturbations in membrane tension and rapidly increase channel activity during hypo-osmotic shock to release vacuolar K^+ [87].

2.4.3. Hypoxia Stress Sensors

O_2 sensors play an important role in plant perception to hypoxia stress (oxygen depletion) induced by flooding. O_2 sensing in plants is mediated by an N-end rule pathway for protein destabilization. In Arabidopsis, the N-end rule pathway of targeted proteolysis acts as severe low oxygen sensor, such as the hypoxia-associated Ethylene Response Factor (ERF) Group VII transcription factors (ERFVIIs) destabilization (Table 3) [88,89]. All ERFVIIs in Arabidopsis contain a conserved amino-terminal amino acid sequence MCGGAIIL to be dedicated to an oxygen-dependent sequence of post-translational modifications. The Met in this conserved sequence is removed by MET AMINO-PEPTIDASE (MetAP), exposing the destabilizing Cys at the N terminus and leading to the initiation of degradation of ERFVIIs under aerobic conditions. During hypoxia, the loss of oxidation of the N terminus of ERFVIIs maintains ERFVIIs stabilization, and then, ERFVIIs migrate to the nucleus and regulate hypoxia-responsive gene expression [88–90]. In addition, the oxidases PLANT CYS OXIDASE 1 (PCO1) and PCO2 are also considered as O_2 sensors because O_2 is the direct ligand of PCO1/2 (Table 3) [91].

3. Conclusions

The mechanisms underlying plant sensors and the sensory systems involved in the detection of environmental abiotic-stress-related stimuli have been explored. In the last decade, major advances have been made in the discovery of abiotic sensors, such as the osmotic stress sensor OSCA1, the temperature stress sensor phyB, the salt stress sensor GIPC, and the oxidation stress sensor HPCA1 (Figure 1). In this review, based on molecular structure, we divide biomolecules participating in abiotic stress sensing into four groups: Ca^{2+}-permeable channels, receptor-like kinases (RLKs), sphingolipids, and other proteins (Figure 1). Our classification analysis is helpful for revealing many unknown stress sensors. For example, most reported abiotic stress sensors are calcium (Ca^{2+}) channels or regulators of Ca^{2+} influx, suggesting the indispensable role of Ca^{2+} in response to environmental stress. Therefore, further experimental investigations of stress sensors could focus on clarifying the functional relevance of the reciprocal bilateral regulation of stress perception by Ca^{2+}.

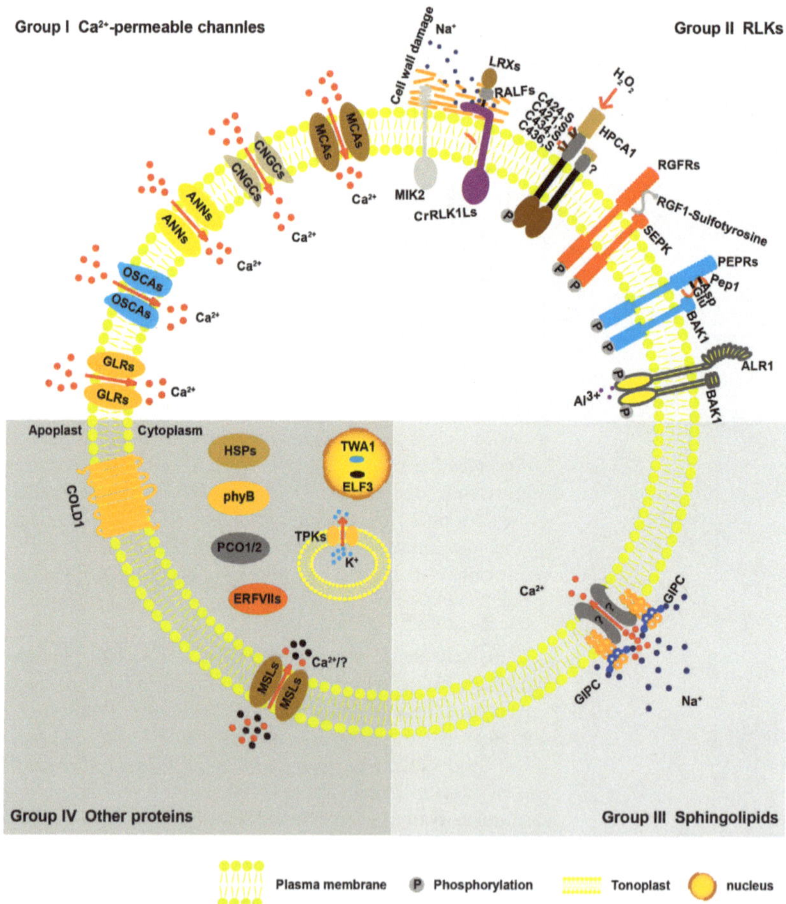

Figure 1. Summary of discussed biomolecule function in abiotic stress sensing. Based on molecular structure, the biomolecule function in abiotic stress sensing can be divided into four groups. Group I: Ca^{2+}-permeable channels, including hyperosmolarity-gated calcium-permeable channel family of proteins (OSCAs), cyclic-nucleotide-gated calcium channels (CNGCs), ANNEXIN proteins (ANNs), glutamate receptor-like proteins (GLRs), and MID1-COMPLEMENTING ACTIVITY proteins (MCAs);

Group II: receptor-like kinases (RLKs), including *Catharanthus roseus* receptor-like kinase 1-like family proteins (CrRLK1Ls), such as FERONIA (FER), THESEUS1 (THE1), HERCULES1 and 2 (HERK1 and 2), MALE DISCOVERER1-INTERACTING RECEPTOR LIKE KINASE 2/LEUCINE-RICH REPEAT KINASE FAMILY PROTEIN INDUCED BY SALT STRESS (MIK2/LRR-KISS), hydrogen-peroxide-induced Ca^{2+} increases 1 (HPCA1), root meristem growth factor receptors and plant elicitor peptide receptors (RGFRs and PEPRs), and the aluminum ion sensor Al Resistance1 (ALR1); Group III: sphingolipids, including glycosylinositol phosphoceramides (GIPCs); Group IV: other proteins, including chilling tolerance divergence 1 (COLD1), EARLY FLOWERING 3 (ELF3), THERMO-WITH ABA-RESPONSE 1 (TWA1), heat shock proteins (HSPs), phytochrome B (phyB), mechanosensitive channel of small conductance (MscS)-like proteins (MSLs), and two-pore potassium family proteins (TPKs), ethylene response factor (ERF) group VII transcription factors (ERFVIIs), and PLANT CYS OXIDASE 1/2 (PCO1/2). Na^+: dark blue dot; K^+: light blue dot; Ca^{2+}: red dot; "?": unknown ions, black dot.

4. Future Perspectives

Many gaps remain in our understanding of plant sensory mechanisms. There are many stress sensors to be identified. Sensing mechanisms of most of the reported stress sensors remain unclear. Crosstalk among these sensors is poorly understood. Future research on abiotic stress sensors must involve two goals. First, we must elucidate the means by which abiotic stress sensors perceive stress. Most reported sensors are also considered putative sensors because of their unclear physiological functions and biochemical sensing mechanisms. Studies to unravel the mode of action of these perception mechanisms are complex, as different sensors in different plant tissues share the same downstream signaling pathway in response to different abiotic stresses and combinations of stresses [7–10,32]. Second, more new abiotic stress sensors need to be identified. It is believed that more sensing mechanisms exist under environmental stress because the known modes of action in abiotic stress perception cannot account for all observed physiological responses. Genetic redundancy and lethality makes it difficult to identify new abiotic stress sensors [32,92,93]. Molecular genetic methods and various bioimaging techniques, including Ca^{2+} imaging-based forward genetic screens and fluorescence-based Ca^{2+} indicators help scientists find new sensors.

The cultivation of stress-resilient crops with improved yield stability is the most effective strategy for overcoming multiple and fluctuating environmental cues. Natural genetic variation in crops, genetic engineering, chemical intervention, and microbial stimulation are usually used in this strategy. In recent years, several success stories about improving crop stress tolerance have been reported. For instance, *HKT1* alleles in rice, wheat, and maize have been identified as major quantitative trait loci regulating salt tolerance and have enabled marker-assisted breeding of wheat with increased yield in saline soils [2,4,94]. The overexpression of the stress-inducible transcription factor *OsDREB2A* enhances dehydration and salt-stress tolerance in rice [1,95]. The application of ABA mimics reduces water loss and promotes drought resistance in plants [96,97]. In addition, approaches combining genetic, chemical, and microbial tactics could provide a promising strategy for cultivating crops with both high stress resistance and high productivity [97,98]. Knowledge of plant sensory mechanisms under abiotic stresses will help us to identify more key molecular targets for engineering stress-resilient crops in the field.

Author Contributions: S.J.: investigation and writing—original draft; M.W. and Y.W.: formal analysis and visualization; Z.J.: conceptualization, writing—review and editing, supervision, and funding acquisition. All authors have read and agreed to the published version of the manuscript.

Funding: This research was funded by the Natural Science Foundation of Guangdong Province (2022B1515020040) and the Research Team Cultivation Program of Shenzhen University (2023DFT005).

Conflicts of Interest: The authors declare no conflicts of interest.

References

1. Cui, M.; Zhang, W.; Zhang, Q.; Xu, Z.; Zhu, Z.; Duan, F.; Wu, R. Induced over-expression of the transcription factor OsDREB2A improves drought tolerance in rice. *Plant Physiol. Biochem.* **2011**, *49*, 1384–1391. [CrossRef] [PubMed]
2. Munns, R.; James, R.A.; Xu, B.; Athman, A.; Conn, S.J.; Jordans, C.; Byrt, C.S.; Hare, R.A.; Tyerman, S.D.; Tester, M.; et al. Wheat grain yield on saline soils is improved by an ancestral Na^+ transporter gene. *Nat. Biotechnol.* **2012**, *30*, 360–364. [CrossRef] [PubMed]
3. Ma, Y.; Dai, X.; Xu, Y.; Luo, W.; Zheng, X.; Zeng, D.; Pan, Y.; Lin, X.; Liu, H.; Zhang, D.; et al. COLD1 confers chilling tolerance in rice. *Cell* **2015**, *160*, 1209–1221. [CrossRef] [PubMed]
4. Zhang, M.; Cao, Y.; Wang, Z.; Wang, Z.q.; Shi, J.; Liang, X.; Song, W.; Chen, Q.; Lai, J.; Jiang, C. A retrotransposon in an HKT1 family sodium transporter causes variation of leaf Na^+ exclusion and salt tolerance in maize. *New Phytol.* **2017**, *217*, 1161–1176. [CrossRef] [PubMed]
5. Alpuerto, J.B.; Hussain, R.M.F.; Fukao, T. The key regulator of submergence tolerance, SUB1A, promotes photosynthetic and metabolic recovery from submergence damage in rice leaves. *Plant Cell Environ.* **2015**, *39*, 672–684. [CrossRef] [PubMed]
6. Akman, M.; Kleine, R.; van Tienderen, P.H.; Schranz, E.M. Identification of the submergence tolerance QTL come quick drowning1 (CQD1) in *Arabidopsis thaliana*. *J. Hered.* **2017**, *108*, 308–317. [CrossRef] [PubMed]
7. Feng, W.; Kita, D.; Peaucelle, A.; Cartwright, H.N.; Doan, V.; Duan, Q.; Liu, M.-C.; Maman, J.; Steinhorst, L.; Schmitz-Thom, I.; et al. The FERONIA receptor kinase maintains cell-wall integrity during salt stress through Ca^{2+} signaling. *Curr. Biol.* **2018**, *28*, 666–675.e665. [CrossRef] [PubMed]
8. Jiang, Z.; Zhou, X.; Tao, M.; Yuan, F.; Liu, L.; Wu, F.; Wu, X.; Xiang, Y.; Niu, Y.; Liu, F.; et al. Plant cell-surface GIPC sphingolipids sense salt to trigger Ca^{2+} influx. *Nature* **2019**, *572*, 341–346. [CrossRef] [PubMed]
9. Yuan, F.; Yang, H.; Xue, Y.; Kong, D.; Ye, R.; Li, C.; Zhang, J.; Theprungsirikul, L.; Shrift, T.; Krichilsky, B.; et al. OSCA1 mediates osmotic-stress-evoked Ca^{2+} increases vital for osmosensing in Arabidopsis. *Nature* **2014**, *514*, 367–371. [CrossRef]
10. Wu, F.; Chi, Y.; Jiang, Z.; Xu, Y.; Xie, L.; Huang, F.; Wan, D.; Ni, J.; Yuan, F.; Wu, X.; et al. Hydrogen peroxide sensor HPCA1 is an LRR receptor kinase in Arabidopsis. *Nature* **2020**, *578*, 577–581. [CrossRef]
11. Knight, H.; Trewavas, A.J.; Knight, M.R. Calcium signalling in *Arabidopsis thaliana* responding to drought and salinity. *Plant J.* **1997**, *12*, 1067–1078. [CrossRef] [PubMed]
12. Knight, M.R.; Campbell, A.K.; Smith, S.M.; Trewavas, A.J. Transgenic plant aequorin reports the effects of touch and cold-shock and elicitors on cytoplasmic calcium. *Nature* **1991**, *352*, 524–526. [CrossRef] [PubMed]
13. Nakagawa, Y.; Katagiri, T.; Shinozaki, K.; Qi, Z.; Tatsumi, H.; Furuichi, T.; Kishigami, A.; Sokabe, M.; Kojima, I.; Sato, S.; et al. Arabidopsis plasma membrane protein crucial for Ca^{2+} influx and touch sensing in roots. *Proc. Natl. Acad. Sci. USA* **2007**, *104*, 3639–3644. [CrossRef] [PubMed]
14. Ward, J.M.; Mäser, P.; Schroeder, J.I. Plant ion channels: Gene families, physiology, and functional genomics analyses. *Annu. Rev. Physiol.* **2009**, *71*, 59–82. [CrossRef]
15. Wang, J.; Ren, Y.; Liu, X.; Luo, S.; Zhang, X.; Liu, X.; Lin, Q.; Zhu, S.; Wan, H.; Yang, Y.; et al. Transcriptional activation and phosphorylation of OsCNGC9 confer enhanced chilling tolerance in rice. *Mol. Plant* **2021**, *14*, 315–329. [CrossRef]
16. Gong, Z.; Xiong, L.; Shi, H.; Yang, S.; Herrera-Estrella, L.R.; Xu, G.; Chao, D.-Y.; Li, J.; Wang, P.-Y.; Qin, F.; et al. Plant abiotic stress response and nutrient use efficiency. *Sci. China Life Sci.* **2020**, *63*, 635–674. [CrossRef] [PubMed]
17. Cui, Y.; Lu, S.; Li, Z.; Cheng, J.; Hu, P.; Zhu, T.; Wang, X.; Jin, M.; Wang, X.; Li, L.; et al. CYCLIC NUCLEOTIDE-GATED ION CHANNELs 14 and 16 promote tolerance to heat and chilling in rice. *Plant Physiol.* **2020**, *183*, 1794–1808. [CrossRef] [PubMed]
18. Finka, A.; Cuendet, A.F.H.; Maathuis, F.J.M.; Saidi, Y.; Goloubinoff, P. Plasma membrane cyclic nucleotide gated calcium channels control land plant thermal sensing and acquired thermotolerance. *Plant Cell* **2012**, *24*, 3333–3348. [CrossRef] [PubMed]
19. Gao, F.; Han, X.; Wu, J.; Zheng, S.; Shang, Z.; Sun, D.; Zhou, R.; Li, B. A heat-activated calcium-permeable channel—Arabidopsis cyclic nucleotide-gated ion channel 6—Is involved in heat shock responses. *Plant J.* **2012**, *70*, 1056–1069. [CrossRef]
20. Mamode Cassim, A.; Gouguet, P.; Gronnier, J.; Laurent, N.; Germain, V.; Grison, M.; Boutté, Y.; Gerbeau-Pissot, P.; Simon-Plas, F.; Mongrand, S. Plant lipids: Key players of plasma membrane organization and function. *Prog. Lipid Res.* **2019**, *73*, 1–27. [CrossRef]
21. Jung, J.-H.; Barbosa, A.D.; Hutin, S.; Kumita, J.R.; Gao, M.; Derwort, D.; Silva, C.S.; Lai, X.; Pierre, E.; Geng, F.; et al. A prion-like domain in ELF3 functions as a thermosensor in Arabidopsis. *Nature* **2020**, *585*, 256–260. [CrossRef]
22. Jiang, B.; Shi, Y.; Peng, Y.; Jia, Y.; Yan, Y.; Dong, X.; Li, H.; Dong, J.; Li, J.; Gong, Z.; et al. Cold-Induced CBF–PIF3 interaction enhances freezing tolerance by stabilizing the phyB thermosensor in Arabidopsis. *Mol. Plant* **2020**, *13*, 894–906. [CrossRef] [PubMed]
23. Jung, J.-H.; Domijan, M.; Klose, C.; Biswas, S.; Ezer, D.; Gao, M.; Khattak, A.K.; Box, M.S.; Charoensawan, V.; Cortijo, S.; et al. Phytochromes function as thermosensors in Arabidopsis. *Science* **2016**, *354*, 886–889. [CrossRef] [PubMed]
24. Zhu, J.-K. Abiotic stress signaling and responses in plants. *Cell* **2016**, *167*, 313–324. [CrossRef] [PubMed]
25. Soltabayeva, A.; Dauletova, N.; Serik, S.; Sandybek, M.; Omondi, J.O.; Kurmanbayeva, A.; Srivastava, S. Receptor-like kinases (LRR-RLKs) in response of plants to biotic and abiotic stresses. *Plants* **2022**, *11*, 2660. [CrossRef] [PubMed]
26. Wang, G.; Zhao, Z.; Zheng, X.; Shan, W.; Fan, J. How a single receptor-like kinase exerts diverse roles: Lessons from FERONIA. *Mol. Hortic.* **2022**, *2*, 25. [CrossRef] [PubMed]
27. Wang, W.; Cheng, H.Y.; Zhou, J.M. New insight into Ca^{2+}-permeable channel in plant immunity. *J. Integr. Plant Biol.* **2024**, *66*, 623–631. [CrossRef] [PubMed]

28. Kang, X.; Zhao, L.; Liu, X. Calcium signaling and the response to heat shock in crop plants. *Int. J. Mol. Sci.* **2023**, *25*, 324. [CrossRef] [PubMed]
29. Nawaz, Z.; Kakar, K.U.; Saand, M.A.; Shu, Q.-Y. Cyclic nucleotide-gated ion channel gene family in rice, identification, characterization and experimental analysis of expression response to plant hormones, biotic and abiotic stresses. *BMC Genom.* **2014**, *15*, 853. [CrossRef]
30. Lee, S.-K.; Lee, S.-M.; Kim, M.-H.; Park, S.-K.; Jung, K.-H. Genome-wide analysis of cyclic nucleotide-gated channel genes related to pollen development in rice. *Plants* **2022**, *11*, 3145. [CrossRef]
31. Liu, X.; Wang, J.; Sun, L. Structure of the hyperosmolality-gated calcium-permeable channel OSCA1.2. *Nat. Commun.* **2018**, *9*, 5060. [CrossRef]
32. Pei, S.; Tao, Q.; Li, W.; Qi, G.; Wang, B.; Wang, Y.; Dai, S.; Shen, Q.; Wang, X.; Wu, X.; et al. Osmosensor-mediated control of Ca^{2+} spiking in pollen germination. *Nature* **2024**, *629*, 1118–1125. [CrossRef] [PubMed]
33. Li, Y.; Zhang, Y.; Li, B.; Hou, L.; Yu, J.; Jia, C.; Wang, Z.; Chen, S.; Zhang, M.; Qin, J.; et al. Preliminary expression analysis of the OSCA gene family in maize and their involvement in temperature stress. *Int. J. Mol. Sci.* **2022**, *23*, 13658. [CrossRef] [PubMed]
34. Li, Y.; Yuan, F.; Wen, Z.; Li, Y.; Wang, F.; Zhu, T.; Zhuo, W.; Jin, X.; Wang, Y.; Zhao, H.; et al. Genome-wide survey and expression analysis of the OSCA gene family in rice. *BMC Plant Biol.* **2015**, *15*, 261. [CrossRef] [PubMed]
35. Hou, C.; Tian, W.; Kleist, T.; He, K.; Garcia, V.; Bai, F.; Hao, Y.; Luan, S.; Li, L. DUF221 proteins are a family of osmosensitive calcium-permeable cation channels conserved across eukaryotes. *Cell Res.* **2014**, *24*, 632–635. [CrossRef]
36. Maity, K.; Heumann, J.M.; McGrath, A.P.; Kopcho, N.J.; Hsu, P.-K.; Lee, C.-W.; Mapes, J.H.; Garza, D.; Krishnan, S.; Morgan, G.P.; et al. Cryo-EM structure of OSCA1.2 from *Oryza sativa* elucidates the mechanical basis of potential membrane hyperosmolality gating. *Proc. Natl. Acad. Sci. USA* **2019**, *116*, 14309–14318. [CrossRef] [PubMed]
37. Jojoa-Cruz, S.; Saotome, K.; Murthy, S.E.; Tsui, C.C.A.; Sansom, M.S.P.; Patapoutian, A.; Ward, A.B. Cryo-EM structure of the mechanically activated ion channel OSCA1.2. *eLife* **2018**, *7*, e41845. [CrossRef] [PubMed]
38. Zhang, M.; Wang, D.; Kang, Y.; Wu, J.-X.; Yao, F.; Pan, C.; Yan, Z.; Song, C.; Chen, L. Structure of the mechanosensitive OSCA channels. *Nat. Struct. Mol. Biol.* **2018**, *25*, 850–858. [CrossRef] [PubMed]
39. Liao, C.; Zheng, Y.; Guo, Y. MYB30 transcription factor regulates oxidative and heat stress responses through ANNEXIN-mediated cytosolic calcium signaling in Arabidopsis. *New Phytol.* **2017**, *216*, 163–177. [CrossRef] [PubMed]
40. Liu, Q.; Ding, Y.; Shi, Y.; Ma, L.; Wang, Y.; Song, C.; Wilkins, K.A.; Davies, J.M.; Knight, H.; Knight, M.R.; et al. The calcium transporter ANNEXIN1 mediates cold-induced calcium signaling and freezing tolerance in plants. *EMBO J.* **2020**, *40*, e104559. [CrossRef]
41. Laohavisit, A.; Richards, S.L.; Shabala, L.; Chen, C.; Colaço, R.D.D.R.; Swarbreck, S.M.; Shaw, E.; Dark, A.; Shabala, S.; Shang, Z.; et al. Salinity-induced calcium signaling and root adaptation in Arabidopsis require the calcium regulatory protein annexin1. *Plant Physiol.* **2013**, *163*, 253–262. [CrossRef]
42. Ma, L.; Ye, J.; Yang, Y.; Lin, H.; Yue, L.; Luo, J.; Long, Y.; Fu, H.; Liu, X.; Zhang, Y.; et al. The SOS2-SCaBP8 complex generates and fine-tunes an AtANN4-dependent calcium signature under salt stress. *Dev. Cell* **2019**, *48*, 697–709.e695. [CrossRef]
43. Qi, Z.; Stephens, N.R.; Spalding, E.P. Calcium entry mediated by GLR3.3, an Arabidopsis glutamate receptor with a broad agonist profile. *Plant Physiol.* **2006**, *142*, 963–971. [CrossRef] [PubMed]
44. Meyerhoff, O.; Müller, K.; Roelfsema, M.R.G.; Latz, A.; Lacombe, B.; Hedrich, R.; Dietrich, P.; Becker, D. *AtGLR3.4*, a glutamate receptor channel-like gene is sensitive to touch and cold. *Planta* **2005**, *222*, 418–427. [CrossRef] [PubMed]
45. Yoshimura, K.; Iida, K.; Iida, H. MCAs in Arabidopsis are Ca^{2+}-permeable mechanosensitive channels inherently sensitive to membrane tension. *Nat. Commun.* **2021**, *12*, 6074. [CrossRef]
46. Hattori, T.; Otomi, Y.; Nakajima, Y.; Soga, K.; Wakabayashi, K.; Iida, H.; Hoson, T. MCA1 and MCA2 are involved in the response to hypergravity in Arabidopsis hypocotyls. *Plants* **2020**, *9*, 590. [CrossRef]
47. Nakano, M.; Furuichi, T.; Sokabe, M.; Iida, H.; Tatsumi, H. The gravistimulation-induced very slow Ca^{2+} increase in Arabidopsis seedlings requires MCA1, a Ca^{2+}-permeable mechanosensitive channel. *Sci. Rep.* **2021**, *11*, 227. [CrossRef]
48. Falcone, D.L.; Ogas, J.P.; Somerville, C.R. Regulation of membrane fatty acid composition by temperature in mutants of Arabidopsis with alterations in membrane lipid composition. *BMC Plant Biol.* **2004**, *4*, 17. [CrossRef] [PubMed]
49. Alfieri, A.; Doccula, F.G.; Pederzoli, R.; Grenzi, M.; Bonza, M.C.; Luoni, L.; Candeo, A.; Romano Armada, N.; Barbiroli, A.; Valentini, G.; et al. The structural bases for agonist diversity in an *Arabidopsis thaliana* glutamate receptor-like channel. *Proc. Natl. Acad. Sci. USA* **2019**, *117*, 752–760. [CrossRef]
50. Toyota, M.; Spencer, D.; Sawai-Toyota, S.; Jiaqi, W.; Zhang, T.; Koo, A.J.; Howe, G.A.; Gilroy, S. Glutamate triggers long-distance, calcium-based plant defense signaling. *Science* **2018**, *361*, 1112–1115. [CrossRef]
51. Lacombe, B.; Becker, D.; Hedrich, R.; DeSalle, R.; Hollmann, M.; Kwak, J.M.; Schroeder, J.I.; Le Novère, N.; Gil Nam, H.; Spalding, E.P.; et al. The identity of plant glutamate receptors. *Science* **2001**, *292*, 1486–1487. [CrossRef] [PubMed]
52. Dennison, K.L.; Spalding, E.P. Glutamate-gated calcium fluxes in Arabidopsis. *Plant Physiol.* **2000**, *124*, 1511–1514. [CrossRef] [PubMed]
53. Dubos, C.; Huggins, D.; Grant, G.H.; Knight, M.R.; Campbell, M.M. A role for glycine in the gating of plant NMDA-like receptors. *Plant J.* **2003**, *35*, 800–810. [CrossRef] [PubMed]
54. Shiu, S.-H.; Bleecker, A.B. Receptor-like kinases from Arabidopsis form a monophyletic gene family related to animal receptor kinases. *Proc. Natl. Acad. Sci. USA* **2001**, *98*, 10763–10768. [CrossRef] [PubMed]

55. Shiu, S.-H.; Bleecker, A.B. Expansion of the receptor-like kinase/pelle gene family and receptor-like proteins in Arabidopsis. *Plant Physiol.* **2003**, *132*, 530–543. [CrossRef] [PubMed]
56. Zhao, C.; Zayed, O.; Yu, Z.; Jiang, W.; Zhu, P.; Hsu, C.-C.; Zhang, L.; Tao, W.A.; Lozano-Durán, R.; Zhu, J.-K. Leucine-rich repeat extensin proteins regulate plant salt tolerance in Arabidopsis. *Proc. Natl. Acad. Sci. USA* **2018**, *115*, 13123–13128. [CrossRef] [PubMed]
57. Hématy, K.; Sado, P.-E.; Van Tuinen, A.; Rochange, S.; Desnos, T.; Balzergue, S.; Pelletier, S.; Renou, J.-P.; Höfte, H. A receptor-like kinase mediates the response of Arabidopsis cells to the inhibition of cellulose synthesis. *Curr. Biol.* **2007**, *17*, 922–931. [CrossRef] [PubMed]
58. Merz, D.; Richter, J.; Gonneau, M.; Sanchez-Rodriguez, C.; Eder, T.; Sormani, R.; Martin, M.; Hématy, K.; Höfte, H.; Hauser, M.-T. T-DNA alleles of the receptor kinase THESEUS1 with opposing effects on cell wall integrity signaling. *J. Exp. Bot.* **2017**, *68*, 4583–4593. [CrossRef] [PubMed]
59. Gonneau, M.; Desprez, T.; Martin, M.; Doblas, V.G.; Bacete, L.; Miart, F.; Sormani, R.; Hématy, K.; Renou, J.; Landrein, B.; et al. Receptor kinase THESEUS1 is a rapid alkalinization factor 34 receptor in Arabidopsis. *Curr. Biol.* **2018**, *28*, 2452–2458.e2454. [CrossRef]
60. Gigli-Bisceglia, N.; van Zelm, E.; Huo, W.; Lamers, J.; Testerink, C. Arabidopsis root responses to salinity depend on pectin modification and cell wall sensing. *Development* **2022**, *149*, dev200363. [CrossRef]
61. Richter, J.; Ploderer, M.; Mongelard, G.; Gutierrez, L.; Hauser, M.-T. Role of CrRLK1L cell wall sensors HERCULES1 and 2, THESEUS1, and FERONIA in growth adaptation triggered by heavy metals and trace elements. *Front. Plant Sci.* **2017**, *8*, 01554. [CrossRef] [PubMed]
62. Van der Does, D.; Boutrot, F.; Engelsdorf, T.; Rhodes, J.; McKenna, J.F.; Vernhettes, S.; Koevoets, I.; Tintor, N.; Veerabagu, M.; Miedes, E.; et al. The Arabidopsis leucine-rich repeat receptor kinase MIK2/LRR-KISS connects cell wall integrity sensing, root growth and response to abiotic and biotic stresses. *PLoS Genet.* **2017**, *13*, e1006832. [CrossRef] [PubMed]
63. Liu, L.; Song, W.; Huang, S.; Jiang, K.; Moriwaki, Y.; Wang, Y.; Men, Y.; Zhang, D.; Wen, X.; Han, Z.; et al. Extracellular pH sensing by plant cell-surface peptide-receptor complexes. *Cell* **2022**, *185*, 3341–3355.e3313. [CrossRef]
64. Ding, Z.J.; Xu, C.; Yan, J.Y.; Wang, Y.X.; Cui, M.Q.; Yuan, J.J.; Wang, Y.N.; Li, G.X.; Wu, J.X.; Wu, Y.R.; et al. The LRR receptor-like kinase ALR1 is a plant aluminum ion sensor. *Cell Res.* **2024**, *34*, 281–294. [CrossRef] [PubMed]
65. Waszczak, C.; Carmody, M.; Kangasjärvi, J. Reactive oxygen species in plant signaling. *Annu. Rev. Plant Biol.* **2018**, *69*, 209–236. [CrossRef] [PubMed]
66. Leshem, Y.; Melamed-Book, N.; Cagnac, O.; Ronen, G.; Nishri, Y.; Solomon, M.; Cohen, G.; Levine, A. Suppression of Arabidopsis vesicle-SNARE expression inhibited fusion of H_2O_2-containing vesicles with tonoplast and increased salt tolerance. *Proc. Natl. Acad. Sci. USA* **2006**, *103*, 18008–18013. [CrossRef]
67. Hao, H.; Fan, L.; Chen, T.; Li, R.; Li, X.; He, Q.; Botella, M.A.; Lin, J. Clathrin and membrane microdomains cooperatively regulate RbohD dynamics and activity in Arabidopsis. *Plant Cell* **2014**, *26*, 1729–1745. [CrossRef]
68. Nagano, M.; Ishikawa, T.; Fujiwara, M.; Fukao, Y.; Kawano, Y.; Kawai-Yamada, M.; Shimamoto, K. Plasma membrane microdomains are essential for Rac1-RbohB/H-mediated immunity in rice. *Plant Cell* **2016**, *28*, 1966–1983. [CrossRef]
69. Hua, D.; Wang, C.; He, J.; Liao, H.; Duan, Y.; Zhu, Z.; Guo, Y.; Chen, Z.; Gong, Z. A plasma membrane receptor kinase, GHR1, mediates abscisic acid- and hydrogen peroxide-regulated stomatal movement in Arabidopsis. *Plant Cell* **2012**, *24*, 2546–2561. [CrossRef]
70. Tsai, H.-H.; Schmidt, W. The enigma of environmental pH sensing in plants. *Nat. Plants* **2021**, *7*, 106–115. [CrossRef]
71. Ma, J.F.; Ryan, P.R.; Delhaize, E. Aluminium tolerance in plants and the complexing role of organic acids. *Trends Plant Sci.* **2001**, *6*, 273–278. [CrossRef] [PubMed]
72. Yang, T.; Chaudhuri, S.; Yang, L.; Du, L.; Poovaiah, B.W. A calcium/calmodulin-regulated member of the receptor-like kinase family confers cold tolerance in plants. *J. Biol. Chem.* **2010**, *285*, 7119–7126. [CrossRef] [PubMed]
73. Zhao, C.; Wang, P.; Si, T.; Hsu, C.-C.; Wang, L.; Zayed, O.; Yu, Z.; Zhu, Y.; Dong, J.; Tao, W.A.; et al. MAP kinase cascades regulate the cold response by modulating ICE1 protein stability. *Dev. Cell* **2017**, *43*, 618–629.e615. [CrossRef] [PubMed]
74. Liu, Z.; Jia, Y.; Ding, Y.; Shi, Y.; Li, Z.; Guo, Y.; Gong, Z.; Yang, S. Plasma membrane CRPK1-mediated phosphorylation of 14-3-3 proteins induces their nuclear import to fine-tune CBF signaling during cold response. *Mol. Cell* **2017**, *66*, 117–128.e115. [CrossRef] [PubMed]
75. Shen, H.; Zhong, X.; Zhao, F.; Wang, Y.; Yan, B.; Li, Q.; Chen, G.; Mao, B.; Wang, J.; Li, Y.; et al. Overexpression of receptor-like kinase ERECTA improves thermotolerance in rice and tomato. *Nat. Biotechnol.* **2015**, *33*, 996–1003. [CrossRef] [PubMed]
76. Yu, J.; Han, J.; Kim, Y.-J.; Song, M.; Yang, Z.; He, Y.; Fu, R.; Luo, Z.; Hu, J.; Liang, W.; et al. Two rice receptor-like kinases maintain male fertility under changing temperatures. *Proc. Natl. Acad. Sci. USA* **2017**, *114*, 12327–12332. [CrossRef] [PubMed]
77. Chen, M.; Han, G.; Dietrich, C.R.; Dunn, T.M.; Cahoon, E.B. The essential nature of sphingolipids in plants as revealed by the functional identification and characterization of the Arabidopsis LCB1 subunit of serine palmitoyltransferase. *Plant Cell* **2006**, *18*, 3576–3593. [CrossRef] [PubMed]
78. Markham, J.E.; Li, J.; Cahoon, E.B.; Jaworski, J.G. Separation and identification of major plant sphingolipid classes from leaves. *J. Biol. Chem.* **2006**, *281*, 22684–22694. [CrossRef] [PubMed]
79. Cacas, J.-L.; Buré, C.; Grosjean, K.; Gerbeau-Pissot, P.; Lherminier, J.; Rombouts, Y.; Maes, E.; Bossard, C.; Gronnier, J.; Furt, F.; et al. Revisiting plant plasma membrane lipids in tobacco: A focus on sphingolipids. *Plant Physiol.* **2016**, *170*, 367–384. [CrossRef]

80. Wang, W.; Yang, X.; Tangchaiburana, S.; Ndeh, R.; Markham, J.E.; Tsegaye, Y.; Dunn, T.M.; Wang, G.-L.; Bellizzi, M.; Parsons, J.F.; et al. An inositolphosphorylceramide synthase is involved in regulation of plant programmed cell death associated with defense in Arabidopsis. *Plant Cell* **2008**, *20*, 3163–3179. [CrossRef]
81. Voxeur, A.; Fry, S.C. Glycosylinositol phosphorylceramides from Rosa cell cultures are boron-bridged in the plasma membrane and form complexes with rhamnogalacturonan II. *Plant J.* **2014**, *79*, 139–149. [CrossRef] [PubMed]
82. Bohn, L.; Huang, J.; Weidig, S.; Yang, Z.; Heidersberger, C.; Genty, B.; Falter-Braun, P.; Christmann, A.; Grill, E. The temperature sensor TWA1 is required for thermotolerance in *Arabidopsis*. *Nature* **2024**, *629*, 1126–1132. [CrossRef] [PubMed]
83. Oroz, J.; Kim, J.H.; Chang, B.J.; Zweckstetter, M. Mechanistic basis for the recognition of a misfolded protein by the molecular chaperone Hsp90. *Nat. Struct. Mol. Biol.* **2017**, *24*, 407–413. [CrossRef] [PubMed]
84. Yamamoto, M.; Takahashi, Y.; Inano, K.; Horigome, T.; Sugano, H. Characterization of the hydrophobic region of heat shock protein 90. *J. Biochem.* **1991**, *110*, 141–145. [CrossRef] [PubMed]
85. Legris, M.; Klose, C.; Burgie, E.S.; Rojas, C.C.R.; Neme, M.; Hiltbrunner, A.; Wigge, P.A.; Schäfer, E.; Vierstra, R.D.; Casal, J.J. Phytochrome B integrates light and temperature signals in Arabidopsis. *Science* **2016**, *354*, 897–900. [CrossRef] [PubMed]
86. Hamilton, E.S.; Haswell, E.S. The tension-sensitive ion transport activity of MSL8 is critical for its function in pollen hydration and germination. *Plant Cell Physiol.* **2017**, *58*, 1222–1237. [CrossRef] [PubMed]
87. Maathuis, F.J.M. Vacuolar two-pore K^+ channels act as vacuolar osmosensors. *New Phytol.* **2011**, *191*, 84–91. [CrossRef] [PubMed]
88. Gibbs, D.J.; Lee, S.C.; Md Isa, N.; Gramuglia, S.; Fukao, T.; Bassel, G.W.; Correia, C.S.; Corbineau, F.; Theodoulou, F.L.; Bailey-Serres, J.; et al. Homeostatic response to hypoxia is regulated by the N-end rule pathway in plants. *Nature* **2011**, *479*, 415–418. [CrossRef]
89. Licausi, F.; Kosmacz, M.; Weits, D.A.; Giuntoli, B.; Giorgi, F.M.; Voesenek, L.A.C.J.; Perata, P.; van Dongen, J.T. Oxygen sensing in plants is mediated by an N-end rule pathway for protein destabilization. *Nature* **2011**, *479*, 419–422. [CrossRef] [PubMed]
90. Gasch, P.; Fundinger, M.; Müller, J.T.; Lee, T.; Bailey-Serres, J.; Mustroph, A. Redundant ERF-VII transcription factors bind to an evolutionarily conserved cis-motif to regulate hypoxia-responsive gene expression in Arabidopsis. *Plant Cell* **2016**, *28*, 160–180. [CrossRef]
91. Weits, D.A.; Giuntoli, B.; Kosmacz, M.; Parlanti, S.; Hubberten, H.-M.; Riegler, H.; Hoefgen, R.; Perata, P.; van Dongen, J.T.; Licausi, F. Plant cysteine oxidases control the oxygen-dependent branch of the N-end-rule pathway. *Nat. Commun.* **2014**, *5*, 3425. [CrossRef]
92. Budziszewski, G.J.; Lewis, S.P.; Glover, L.W.; Reineke, J.; Jones, G.; Ziemnik, L.S.; Lonowski, J.; Nyfeler, B.; Aux, G.; Zhou, Q.; et al. Arabidopsis genes essential for seedling viability: Isolation of insertional mutants and molecular cloning. *Genetics* **2001**, *159*, 1765–1778. [CrossRef] [PubMed]
93. Dai, X.; Zhang, Y.; Zhang, D.; Chen, J.; Gao, X.; Estelle, M.; Zhao, Y. Embryonic lethality of Arabidopsis *abp1-1* is caused by deletion of the adjacent BSM gene. *Nat. Plants* **2015**, *1*, 15183. [CrossRef] [PubMed]
94. Ren, Z.-H.; Gao, J.-P.; Li, L.-G.; Cai, X.-L.; Huang, W.; Chao, D.-Y.; Zhu, M.-Z.; Wang, Z.-Y.; Luan, S.; Lin, H.-X. A rice quantitative trait locus for salt tolerance encodes a sodium transporter. *Nat. Genet.* **2005**, *37*, 1141–1146. [CrossRef] [PubMed]
95. Mallikarjuna, G.; Mallikarjuna, K.; Reddy, M.K.; Kaul, T. Expression of OsDREB2A transcription factor confers enhanced dehydration and salt stress tolerance in rice (*Oryza sativa* L.). *Biotechnol. Lett.* **2011**, *33*, 1689–1697. [CrossRef] [PubMed]
96. Cao, M.; Liu, X.; Zhang, Y.; Xue, X.; Zhou, X.E.; Melcher, K.; Gao, P.; Wang, F.; Zeng, L.; Zhao, Y.; et al. An ABA-mimicking ligand that reduces water loss and promotes drought resistance in plants. *Cell Res.* **2013**, *23*, 1043–1054. [CrossRef] [PubMed]
97. Cao, M.-J.; Zhang, Y.-L.; Liu, X.; Huang, H.; Zhou, X.E.; Wang, W.-L.; Zeng, A.; Zhao, C.-Z.; Si, T.; Du, J.; et al. Combining chemical and genetic approaches to increase drought resistance in plants. *Nat. Commun.* **2017**, *8*, 1183. [CrossRef]
98. Gallegos-Cedillo, V.M.; Nájera, C.; Signore, A.; Ochoa, J.; Gallegos, J.; Egea-Gilabert, C.; Gruda, N.S.; Fernández, J.A. Analysis of global research on vegetable seedlings and transplants and their impacts on product quality. *J. Sci. Food Agric.* **2024**, *104*, 4950–4965. [CrossRef]

MDPI AG
Grosspeteranlage 5
4052 Basel
Switzerland
Tel.: +41 61 683 77 34

Plants Editorial Office
E-mail: plants@mdpi.com
www.mdpi.com/journal/plants

www.ingramcontent.com/pod-product-compliance
Lightning Source LLC
Chambersburg PA
CBHW042010260926
51850CB00020B/3007
* 9 7 8 3 7 2 5 8 3 8 8 9 9 *